Study Topics in Physics

Revision and Workbook

- Revision Notes
- Questions with Explanatory Answers
- Questions from Past Examination Papers

W Bolton

Butterworths
London Boston Sydney Wellington Toronto Durban

First published 1982

British Library Cataloguing in Publication Data

Bolton, W.
Study topics in physics; revision and workbook
1. Physics
I. Title
530 QC23

ISBN 0-408-10829-0

Typeset by Scribe Design, Gillingham, Kent

Printed in England by Page Bros (Norwich) Ltd

Preface

This book is designed primarily for use as a workbook and revision guide for students preparing for the GCE Advanced Level Physics examinations. The format involves:

- concise notes on the main aspects of physics;
- worked examples of the various types of questions that can be posed;
- further questions, with hints as to methods of solution and answers;
- questions from A-level papers, with hints as to solutions and answers.

The aim has been to provide a book that teachers may wish to adopt for classroom use, as well as a book that students may buy and use for their own revision for the A-level examination.

The sequence adopted for the book is that used in the series Study Topics in Physics, chapter 1 of this book equating to Book 1, chapter 2 to Book 2, etc. This enables easy cross-reference between this book and that series.

I am grateful to the GCE examination boards (AEB, Joint Matriculation Board, Oxford Local Examinations, Southern Universities, University of Cambridge, and University of London) for permission to reproduce questions from their examinations. The hints to solutions for these questions are, however, my interpretation and not that of the boards. I must also point out that I have only given hints as to methods of solution and not all the detail that would be needed for adequate answers to the board questions. I am indebted to Norman Riches who worked bravely through all the questions in the book.

W. Bolton

Contents

Each chapter subsection contains revision notes, worked examples and further problems. N.B. *The answers are given to two or three significant figures only*

1 Motion and force

Describing motion

Notes

Average speed

$$= \frac{\text{distance covered}}{\text{time taken}}$$

Average velocity in a given direction

$$= \frac{\text{distance covered in that direction}}{\text{time taken}}$$

Average acceleration

$$= \frac{\text{change in velocity}}{\text{time taken for the change}}$$

The smaller the time intervals considered the more closely the average quantities defined above approximate to the instantaneous values.

Displacement differs from distance in that distance is measured along the actual path taken by the moving object whereas displacement is the straight line distance between the positions at the beginning and at the end of the time interval, regardless of what path the object moved along within that time. Distance and speed are scalar quantities, displacement and velocity are vector quantities.

For linear motion with constant (uniform) acceleration a, if u is the initial velocity at time $t = 0$ and v is the velocity after time t, then the definitions above lead to

Acceleration, $a = (v - u)/t$

Hence

$$v = u + at$$

As

Displacement = (average velocity) × (time)

Displacement, $s = \frac{1}{2}(u + v)t$

This equation can be combined with $v = u + at$ to give

$$s = ut + \frac{1}{2}at^2$$

and

$$v^2 = u^2 + 2as$$

All objects falling freely in a vacuum fall with a constant acceleration, this acceleration being known as the acceleration due to gravity (g). At the Earth's surface it has a value of about 9.8 m/s^2.

Units: Distance – m; displacement – m; time – s, speed – m/s or m s^{-1}, velocity – m/s or m s^{-1}, acceleration – m/s^2 or m s^{-2}.

Examples

Take the acceleration due to gravity, g, to be 9.8 m/s^2.

1.1 A bullet strikes and penetrates a bank of earth. If the velocity of the bullet on entering the earth is 600 m/s and it penetrates 3.0 m before coming to rest with uniform retardation in the earth, what is the retardation?

Ans

$$v^2 = u^2 + 2as$$

$$0 = 600^2 + 2 \times a \times 3.0$$

Hence

$$a = -60\,000 \text{ m/s}^2$$

and so retardation = 60 000 m/s^2

1.2 A uniformly accelerated object is seen to cover a distance of 14 m in 3.0 s and then a further 26 m in the next 3.0 s. What is the acceleration?

Ans $s = ut + \frac{1}{2}at^2$

For the first 3.0 s,

$$14 = 3u + \frac{1}{2} \times a \times 3.0^2$$

For the first 6.0 s,

$$40 = 6u + \frac{1}{2} \times a \times 6.0^2$$

Eliminating u between these two equations gives

$$a = 1.3 \text{ m/s}^2$$

1.3 A motorist starts his car from rest and accelerates uniformly at 0.5 m/s^2 for 30 s. The car then travels at constant velocity for 40 s. The brakes are then applied, producing a uniform retardation of 3.0 m/s^2 and bringing the car to rest. What is (a) the maximum velocity of the car and (b) the distance travelled during the entire journey?

Ans (a) During the first 30 s

$$v = u + at$$
$$= 0 + 0.5 \times 30$$
$$= 15 \text{ m/s}$$

This is the maximum velocity.
(b) During the first 30 s

$$s = ut + \frac{1}{2}at^2$$
$$= 0 + \frac{1}{2} \times 0.5 \times 30^2$$
$$= 225 \text{ m}$$

During the next 40 s

$$s = ut$$
$$= 15 \times 40$$
$$= 600 \text{ m}$$

During the retardation

$$v^2 = u^2 + 2as$$
$$0 = 15^2 - 2 \times 3.0 \times s$$

Hence

$$s = 37.5 \text{ m}$$

Thus

Total distance travelled
$= 225 + 420 + 37.5$
$= 682.5$ m

1.4 A ball freely falls from rest with the acceleration due to gravity of 9.8 m/s^2. How far will the ball fall in (a) the first 1 s, (b) the first 2 s?

Ans (a) $s = ut + \frac{1}{2}at^2$
$= 0 + \frac{1}{2} \times 9.8 \times 1^2$
$= 4.9$ m

(b) $s = ut + \frac{1}{2}at^2$
$= 0 + \frac{1}{2} \times 9.8 \times 2^2$
$= 19.6$ m

1.5 An object is thrown vertically upwards. How high is the object thrown if it takes 2.0 s to reach its maximum height

Ans At the maximum height the velocity is zero. Hence, as there is a retardation of 9.8 m/s^2 while the ball is moving upwards

$$s = ut + \frac{1}{2}at^2$$
$$s = 2.0u - \frac{1}{2} \times 9.8 \times 2.0^2$$

Also

$$v = u + at$$
$$0 = u - 9.8 \times 2.0$$
$$u = 19.6 \text{ m/s}$$

Hence

$$s = 2.0 \times 19.6 - \frac{1}{2} \times 9.8 \times 2.0^2$$
$$= 19.6 \text{ m}$$

This is the greatest height reached.

Further problems

Take g to be 9.8 m/s^2.

1.6 A ball starts rolling from rest down a constant slope with a constant acceleration of 4.9 m/s. If the slope has a length of 1.2 m how long does the ball take to travel down the slope? What will be the speed of the ball when it reaches the bottom of the slope?

1.7 An arrow, in being fired from a bow, is accelerated over a distance of 0.40 m and leaves the bow with a velocity of 40 m/s. What is the average acceleration of the arrow while being fired?

1.8 A bullet passes through a wall 200 mm thick. What is the average retardation of the bullet if the velocity is reduced from 400 m/s to 250 m/s in traversing the wall? What is the time taken for the bullet to traverse the wall?

1.9 An underground train starts from rest with a constant acceleration of 1.4 m/s^2 for 10 s, before moving with constant speed for a further 20 s. The brakes are then applied and bring the train to a halt with a uniform retardation of 2.8 m/s^2. What is the total distance covered by the train?

1.10 A car, moving with a constant acceleration, is observed to cover the distance between two points 50 m apart in 2.5 s. If the car passes the first of the two points with a velocity of 12 m/s, what is its velocity on passing the second point and the acceleration between the two points?

1.11 A ball falls freely from rest and accelerates with a constant acceleration of 9.8 m/s^2 What is the distance covered in (a) the first 2 s of fall, (b) the first 4 s of fall?

1.12 With what speed should a ball be thrown vertically upwards if it is to reach a maximum height of 10 m? What time will it take to reach this maximum height?

1.13 Object A is allowed to drop and then 1 s later object B is allowed to drop. If both objects start from rest and fall freely with the acceleration due to gravity, what will be the distance apart of the two objects 3 s after object A started to fall?

1.14 Describe a method that could be used in a school laboratory for the measurement of the acceleration due to gravity.

Projectiles

Notes

The motion of a projectile, if there is negligible air resistance, can be considered to be made up of two independent motions – vertical, with an acceleration due to gravity, and horizontal, with a constant velocity.

Examples

Take g to be 9.8 m/s^2.

1.15 A ball is rolled along the horizontal surface of a table and leaves the edge of the table with a horizontal velocity of 4 m/s. If the table top is 1 m above the horizontal floor what will be (a) the time taken for the ball, on leaving the table top, to reach the floor and (b) the horizontal distance it will travel from the edge of the table top before it reaches the floor?

Ans For the horizontal motion

$$s = ut$$

where s is the horizontal distance travelled in time t, t is the time taken for the ball to reach the floor and u is the horizontal velocity of the ball. Hence

$$s = 4t$$

For the vertical motion

$$h = \tfrac{1}{2}\,at^2$$

where h is the height of the table top above the floor and a the acceleration due to gravity. There is zero initial vertical velocity. Hence

$$1 = \tfrac{1}{2} \times 9.8 \times t^2$$

(a) The vertical motion equation enables t to be calculated.

$$t = 0.45 \text{ s}$$

(b) Hence

$$s = 4 \times 0.45$$
$$= 1.8 \text{ m}$$

1.16 A projectile starts with an initial velocity of 10 m/s at an angle of 30° to the horizontal. What will be the greatest height reached by the projectile and the horizontal distance it travels before hitting the horizontal plane from which it starts? (This horizontal distance is called the range.)

Ans The initial horizontal velocity is $u \cos \theta$, where u is the initial velocity and θ is the angle between this velocity and the horizontal plane. Thus, the horizontal velocity is $10 \cos 30°$, i.e. 8.66 m/s. The initial vertical velocity is $u \sin \theta$, i.e.

$$10 \sin 30° = 5.0 \text{ m/s}$$

For the vertical motion, $v = 0$ at the greatest height h, hence

$$v^2 = u^2 + 2as$$
$$0 = 5.0^2 - 2 \times 9.8 \times h$$

The acceleration due to gravity is in the opposite direction to the upwardly directed velocity component. Hence

$$h = 1.3 \text{ m}$$

The greatest height is thus 1.3 m (to two significant figures). The time taken to reach the greatest height can be obtained by using the equation

$$v = u + at$$
$$0 = 5.0 - 9.8 \times t$$

Hence

$$t = 0.51 \text{ s}$$

The time taken for the projectile to cover its range is twice the time taken for the projectile to reach its greatest height, i.e.

$$t = 2 \times 0.51 = 1.02 \text{ s}$$

For the horizontal motion

$$s = ut$$
$$s = 8.66 \times 1.02$$
$$= 8.83 \text{ m}$$

Thus, the greatest height reached is 1.3 m and the range is 8.83 m.

Further problems

Take g to be 9.8 m/s^2.

1.17 A bullet is fired horizontally from a gun held 1.4 m above the ground. If the bullet leaves the gun with a velocity of 300 m/s at what distance from the gun will the bullet strike the ground?

1.18 A projectile is given an initial velocity of 12 m/s at an elevation of 40° above the horizontal. What will be the greatest height reached above the starting point and the time taken to reach that height?

1.19 A football is kicked while on the ground and given a velocity of 15 m/s at an angle of 30° to the horizontal. What will be (a) the greatest height reached and (b) the distance the ball will travel before hitting the ground?

1.20 A rifle is aimed so that it lines up horizontally with a target 30 m away. When the rifle is fired the bullet hits the target 800 mm below the aiming point. What is the velocity with which the bullet left the rifle?

Newton's laws

Notes

Newton's laws can be expressed as:

First law. When no net force acts on an object it remains at rest or moves with a uniform velocity.

Second law. When a net force is acting on an object, the object moves with an acceleration that is proportional to the size of the force and which is in the same direction as the force (see the notes on momentum in this chapter for an alternative statement)

$$F = ma$$

Third law. If an object exerts a force on a second object, the second object exerts an equal and opposite force on the first (or action equals reaction).

Mass is a scalar quantity. It represents the resistance of a body to acceleration and is independent of the place in which a body is located. Weight is a force, i.e. a vector quantity, and can be defined as the downward force exerted by a body on a support which holds it at rest relative to the Earth, or some other planetary body. A body is weightless if falling freely with acceleration due to gravity. The weight of a body when at rest is mg, where g is the value of the acceleration due to gravity at that place.

Units: acceleration – m/s^2, mass – kg, force – N (newton).

Examples

Take g to be $9.8\ m/s^2$

1.21 A man stands in a lift which is being accelerated upwards at $2.0\ m/s^2$. If the man has a mass of 60 kg what is the force exerted on the man by the floor of the lift?

Ans If the lift had been at rest the force exerted by the lift on the man would have been 60×9.8 N, i.e. mg where g is the acceleration due to gravity. Because the man in the accelerating lift is being accelerated there must be a net force acting on him.

$$\text{Net force} = ma$$

where m is the mass of the man and a his acceleration. The net force is the force F exerted by the floor of the lift minus the gravitational force acting on the man. Thus

$$F - 60 \times 9.8 = ma$$

Hence

$$F = 60 \times 2.0 + 60 \times 9.8$$
$$= 708\ \text{N}$$

1.22 A lift, together with the passengers in the lift, has a mass of 500 kg. What is the tension in the cable supporting the lift when it is (a) at rest, (b) accelerating upwards with an acceleration of $1.5\ m/s^2$?

Ans (a) Tension $= mg$
$= 500 \times 9.8$
$= 4900$ N

(b) When the lift is accelerating upwards the net force needed to accelerate the lift plus passengers is 500×1.5 N. This net force is given by the force T exerted by the cable minus the gravitational force mg acting on the lift plus passengers, i.e.

$$T - mg = ma$$
$$T - 500 \times 9.8 = 500 \times 1.5$$
$$T = 5650\ \text{N}$$

This is the tension in the cable.

1.23 A light string joins two objects of masses 5 kg and 10 kg and the string passes over a frictionless pulley. What is (a) the acceleration of the objects, (b) the tension in the string?

Ans The gravitational forces acting on the objects are 5×9.8 N and 10×9.8 N. The total force acting on the 10 kg object is 10×9.8 N minus the tension T in the string. Thus

$$10 \times 9.8 - T = ma$$
$$= 10 \times a$$

where a is the acceleration of the 10 kg object. The total force acting on the 5 kg object is, in the direction of the acceleration,

$$T - 5 \times 9.8 = 5 \times a$$

where a is the acceleration of the 5 kg object. This must, however, be the same as the acceleration of the 10 kg object as they are connected. These two equations can be solved to give the acceleration a as $3.2\ m/s^2$ and the tension T as 65 N.

Further problems

Take g to be $9.8\ m/s^2$.

1.24 A car of mass 1200 kg is travelling at a velocity of 20 m/s when the brakes are applied and the car is brought to rest in a distance of 60 m. What is the deceleration, if it is assumed to be uniform during the braking, and the braking force?

1.25 A lift, together with the passengers in the lift, has a mass of 400 kg. What is the tension in the cable supporting the lift when it is accelerating downwards with an acceleration of $1.4\ m/s^2$?

1.26 A light string joins two objects of masses 2 kg and 4 kg. The object of mass 2 kg is on a smooth horizontal table and the 4 kg object hangs over the edge of the table. What is the tension in the string and the acceleration of the objects?

1.27 An object of mass 2.0 kg is suspended from a spring balance, calibrated in newtons, attached to the ceiling of a lift. What is the reading on the balance when the lift is (a) at rest, (b) moving with a constant velocity of 2 m/s upwards, (c) accelerating downwards with an acceleration of 2 m/s^2?

1.28 Two objects with masses of 1.0 kg and 3.0 kg are connected by a light, taut string and rest on a smooth horizontal table. If a force of 20 N is applied along the line of the string to the 3.0 kg object in such a way that the two objects both accelerate along the table with the string remaining taut, what will be the acceleration of the objects and the tension in the string?

1.29 What is the initial acceleration of a rocket of mass 1.2×10^4 kg when it is fired vertically upwards with an initial thrust of 2.4×10^5 N?

Friction

Notes

Frictional forces act along the common surface between two bodies in contact and are in such a direction as to oppose the relative motion of the bodies. Thus, for an object on a horizontal surface the forces may be as shown in *Figure 1.1*. For

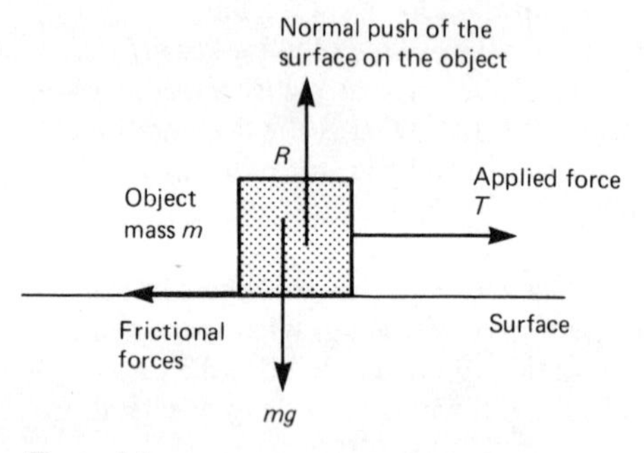

Figure 1.1

small values of the applied force there may be no motion of the object. At some particular value of the applied force the object begins to slide. Up to this point there must have been no net force acting on the object and so the frictional force must have been opposite and equal to the applied force. When the object begins to slide the applied force has exceeded the frictional force. The maximum frictional force between the two surfaces is called the limiting frictional force (F). Thus, when the object is accelerating along the surface the net force is

$$T - F = ma$$

where m is the mass of the sliding object and a its acceleration.

The value of the limiting frictional force is determined by the nature of the two surfaces concerned in the interaction and is proportional to the normal force R between the surface and the object.

$$F = \mu R$$

where μ is the coefficient of limiting friction. The coefficient depends on the nature of the surfaces concerned. It is called the static coefficient when the limiting frictional force concerned is that involved in the start of relative motion. When the object is sliding a force is needed to keep the object in motion with a constant velocity. The frictional force involved in this situation is given by

$$F = \mu_d R$$

where the coefficient μ_d is termed the coefficient of dynamic friction.

Examples

Take g to be 9.8 m/s^2.

1.30 What force is needed to start a block of steel of mass 1.2 kg sliding along a horizontal surface if the coefficient of static friction is 0.6?

Ans Since there is no vertical acceleration, the normal force R is opposite and equal to the gravitational force mg acting on the object. Hence

$$F = \mu R$$
$$= 0.6 \times 1.2 \times 9.8$$
$$= 7.1 \text{ N} \quad \text{(to two significant figures)}$$

1.31 An object of mass 2 kg is acted on by a horizontal force which pushes it along a horizontal surface with an acceleration of 0.2 m/s^2. If the coefficient of friction between the object and the surface is 0.5 what are the values of the applied force and the limiting frictional force?

Ans $F = \mu R$
$= 0.5 \times 2 \times 9.8$
$= 9.8$ N

The net force accelerating the object is the applied force T minus the frictional force F, hence

$T - F = ma$
$T - 9.8 = 2 \times 0.2$
$T = 10.2$ N

The applied force is 10.2 N.

Further problems

Take g to be 9.8 m/s^2.

1.32 What is the horizontal force that is needed to start sliding a 2 kg block of steel along a horizontal table if the coefficient of static friction between the block and the table is 0.6?

1.33 What is the horizontal force that is needed to accelerate a block of mass 3 kg at 2 m/s^2 along a horizontal table if the coefficient of dynamic friction is 0.5?

1.34 What is the acceleration produced when a horizontal force of 60 N acts on a block of mass 2.5 kg resting on a horizontal surface where the coefficient of dynamic friction is 0.6?

1.35 A block of mass 2.0 kg rests on a horizontal surface. If the coefficient of friction between the block and the surface is 0.5 what will be the size of the frictional force when (a) there is no horizontal force acting on the block, (b) a horizontal force of 5.0 N acts on the block?

Relative velocity

Notes

Velocity is a vector quantity. The velocity of some object A relative to another object B is the velocity that A appears to have to an observer moving with object B. Similarly, the velocity of object B relative to object A is the velocity that appears to have to an observer moving with object A. The velocity of A relative to B is opposite and equal to the velocity of B relative to A.

To determine the velocity of object A relative to B the following procedure can be followed:

(i) A velocity opposite and equal to that of B is superimposed on both the velocities of A and B.
(ii) In effect this superimposition brings B to rest.
(iii) The velocity of A relative to B can then be found by 'summing' the two velocities now acting on A, taking into account both the directions and magnitudes of the velocities.

The two velocities acting on A can be combined by what is termed the parallelogram law for the summation of vector quantities. The two velocities are represented in magnitude and direction by arrowed lines, the arrows indicating the directions and the lengths of the lines indicating the magnitudes. When the two arrowed lines form the adjacent sides of a parallelogram the diagonal of that parallelogram, passing through the intersection of the two arrowed lines, represents in magnitude and direction the single velocity which could replace the two separate velocities. It is the velocity of object A relative to object B (*Figure 1.2*).

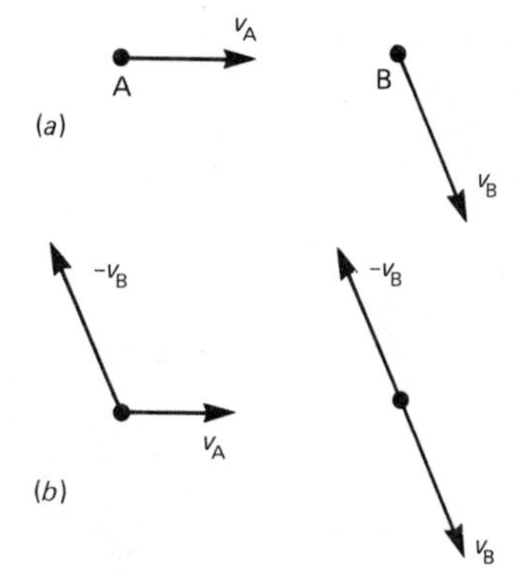

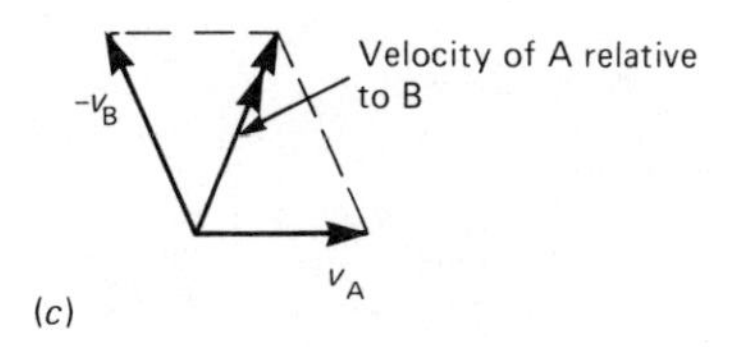

Figure 1.2 (a) The situation; (b) superimposing $-v_B$; *(c) the parallelogram*

Examples

1.36 A cyclist observes that when he is stationary the wind blows from the North. When he is pedalling due West he observes that the wind appears to blow from the North West. If the cyclist is pedalling at 6 m/s what is the velocity of the wind relative to the Earth?

Ans *Figure 1.3* shows the vector diagram. The wind velocity relative to the Earth is 6 m/s from the North.

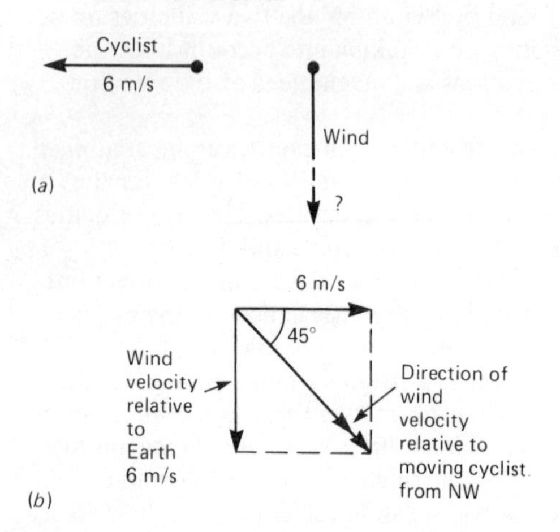

Figure 1.3 (a) The situation; (b) the parallelogram

1.37 Rain falls vertically, relative to the ground, with a velocity of 6.0 m/s. What is the velocity of a car if the raindrops make tracks on a side window at an angle of 60° from the vertical?

Ans *Figure 1.4* shows the vector diagram. The velocity of the car relative to the ground is 10 m/s.

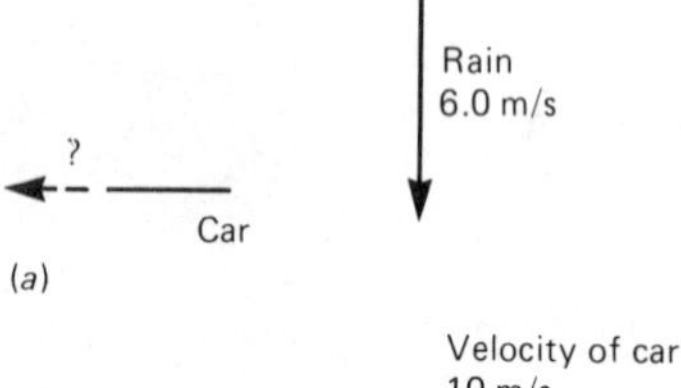

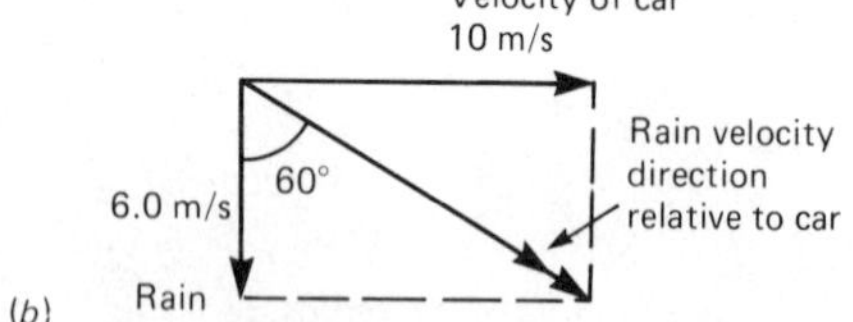

Figure 1.4 (a) The situation; (b) the parallelogram

Further problems

1.38 A man walking due West at 2.5 m/s feels the wind blowing from the North. When he increases his velocity to 3.5 m/s due West he feels the wind blowing from the North West. If the wind velocity is unchanged relative to the ground what is its value?

1.39 A woman observes that when she is stationary the snow appears to be falling vertically at 6.0 m/s. When she gets into a car and it is in motion with a constant velocity she observes that the snow flakes appear to be falling at an angle of 40° from the vertical. With what velocity is the car moving?

1.40 An aircraft A is flying due West at 1000 km/h when the pilot observes another aircraft B and is informed by the pilot of B that aircraft B is flying a true course 1200 km/h due South. What is the relative velocity of aircraft B as perceived by the pilot of aircraft A?

Force as a vector quantity

Notes

Force is a vector quantity. If two forces acting at a point are represented in magnitude and size by the adjacent sides of a parallelogram then the resultant force is represented by the diagonal of the parallelogram drawn from the junction of the two adjacent sides.

A single force can be resolved into two components at right angles to each other. Thus, a force F can be replaced by the two components $F \cos \theta$ and $F \sin \theta$, where θ is the angle between the force and the $F \cos \theta$ component.

Examples

Take g to be 9.8 m/s^2.

1.41 An object of mass 3 kg rests on an incline which is 30° to the horizontal. What is the component of the gravitational force acting on the object parallel to the inclined plane?

Ans The component parallel to the plane is

$$3 \times 9.8 \times \sin 30° = 14.7 \text{ N}$$

1.42 An object of mass 2.0 kg rests on a smooth inclined plane and is connected by a light string over a frictionless pulley at the top of the plane to a vertically hanging object of mass 0.5 kg. If the incline is at 30° to the horizontal and the string passes up the slope along the line of greatest slope, what is the acceleration of the objects and the tension in the string?

Ans The component of the 2.0×9.8 N force acting down the plane is $2.0 \times 9.8 \times \sin 30°$

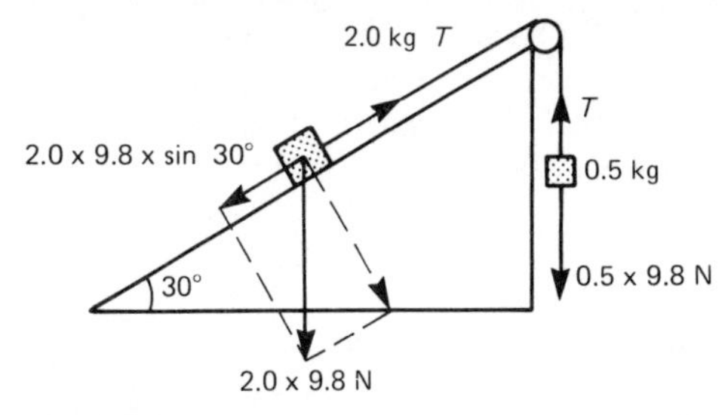

Figure 1.5

(*Figure 1.5*). The force acting on the 2.0 kg object is thus

$$2.0 \times 9.8 \times \sin 30° - T = ma$$
$$= 2.0 \times a$$

where T is the tension in the string and a the acceleration. The force acting on the 0.5 kg object is

$$T - 0.5 \times 9.8 = 0.5a$$

Eliminating T from the two equations gives the acceleration a as 5.9 m/s² (to two significant figures). Eliminating a from the two equations gives the tension T as 7.9 N (to two significant figures).

1.43 An object of mass 2.0 kg lies on an inclined plane. When the angle of inclination of the plane is slowly increased from zero no motion occurs until the plane is at an angle of 30° to the horizontal. What is the coefficient of static friction between the object and the surface of the plane?

Ans The gravitational force acting on the object is 2.0×9.8 N. The component of this acting down the plane is $2.0 \times 9.8 \sin 30°$ and the component at right angles to the plane is $2.0 \times 9.8 \cos 30°$. When the object just begins to slide the frictional force is opposite and equal to the gravitational field component acting down the plane. The normal force acting on the object is opposite and equal to the gravitational force component at right angles to the plane. Hence

$$F = \mu R$$
$$2.0 \times 9.8 \sin 30° = \mu \times 2.0 \times 9.8 \cos 30°$$
$$\mu = \tan 30°$$
$$= 0.58$$

Note that the calculation does not need the mass of the object.

Further problems

Take g to be 9.8 m/s².

1.44 An object of mass 1.2 kg rests on a smooth, inclined plane and is connected by a light string over a frictionless pulley at the top of the plane to a vertically hanging object of mass 3.0 kg. If the incline is at 30° to the horizontal and the string passes up the slope along the line of greatest slope, what is the initial acceleration of the objects and the tension in the string?

1.45 An object rests on a plane. When the angle of the plane to the horizontal is increased no motion of the object occurs until the angle reaches 40°. What is the static coefficient of friction?

1.46 A block of mass 2.0 kg slides down an inclined plane which is at 50° to the horizontal. If the dynamic coefficient of friction between the block and the plane is 0.4, what will be the acceleration of the object?

1.47 An object of mass 3.0 kg rests on an inclined plane and is connected by a light string over a frictionless pulley at the top of the plane to a vertically hanging object of mass 0.5 kg. If the incline is at 50° to the horizontal and the static coefficient of friction between the object and the plane is 0.5, what is the initial acceleration of the objects and the tension in the string?

Forces in equilibrium

Notes

An object when acted on by a net force will accelerate in the direction of the force. If, however, forces act on the object and there is no acceleration (or rotation) then the forces are said to be in equilibrium. For three forces to be in equilibrium the following conditions are necessary.

(i) They must all be in the same plane, i.e. coplanar.
(ii) They must have all their lines of action passing through one point, i.e. concurrent.
(iii) If the forces are represented in magnitude and direction by arrowed straight lines then these lines, when taken in the sequence of the forces, must form a triangle, known as the triangle of forces, i.e. there is no resultant force.

If more than three forces act at a point then the triangle of forces is extended to become a polygon of forces. If the forces are coplanar and concurrent and if each force is represented in magnitude and direction by arrowed straight lines then these lines, when taken in the order of the forces, must form a closed shape, i.e. a polygon. If the system of forces is not in equilibrium then a closed shape is not produced.

Examples

1.48 *Figure 1.6* shows an object supported by two wires. If the object is in equilibrium what are the forces in each wire at their points of connection to the object?

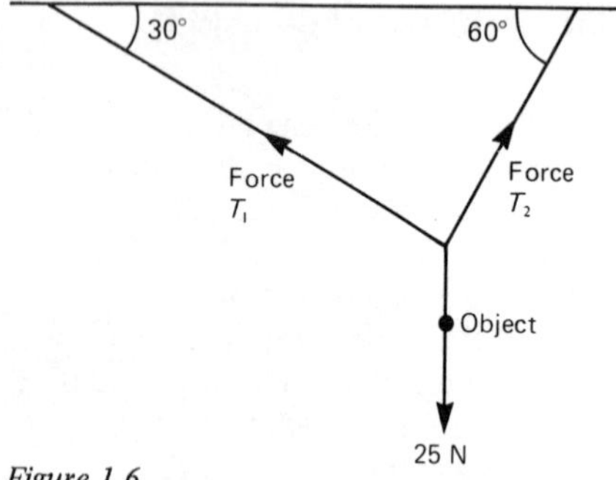

Figure 1.6

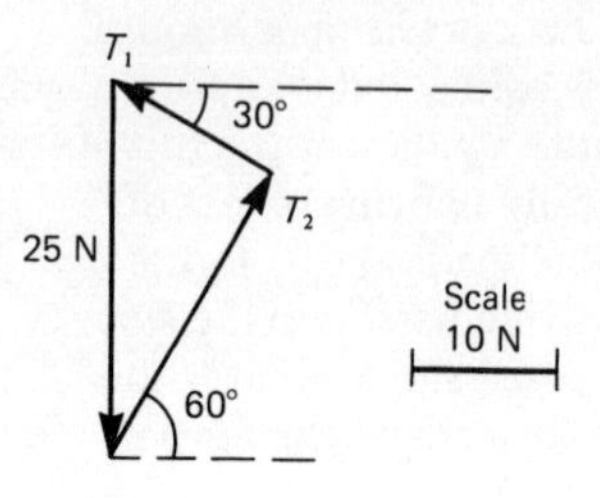

Figure 1.7

Ans *Figure 1.7* shows a scale diagram for the triangle of forces. The line representing T_1 has a length of 1.2 cm and thus T_1 is 12 N; the line for T_2 has a length of 2.2 cm and thus T_2 is 22 N.

1.49 Determine the forces F_1 and F_2 for the equilibrium shown in *Figure 1.8.*

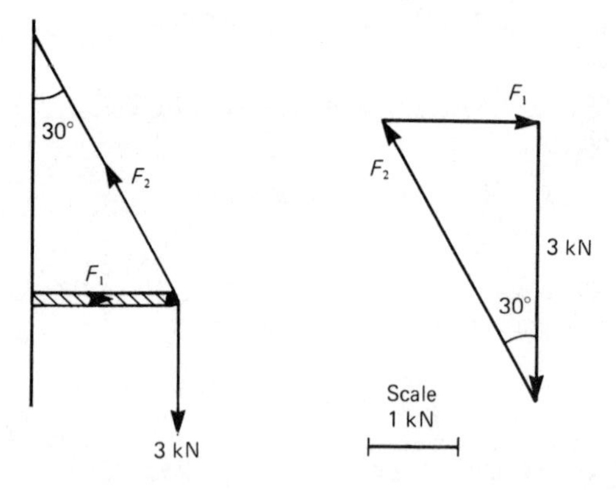

Figure 1.8 *Figure 1.9*

Ans *Figure 1.9* shows a scale diagram for the triangle of forces. The line representing F_1 has a length of 1.7 cm and thus F_1 is 1.7 kN the line for F_2 has a length of 3.4 cm and thus F_2 is 3.4 kN.

Further problems

1.50 What force must be applied to the rope in *Figure 1.10* in order for the box, of mass 200 kg, to be held in the position shown?

1.51 Find the tensions in the strings supporting the objects shown in *Figure 1.11.*

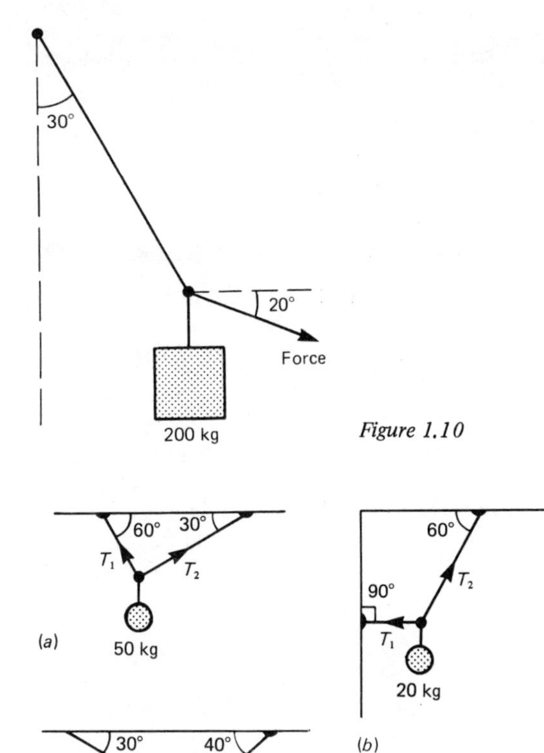

Figure 1.10

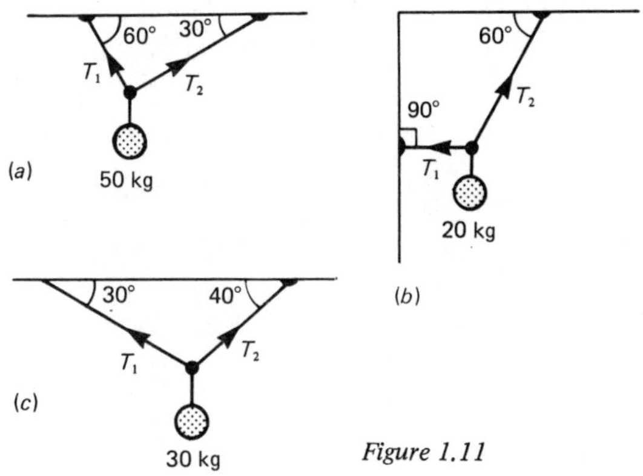

Figure 1.11

Circular motion

Notes

Figure 1.12 shows an object moving in a circular path with a constant speed, with the position and velocity of the object indicated at t_1 and t_2, time t apart. For the object moving from A to B there is

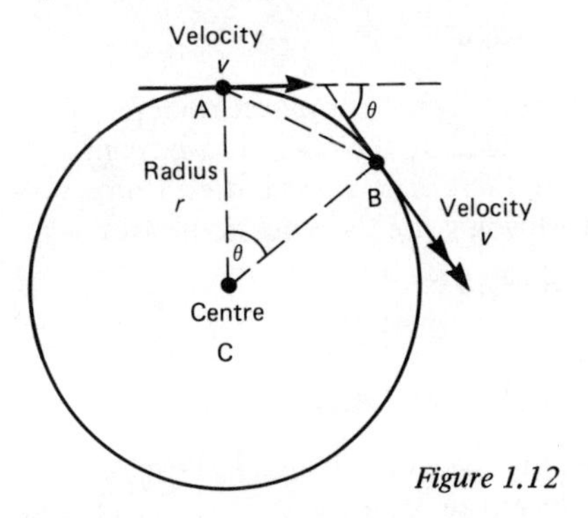

Figure 1.12

a change in velocity of Δv, where Δv is given by the parallelogram of velocities shown in *Figure 1.13*. Thus Δv is the change in velocity that occurs in time t. The average acceleration a during this time interval is thus

Average acceleration, $a = \Delta v/t$

The object is moving with a constant speed, the magnitude of which is v. In time t it covers the

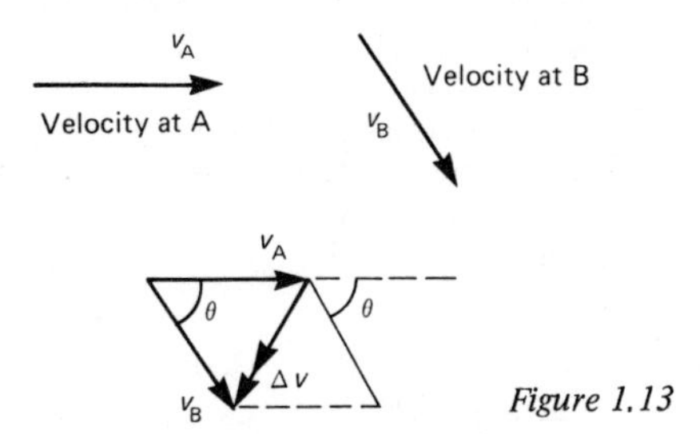

Figure 1.13

arc AB. If only small intervals of time have been considered this arc length is approximately equal to the distance AB. Thus

$v = \text{AB}/t$

Triangles CAB and SPQ are similar triangles, thus

$$\frac{\text{CA}}{\text{AB}} = \frac{\text{SP}}{\text{PQ}}$$

Hence

$r/\text{AB} = v/\Delta v$

Thus, the average acceleration a is given by

$$a = \Delta v/t$$
$$= \text{AB}\ v/tr$$
$$= v^2 r$$

This is called the centripetal acceleration because it is directed towards the centre of the circle in which the object is moving. The centripetal force is given by

$F = ma$

hence

$F = mv^2/r$

Examples

Take g to be 9.8 m/s^2.

1.52 A stone of mass 0.20 kg held on the end of a piece of string is whirled round in a horizontal circle of radius 1.0 m with a constant speed of 2.4 m/s. What is the tension in the string?

Ans $F = mv^2/r$
$= 0.20 \times 2.4^2/1$
$= 1.2$ N (to two significant figures)

This is the tension in the string, the centripetal force needed to pull the stone into its circular motion.

1.53 A wheel rotates at a constant rate of 10 revolutions per second. What is the centripetal acceleration at a distance of 0.80 m from the centre of the wheel?

Ans In one revolution a point 0.80 m from the centre covers a total distance equal to the circumference of a circle of radius 0.80 m, i.e. $2\pi \times 0.80$ m. In one second the total distance covered is $10 \times 2\pi \times 0.80$ m. Hence

$$v = 10 \times 2\pi \times 0.80/1$$

Thus

Centripetal acceleration,
$a = v^2/r$
$= (10 \times 2\pi \times 0.80)^2/0.80$
$= 3160 \text{ m/s}^2$

1.54 An object rests on the horizontal turntable of a record player. If the turntable is rotating at 33 revolutions per minute and the coefficient of friction between the object and the turntable is 0.30, what is the greatest distance from the axis of rotation at which the object can be placed without it sliding off?

Ans $v = 2\pi r/t$

where r is the distance from the axis and t is the time for one revolution. This is 60/33 s. Hence

$$v = 2\pi r \times 33/60$$

The limiting frictional force F is given by

$$F = \mu R$$

where R is the normal force. This is mg, where m is the mass of the object and g the acceleration due to gravity. Hence

$$F = \mu mg$$

This force would keep a body moving in a circular path of radius r where

$$F = mv^2/r$$

Thus

$$\mu mg = \frac{m}{r} \times \left(\frac{2\pi r \times 33}{60}\right)^2$$

Hence r is 0.25 m to two significant figures. This is the greatest distance from the axis at which an object can be placed without sliding off.

1.55 What is the angle of banking required for a race-track bend of radius 80 m so that cars can travel round the bend at 25 m/s without requiring a centripetal frictional force?

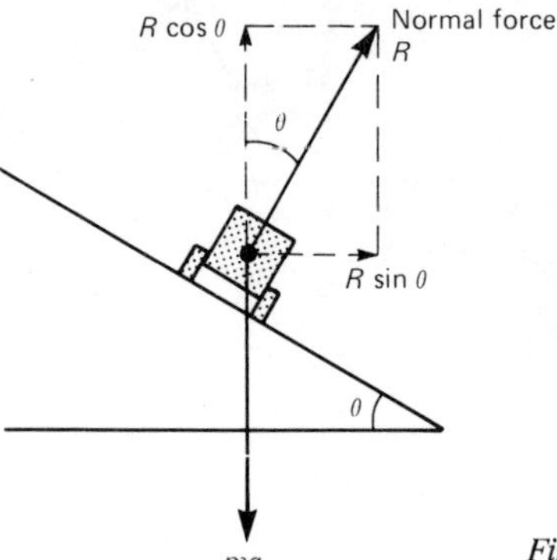

Figure 1.14

Ans *Figure 1.14* shows the forces acting on a car on the banked bend. There is no net force in the vertical direction, so

$$R \cos\theta = mg$$

The force in the horizontal direction is $R \sin\theta$ and this provides the centripetal force. Thus

$$R \sin\theta = mv^2/r$$

This assumes that the entire centripetal force is provided by the normal force component and there is no contribution from friction. Dividing the two equations gives

$$\tan\theta = v^2/rg$$

Hence

$$\tan\theta = 25^2/(80 \times 9.8)$$

and so θ, the angle of banking, is 39° (to two significant figures).

1.56 A small body of mass 50 g is swung round in a vertical circle at the end of a string of length 1.0 m. What is the tension in the string when the body is at its lowest point in the circular path if the speed at that point is 2.0 m/s?

Ans The forces acting on the body at the lowest point are the gravitational force mg due to its mass and the tension T in the string. These forces are in opposite directions with the net force providing the centripetal force. Thus

$$T - mg = mv^2/r$$

Hence

$$T = (0.050 \times 2.0^2)/1 + 0.050 \times 9.8$$
$$= 0.69 \text{ N}$$

Further problems

Take g to be 9.8 m/s^2.

1.57 A satellite moves in a circular orbit 640 km above the surface of the Earth. If it completes one orbit in 5880 s, what is its centripetal acceleration? The radius of the Earth is 6400 km.

1.58 What is the centripetal acceleration at a distance of 0.20 m from the centre of a record player turntable rotating at 33 revolutions per minute?

1.59 A car racing-track is banked at an angle of 40° on a bend of radius 200 m. With what speed can a car travel round the bend without relying on a centripetal frictional force?

1.60 It is proposed that a space station should be built in the form of a rotating wheel of radius 60 m. How many revolutions per second should the wheel make if the centripetal acceleration at the rim of the wheel is to be the same as the acceleration due to gravity, i.e. 9.8 m/s^2?

1.61 A small object of mass 80 g is swung round in a vertical circle at the end of a string of length 1.2 m. What is the tension in the string when the object is at its highest point in the circular path if the speed at that point is 4.8 m/s?

1.62 A car travels round a corner of radius 30 m. If the coefficient of friction between the car and the road surface is 0.6, what is the greatest speed for the car without skidding?

Angular motion

Notes

The measure of an angle θ in radians is given by the arc length s which has swept out the angle divided by the radius r of the rotation.

$$\theta = s/r$$

An angle in radians can be considered to be without units in that it is a length divided by a length.

Angular velocity ω is the rate at which an angle is covered or swept out. It has units of radians per second or just s^{-1}.

Average angular velocity, ω

$$= \frac{\text{angle rotated through}}{\text{time taken for that rotation}}$$

The frequency of a rotation is the number of rotations completed per second, one rotation per second being a frequency of 1 Hz (hertz). One revolution per second means a rotation through 2π radians. Thus a frequency f means a rotation through $2\pi f$ radians per second. This is the angular velocity ω. Thus

$$\omega = 2\pi f$$

The angular acceleration α is the rate at which angular velocity changes.

Average angular acceleration, α

$$= \frac{\text{change in angular velocity}}{\text{time taken for that change}}$$

Angular acceleration has units of radians/s^2 or just s^{-2}.

If a point on some object moves with a constant angular velocity, then in time t it will sweep out an angle θ, where

$$\omega = \theta/t$$

As the arc length s swept out by the point in time t is given by

$$s = r\theta$$

then

$$s = r\omega t$$

and so

$$s/t = r\omega$$

But s/t is the distance covered by the point divided by the time taken, i.e. the linear speed v. With small time intervals this approximates to the velocity

$$v = r\omega$$

If the point had an initial linear velocity of u and an initial angular velocity of ω_0 then

$$u = r\omega_0$$

If, after some time interval, t, the linear velocity is v and the angular velocity ω, the average angular acceleration α is given by

$$\alpha = (\omega - \omega_0)/t$$

and the average linear acceleration a by

$$a = (v - u)/t$$

Hence

$$a = (r\omega - r\omega_0)/t$$

and

$$a = r\alpha$$

Units: angle – radians (a ratio), angular velocity – radians/second or s^{-1}, angular acceleration – radians/second^2 or s^{-2}, frequency – s^{-1} or Hz, where 1 Hz = 1 s^{-1}.

Examples

1.63 A wheel starts from rest and accelerates uniformly with an angular acceleration of 3 s^{-2}. What will be its angular velocity after 4 s?

Ans $$\alpha = \frac{\text{change in angular velocity}}{\text{time taken for change}}$$

$$= (\omega - \omega_0)/t$$

where ω_0 is the initial angular velocity and ω the angular velocity after time t, when the uniform angular acceleration is α. Hence, as ω_0 is zero

$$\alpha = \omega/t$$

and so

$$\omega = 3 \times 4$$
$$= 12\ s^{-1}$$

1.64 A wheel has a radius of 0.2 m and is rotating with an angular velocity of 4 s^{-1}. What is the linear velocity of a point on the rim of the wheel?

Ans $$v = r\omega$$
$$= 0.2 \times 4$$
$$= 0.8\ \text{m/s}$$

Further problems

1.65 A wheel rotates with an angular velocity of 8.0 s^{-1} for 3.0 s. Through what angle has a point on the rim of the wheel rotated in that time?

1.66 A flywheel rotating at 3 revolutions per second accelerates at a constant rate to 5 revolutions per second in 40 s. What is the angular acceleration?

1.67 A wheel starts from rest and accelerates uniformly with an angular acceleration of 2.0 s^{-2}. What will be its angular velocity after 4.0 s?

1.68 A car has wheels of diameter 0.80 m. If the linear velocity of the car is 40 km/h what is the angular velocity of the wheels?

1.69 A flywheel accelerates from 1.2 to 2.8 revolutions per second in 15 s. If the wheel has a radius of 1.4 m what is (a) the angular acceleration, (b) the linear acceleration of a point on the rim of the wheel?

Torque and angular acceleration

Notes

A force F applied to a pivoted object is said to exert a torque T of Fr where r is the distance from the point of application of the force F to the pivot axis.

$$\text{Torque, } T = Fr$$

The angular acceleration α resulting from this torque is related to the torque by

$$T = I\alpha$$

where I is the moment of inertia.

For a small object of mass m at the end of a pivoted arm of radius r the moment of inertia is mr^2; for a sphere about its diameter

$$I = \frac{2}{3} mr^2$$

For a disc about its axis

$$I = \frac{1}{2} mr^2$$

Units: force – N, radius – m, torque – N m; angular acceleration – s^{-2}, moment of inertia – kg m^2.

Examples

1.70 What torque is required to give a pulley wheel an angular acceleration of 2 s^{-2} if it has a moment of inertia of 200 kg m^2?

Ans $T = I\alpha$
$= 200 \times 2$
$= 400$ N m

1.71 An electric motor has a rotor with a moment of inertia of 3.9 kg m^2. What is the torque required to give it an angular acceleration of 0.40 s^{-2}?

Ans $T = I\alpha$
$= 3.9 \times 0.40$
$= 1.6$ N m

Further problems

1.72 What torque is required to give a wheel an angular acceleration of 1.6 s^{-2} if it has a moment of inertia of 120 kg m^2?

1.73 A flywheel has a moment of inertia of 100 kg m^2. What is the retardation produced when a constant braking torque of 200 N m is applied to it?

1.74 A wheel has a moment of inertia of 0.80 kg m^2. If the wheel has a radius of 0.40 m and a force of 6.0 N acts tangentially on the rim of the wheel, what will be the angular acceleration?

1.75 A grindstone of mass 45 kg and radius 1.4 m is rotating at 12 revolutions per second. When a tool is pressed against the rim of the grindstone it decelerates uniformly and comes to rest in 10 s. What is (a) the torque applied by the tool, (b) the coefficient of friction between the tool and the grindstone if the tool is pressed normally against the rim with a force of 280 N?

1.76 A cord is wrapped round the rim of a flywheel and a steady pull of 5.0 N is applied to the cord, causing it to unwind and the flywheel to accelerate. If the flywheel has a radius of 0.70 m and a moment of inertia of 0.80 kg m^2, and is mounted on frictionless bearings, what will be the angular acceleration produced?

Linear momentum

Notes

The momentum of an object of mass m moving with a velocity v is defined as the product of the mass and velocity.

$$\text{Momentum} = mv$$

Momentum is a vector quantity. **Newton's second law of motion** can be expressed as:

> when a net force acts on an object the rate of change of momentum equals the force.

Force = rate of change of momentum
$F = d(mv)/dt$

This statement is equivalent to

$$F = ma$$

if m is constant.

$$\text{Average force} = \frac{\text{change in } (mv)}{\text{change in } t} = m \times \frac{\text{change in } v}{\text{change in } t} = ma$$

The product of force and the time for which the force is acting is called the impulse.

Impulse = force × time interval

But

Force × time interval = change in momentum

Hence

Impulse = change in momentum

In any closed system momentum is conserved, i.e. the total momentum remains constant. This is known as the **law of conservation of momentum.** This means that in some event, such as a collision, the change in momentum of one body must be exactly equal to and opposite to the change in momentum of the other body. This means that the impulse acting on one body must be opposite and equal to the impulse acting on the other body. In a collision the time for which the force acts on one body is equal to the time for which the other force acts on the other body. Thus, the force acting on one body must be equal and opposite to the force acting on the other body. This is **Newton's third law of motion.**

Units: mass – kg, velocity – m/s, momentum – kg m s^{-1}.

Examples

1.77 A bullet of mass 2 g emerges from the barrel of a gun with a velocity of 400 m/s. What is (a) the momentum of the bullet and (b) the recoil momentum of the gun?

Ans (a) Momentum $= mv$
$= 2 \times 10^{-3} \times 400$
$= 0.8$ kg m s^{-1}

(b) The gun must recoil with a momentum of –0.8 kg m s^{-1} if momentum is to be conserved.

1.78 What is the average recoil force experienced by a machine gun which is firing 120 bullets per minute, each bullet having a mass of 2 g and a muzzle velocity of 400 m/s?

Ans Momentum of one bullet $= mv$
$= 2 \times 10^{-3} \times 400$
$= 0.8$ kg m s^{-1}

Momentum of 120 bullets $= 120 \times 0.8$
$= 96$ kg m s^{-1}

This momentum is given to the bullets in 60 s. Thus, over that period the average change in momentum per second is 96/60 kg m s^{-2}. The average force is the rate of change of momentum, thus the average force is 1.6 N.

1.79 A truck with a mass of 50 kg moving with a velocity of 3 m/s collides with a stationary truck of mass 30 kg. When the trucks collide they lock together. What will be their velocity after the collision?

Ans Momentum before the collision $= 50 \times 3$
$= 150$ kg m s^{-1}
Momentum after the collision $= (50 + 30)\ v$

where v is the velocity of the two trucks when locked together. Hence, as momentum is conserved,

$$80v = 150$$
$$v = 1.9 \text{ m/s}$$

1.80 A billiard cue hits a stationary billiard ball of mass 0.20 kg and applies an average force of 40 N over a time of 0.009 s. What is the velocity of the billiard ball after the impact?

Ans Impulse $= 40 \times 0.009$
$= 0.36$ N s

Hence

Change in momentum = 0.36 N s

The velocity is given by

$$mv = 0.36$$

and so v is 1.8 m/s

1.81 A jet of water 40 mm in diameter and having a velocity of 30 m/s, in striking a flat plate fixed at right angles to the jet, loses all its velocity on impact. What is the force acting on the plate (take the density of water to be 1000 kg/m^3)?

Ans The volume of water hitting the plate per second is Av, where A is the cross-sectional area of the jet and v its velocity. If ρ is the density of the water, then

Mass of water hitting the plate per second
$= Av\rho$

and thus, as this mass of water has the momentum $Av\rho v$, the rate of change of momentum is $Av^2\rho$. Hence

$$\text{Force on the plate} = Av^2\rho$$
$$= \frac{\pi}{4}d^2v^2\rho$$

where d is the diameter of the jet. Hence

$$\text{Force} = \pi/4 \times 0.040^2 \times 30^2 \times 1000$$
$$= 1.1 \text{ kN}$$

1.82 A rocket in flight explodes and breaks into two pieces. One of the pieces, of mass 10 kg, moves off in the same direction as that of the rocket with a velocity of 1500 m/s while the other piece, of mass 4 kg, moves off in the opposite direction with a velocity of 50 m/s. What was the velocity of the rocket before the explosion?

Ans The momentum of the rocket before the explosion was $(10 + 4)v$, where v was the velocity. After the explosion

$$\text{Momentum} = 10 \times 1500 - 4 \times 50$$
$$= 14\,800 \text{ kg m s}^{-1}$$

Hence, as momentum is conserved,

$$14v = 14\,800$$
$$v = 1060 \text{ m/s}$$

in the direction of the velocity of the 10 kg piece.

1.83 A firework rocket is discharging gases at the rate of 5 g/s and the gas emerges from the rocket with a velocity of 300 m/s. What is the force acting on the rocket due to the discharge?

Ans Mass of gas ejected per second $= 5 \times 10^{-3}$ kg

Thus

Rate of change of momentum
$= 5 \times 10^{-3} \times 300 \text{ kg m s}^{-1}$

Hence

Resulting force = 1.5 N

1.84 A space rocket is in the process of being fired from its stand on the Earth. The mass of the rocket, plus fuel, is 5000 kg and the exhaust gas emerges at 800 m/s. What will be the thrust required for the rocket to overcome the gravitational force acting on it and begin to rise vertically? What will be the rate at which the exhaust gas has to emerge for this thrust to be obtained?

Ans The thrust has to overcome the gravitational force of mg, where m is the mass of the rocket and g the acceleration due to gravity. Hence

$$\text{Required thrust} = 5000 \times 9.8$$
$$= 49\,000 \text{ N}$$

This thrust is obtained from the rate of change of momentum resulting from the exhaust gases. Hence

$$49\,000 = M \times 800$$

where M is the mass of gas leaving the rocket per second.

$$M = 61 \text{ kg/s}$$

Further problems

Take water to have a density of 1000 kg/m^3.

1.85 A shell is fired from a gun with a muzzle velocity of 500 m/s. If the shell has a mass of 6.0 kg what is (a) the momentum of the shell as it leaves the gun, (b) the recoil momentum of the gun?

1.86 A bullet of mass 2.0 g is fired from a rifle with a muzzle velocity of 250 m/s. If the rifle has a mass of 3.5 kg what is its recoil velocity?

1.87 A rocket burns fuel at the rate of 60 g per second and ejects the spent fuel at a velocity of 4000 m/s. What is the force acting on the rocket?

1.88 A machine gun fires bullets of mass 2 g with a muzzle velocity of 450 m/s. What is the maximum number of bullets that can be fired per minute if the man holding the gun can exert a force of 200 N against the gun?

1.89 Describe an experiment by which you could demonstrate, in the school laboratory, that momentum is conserved.

1.90 A bullet of mass 2 g and moving horizontally with a velocity of 400 m/s strikes a wooden block of mass 2.0 kg resting on a smooth, horizontal surface. What will be the velocity of the block after it has absorbed the bullet if frictional forces are neglected?

1.91 A car with its occupants has a mass of 2000 kg and is moving at a velocity of 45 km/h when it strikes, head on, a smaller car having a mass of 1200 kg and a velocity of 60 km/h. What is the velocity of the cars after the impact if they become locked together?

1.92 A ball having a mass of 100 g has a velocity of 30 m/s when it is hit by a bat. After the collision the ball reverses its direction of motion and has a velocity of 50 m/s. What was the impulse given to the ball by the bat? If the ball was in contact with the bat for 4.0 ms (milliseconds) what was the average force acting on the ball?

1.93 A jet of water of diameter 12 mm has a velocity of 12 m/s. What is the force experienced by a stationary plate which is hit by the jet if the jet is directed along the normal to the plate and loses all its momentum on impact with the plate?

1.94 A jet of water with a diameter of 300 mm and moving with a velocity of 6.0 m/s strikes a stationary vane of a water wheel along a line at right angles to the vane. If the water loses all its momentum when hitting the vane, what will be the force exerted on the vane by the jet?

1.95 A rocket is to be launched vertically. If it has a total mass of 8000 kg and the gases are ejected from the rear of the rocket with a speed of 2000 m/s what must be the mass of gas ejected per second to overcome the weight of the rocket plus fuel and enable 'lift off' to occur? What must be the mass of gas ejected per second if the rocket is to be given an initial acceleration of 20 m/s^2?

Examination questions

1.96 (a) A ball is thrown vertically upwards from the surface of the Earth with an initial velocity u. Neglecting frictional forces, sketch a graph to show the variation of the velocity v of the ball with time t as the ball rises and then falls back to Earth. What information contained in the graph enables you to determine

(i) the gravitational acceleration,
(ii) the maximum height to which the ball rises?

(b) If the frictional forces in the air were not negligible, how, in the above situation, would

(i) the initial deceleration of the ball,
(ii) the maximum height reached by the ball, be affected?

(AEB)

1.97 Mass, length and time are base quantities whereas momentum, acceleration and force are derived ones. Define each of the derived quantities and express each of them in terms of the base ones.

In a collision linear momentum is conserved. Outline an experiment to illustrate this conservation law. Give two further examples of conservation laws.

By considering a collision as one body exerting a constant force F on another for a time t, show that the conservation of linear momentum is a direct result of Newton's laws of motion.

A rocket with its motors burning is moving in space in a straight line with constant acceleration. Sketch graphs showing how (a) acceleration, (b) velocity, (c) momentum and (d) distance moved change with time.

(University of London)

1.98 A hose with a nozzle 80 mm in diameter ejects a horizontal stream of water at a rate of 0.044 m^3/s. With what velocity will the water leave the nozzle? What will be the force exerted on a vertical wall situated close to the nozzle and at right angles to the stream of water if, after hitting the wall,

(a) the water falls vertically to the ground,
(b) the water rebounds horizontally?

Density of water = 1000 kg m^3

(AEB)

1.99 (a) State the laws of conservation of linear and angular momentum. Discuss briefly the application of momentum conservation to the motion of (i) a jet aircraft, (ii) a space rocket and (iii) a helicopter.
(b) A wooden block of mass 1.6 kg is at rest on a wooden plane inclined at 30° to the horizontal.

(i) Write down the component of its weight acting parallel to the plane, the component of its weight acting at right angles to the plane, and the frictional force between the block and the plane.
(ii) When the block is given a sharp tap directed down the plane it begins to move with a speed of 0.3 m/s and thereafter slides down the plane with a uniform acceleration of 0.2 m/s^2. Calculate the impulse applied to the block and the frictional resistance to the ensuing motion.

(Oxford Local Examinations)

1.100 (a) A solid metal ball released from rest rolls down a curved ramp which has been adjusted so that the ball leaves it horizontally and subsequently falls through the air to the ground. Describe, giving experimental details, how you would determine experimentally the trajectory of the ball after it has left the ramp.
(b) If the horizontal speed of the ball may be assumed to be constant after it has left the ramp, explain how the vertical motion of the ball can be shown to be uniformly accelerated.
(c) What would be the effect on the trajectory of (i) releasing the ball from a different point on the ramp, (ii) using solid balls, equal in size but having different masses, released from the same point on the ramp?
(d) A solid metal ball rolls off a horizontal table top of height h with a speed v in a direction perpendicular to the edge and hits the floor a horizontal distance d from the table. In terms of h, d and g (the acceleration of free fall), and assuming no air resistance, calculate v, and hence calculate the magnitude and direction of the velocity with which the ball hits the floor.

(JMB)

1.101 A special prototype model aeroplane of mass 400 g has a control wire 8 m long attached to its body. The other end of the control wire is attached to a fixed point. When the aeroplane flies with its wings horizontal in a horizontal circle, making one revolution every 4 s, the control wire is elevated 30° above the horizontal. Draw a diagram showing the forces exerted on the plane and determine

(a) the tension in the control wire,
(b) the lift on the plane.

Assume that the acceleration of free fall $g = 10\ m/s^2$ and $\pi^2 = 10$.

(AEB)

Answers

1.6 Ans $s = ut + \frac{1}{2}at^2$, hence $t = 0.70$ s
$v^2 = u^2 + 2as$; hence $v = 3.4$ m/s

1.7 Ans $v^2 = u^2 + 2as$, hence $a = 2000\ m/s^2$

1.8 Ans $v^2 = u^2 + 2as$, hence $a = -2.4 \times 10^5\ m/s^2$ and so retardation = $2.4 \times 10^5\ m/s^2$.
$v = u + at$, hence $t = 6.3 \times 10^{-4}$ s

1.9 Ans $s_1 = ut + \frac{1}{2}at^2 = 70$ m;
$v_1 = u + at = 14$ m/s;
$s_2 = 14 \times 20 = 280$ m; $v^2 = v_1^2 + 2as_3$
hence $s_3 = 35$ m and hence total distance = 385 m

1.10 Ans $s = ut + \frac{1}{2}at^2$, hence $a = 6.4$ m/s^2;
$v = u + at = 28$ m/s

1.11 Ans (a) $s = \frac{1}{2}at^2 = 19.6$ m
(b) 78.4 m

1.12 Ans $v^2 = u^2 + 2as = 0$, hence $u = 14$ m/s;
$v = u + at = 0$, hence $t = 1.4$ s

1.13 Ans $s = \frac{1}{2}at^2$, hence $s_1 = \frac{1}{2} \times 9.8 \times 9 = 44.1$ m;
$s_2 = \frac{1}{2} \times 9.8 \times 4 = 19.6$ m. Hence the distance apart = 24.5 m

1.14 Ans You could describe a large mass falling and pulling ticker tape through a vibrator, or a stroboscopic photograph being taken of a falling object, or measurement (with an electric stopclock for high accuracy) of the time taken for a steel ball to fall from rest through a measured distance

1.17 Ans For the vertical motion $s = \frac{1}{2}gt^2$, hence $t = 0.76$ s. In this time the horizontal motion is $s = ut = 300 \times 0.76 = 227$ m

1.18 Ans Vertical velocity = 12 sin 40°; for the vertical motion $v^2 = u^2 + 2as = 0$,
hence $s = u^2/2g = 3.0$ m; $v = u + at$,
hence $t = u/g = 0.79$ s

1.19 Ans (a) Vertical velocity = 15 sin 30°.
For vertical motion $v^2 = u^2 + 2as$, hence
$s = u^2/2g = 2.87$ m; $v = u + at$, hence
$t = u/g = 0.765$ s

(b) For horizontal motion,
velocity = 15 cos 30°, hence $s = ut$ =
9.94 m and the total distance = 19.9 m

1.20 Ans For vertical motion $s = \frac{1}{2}at^2$, hence
$t = 0.404$ s. For horizontal motion
$s = ut$, hence $u = 30/0.404 = 74.2$ m/s

1.24 Ans $v^2 = u^2 + 2as = 0$, hence $a = 4.1$ m/s^2
$F = ma = 4.9$ kN

1.25 Ans $T + 400 \times 9.8 = 400 \times 1.4$, hence
$T = -3.4$ kN. The force in the cable is in the opposite direction to the acceleration.

1.26 Ans Force acting on the 2.0 kg mass =
$T = 2.0a$. Force acting on the 4.0 kg mass = $4.0 \times 9.8 - T = 4.0a$.
Hence $a = 6.5$ m/s^2 and $T = 13$ N

1.27 Ans (a) 19.6 N
(b) 19.6 N
(c) Force = $mg - ma = 15.6$ N

1.28 Ans Force of 3.0 kg mass = $20 - T = 3.0a$.
Force on 1.0 kg mass = $T = 1.0a$. Hence
$a = 5.0$ m/s and $T = 5.0$ N

1.29 Ans $F - mg = ma$, hence $a = 10.2$ m/s^2

1.32 Ans $F = \mu R = \mu mg = 12$ N

1.33 Ans Net force = $F - \mu R = ma$, hence
$F = 21$ N

1.34 Ans Net force = $F - \mu R = ma$, hence
$a = 18$ m/s^2

1.35 Ans (a) Zero
(b) $F = \mu R = \mu mg = 9.8$ N. This is greater than the 5.0 N force and so the frictional force will be 5.0 N

1.38 Ans 2.7 m/s from N 68° E (*Figure 1.15*)

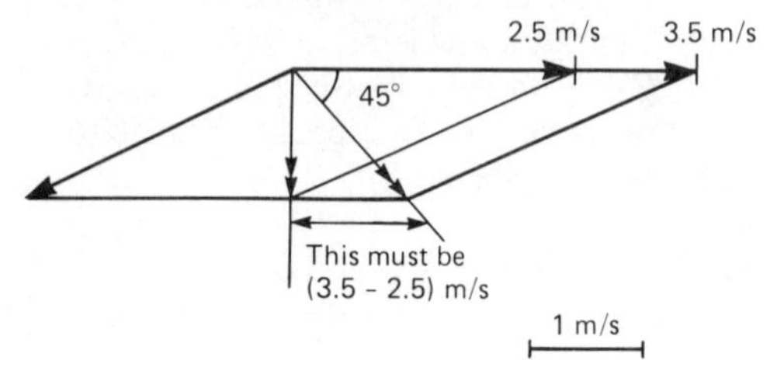

Figure 1.15

1.39 Ans See **1.37**; 5.0 m/s

1.40 Ans 1560 km/h at 40° W of S (*Figure 1.16*)

1.44 Ans For the 3.0 kg object, $3.0 \times 9.8 - T = 3.0a$. For the 1.2 kg object
$T - 1.2 \times 9.8 \times \sin 30° = 1.2a$;
hence $a = 5.6$ m/s^2 and $T = 13$ N

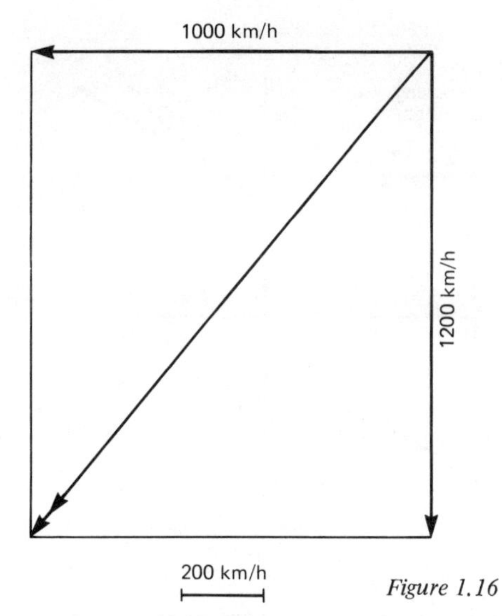

Figure 1.16

1.45 Ans $F = \mu R$, hence $mg \sin 40° = \mu mg \times \cos 40°$ and so $\mu = \tan 40° = 0.84$

1.46 Ans $2.0 \times 9.8 \sin 50° - 0.4 \times 2.0 \times 9.8 \times \cos 50° = 2.0a$, hence $a = 5.0$ m/s^2

1.47 Ans For the 3.0 kg object, $3.0 \times 9.8 \sin 50° - 0.5 \times 3.0 \times 9.8 \times \cos 50° - T = 3.0a$. For the 0.5 kg object, $T - 0.5 \times 9.8 = 0.5a$, hence $a = 2.33$ m/s^2, i.e. the object is moving down the plane and $T = 6.06$ N

1.50 Ans 1.5 kN

1.51 Ans (a) $T_1 = 421$ N, $T_2 = 245$ N;
(b) $T_1 = 113$ N, $T_2 = 226$ N;
(c) $T_1 = 240$ N, $T_2 = 272$ N

1.57 Ans $a = v^2/r = 4\pi^2 r/T^2 = 8.0$ m/s^2

1.58 Ans $a = v^2/r = 4\pi^2 rf^2 = 2.4$ m/s^2

1.59 Ans Tan $\theta = v^2/rg$, hence $v = 40.6$ m/s

1.60 Ans $a = v^2/r = 4\pi^2 rf^2$, hence $f = 0.064$ revolutions per second

1.61 Ans $T + mg = mv^2/r$, hence $T = 0.75$ N

1.62 Ans $mv^2/r = \mu mg$, hence $v = 13$ m/s

1.65 Ans $\theta = \omega t = 24$ radians

1.66 Ans $\alpha = 0.31$ s^{-2}

1.67 Ans $\omega = \alpha t = 8.0$ s^{-1}

1.68 Ans $\omega = v/r = 50 \times 10^3$ h^{-1} $= 14$ s^{-1}

1.69 Ans (a) $\alpha = 0.67$ s^{-2}
(b) $a = r\alpha = 0.94$ m/s^2

1.72 Ans $T = I\alpha = 192$ N m

1.73 Ans $T = I\alpha$, hence $\alpha = 2$ s^{-2}

1.74 Ans $T = Fr = I\alpha$, hence $\alpha = 3.0$ s^{-2}

1.75 Ans (a) $I = \frac{1}{2}mr^2 = 44$ kg m^2, $\alpha = 2\pi \times 12/10$, $T = I\alpha = 333$ N m
(b) $F = T/r = 236$ N, $F = \mu R$, hence $\mu = 0.85$

1.76 Ans $T = Fr = I\alpha$, hence $\alpha = 4.4$ s^{-2}

1.85 Ans (a) $mv = 3000$ kg m s^{-1}
(b) -3000 kg m s^{-1}

1.86 Ans $mv = MV$, hence $V = 0.14$ m/s

1.87 Ans Force = rate of change of momentum = 240 N

1.88 Ans Force = rate of change of momentum, hence number per minute = 13 000

1.89 Ans You could describe an experiment involving trolleys and ticker tape, or gliders on an air track with electrical timing, or a ballistic pendulum

1.90 Ans $mv = (M + m)V$, hence $V = 0.40$ m/s

1.91 Ans $mu - m'u' = (m' + m)V$, hence $V = 5.6$ km/h in the direction of the initial velocity of the 2000 kg car

1.92 Ans Impulse = change in momentum $= 100 \times 10^{-3} \times 30 - (-100 \times 10^{-3} \times 50) = 8.0$ N s; average force = 2000 N

1.93 Ans See **1.81**.

Force $= \frac{\pi}{4} d^2 v^2 \rho = 16$ N

1.94 Ans See **1.81**.

$$\text{Force} = \frac{\pi}{4} d^2 v^2 \rho = 2.5 \text{ kN}$$

1.95 Ans See **1.84**. Thrust = 8000 × 9.8 N. This equals the rate of change of momentum of $M \times 2000$. Hence the mass ejected per second, $M = 39.2$ kg/s. For an acceleration of 20 m/s^2, thrust = 8000(9.8 + 20), hence $M = 119.2$ kg/s

1.96 Ans (a) $v = u - gt$ (see *Figure 1.17*).
(i) The slope of the graph is $-g$
(ii) The area A is the greatest height
(b) (i) The initial deceleration would be increased to $g + F/m$, where F is the frictional force and m the mass of the object

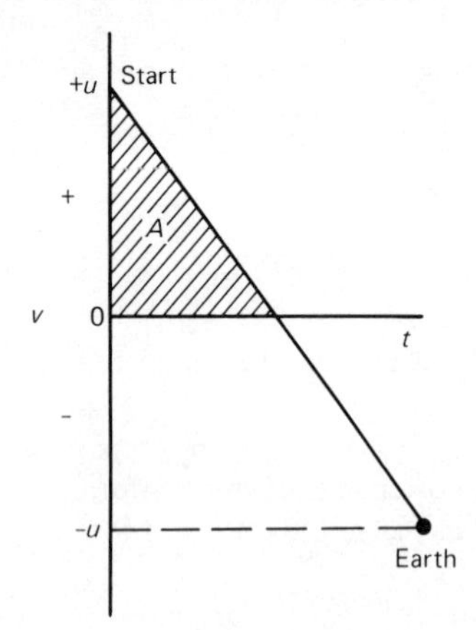

Figure 1.17

(ii) The increase in the retardation would reduce the area A and hence the greatest height reached

1.97 Ans See the notes in this chapter for the definitions. Momentum, MLT^{-1}; acceleration, LT^{-2}; force, MLT^{-2} (M, mass; L, length; T, time). See **1.89** for the experiment. Other conservation laws are for charge and energy.

Action equals reaction means $F_1 = F_2$. Hence as force is the rate of change of momentum, the rate of change of momentum of object 1 equals the rate of change of momentum of object 2. But the time t for which the two objects are in contact is the same, therefore the change in momentum of one object equals the change of momentum of the other. The total change in momentum is thus zero, i.e. constant

See *Figure 1.18* for the graphs. The graphs assume that there is no significant change in the mass of the rocket during the time considered. A decrease in mass would, however, mean an increase in acceleration

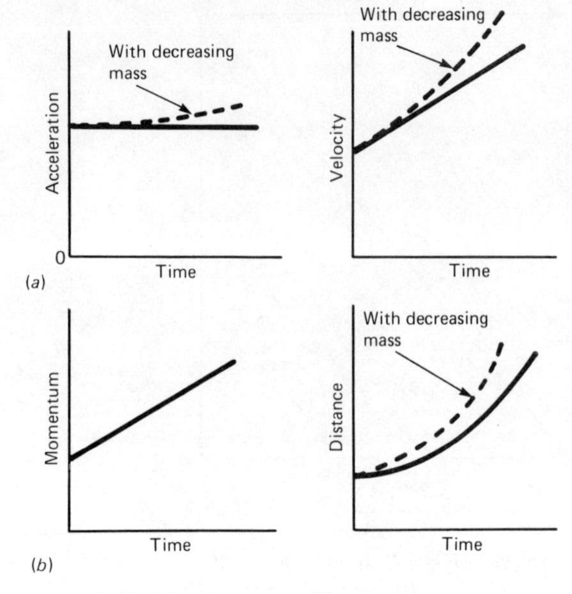

Figure 1.18 (a) a = constant; (b) $v = u + at$; (c) $mv = m(u + at)$; (d) $s = ut + \frac{1}{2}at^2$

1.98 Ans See **1.81**. Velocity of water = $0.044/(\pi 0.040^2) = 8.75$ m/s

(a) $\text{Force} = \frac{\pi}{4} d^2 v^2 \rho = 385$ N

(b) Twice the momentum change occurs and so force = 2 × 385 = 770 N

1.99 Ans (a) See the notes in this chapter for the definitions. (i) Air is sucked in to the engine and ejected from the rear at high speed; the momentum gives rise to thrust on the engine. (ii) Hot gas is ejected from the rear as a result of burning the fuel and so conservation of momentum leads to a forward motion of the rocket. (iii) The rotating blades drive air downwards and so themselves experience an upward force as a result of momentum conservation

(b) (i) Weight = 1.6 × 9.8 sin 30° = 7.84 N parallel to the plane. At right angles to plane weight = 1.6 × 9.8 cos × 30° = 13.6 N. Limiting frictional force has not been reached, thus the frictional force = 7.84 N as the object is at rest.
(ii) Impulse = change in momentum = 1.6 × 0.3 = 0.48 N s. Frictional force F is given by $7.84 - F = ma$, hence $F = 7.52$ N

1.100 Ans (a) You could use stroboscopic photography or repeat the experiment a number of times and allow the ball to hit a screen placed different distances from the ramp. The screen could be marked by the impact if a piece of carbon paper covering white paper is used
(b) Equal horizontal displacements mean equal intervals of time. Thus the vertical distance fallen in 1 unit, 2 units, 3 units, etc. of time from rest can be computed. They should fit the equation $s = \frac{1}{2}gt^2$
(c) (i) A different horizontal velocity would be produced and thus the horizontal distances covered in the same time intervals would be changed. There would be no effect on the vertical motion.
(ii) No change.
(d) $h = \frac{1}{2}gt^2$ and $d = vt$, hence $v = dg^{1/2}/(2h)^{1/2}$. Horizontal velocity at impact = v. Vertical velocity at impact = $(2gh)^{1/2}$ (from $v^2 = u^2 + 2as$). Thus the resultant velocity $V^2 = (d^2g/2h) + 2gh$. The angle to the vertical is given by $\tan\theta = (2gh)^{1/2}/(d^2g/2h)^{1/2} = 2h/d$

1.101 Ans (a) See *Figure 1.19*. $T \cos 30° = mv^2/r$, where $r = 8 \cos 30°$. This is the radius of the horizontal circle. Hence $T = 0.400 \times 4\pi^2 \times 8/4^2 = 8.0$ N.
(b) Lift = $mg + T \sin 30° = 8.0$ N

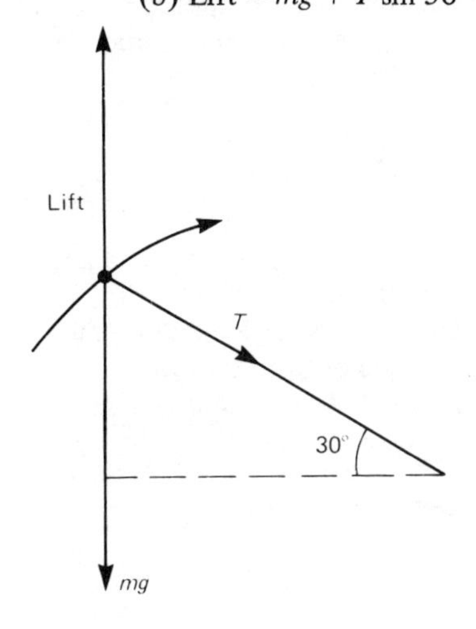

Figure 1.19

2 Materials

Elasticity

Notes

For many materials, when they are pulled by a pair of forces the extension is directly proportional to the applied force. This relationship is called **Hooke's law**. It can be represented by the equation

$$F = kx$$

where F is the applied force, x the extension and k a constant of proportionality called the force constant. This relationship is not generally valid for all forces, but is valid only up to some particular value which is the limit of proportionality (*Figure 2.1*).

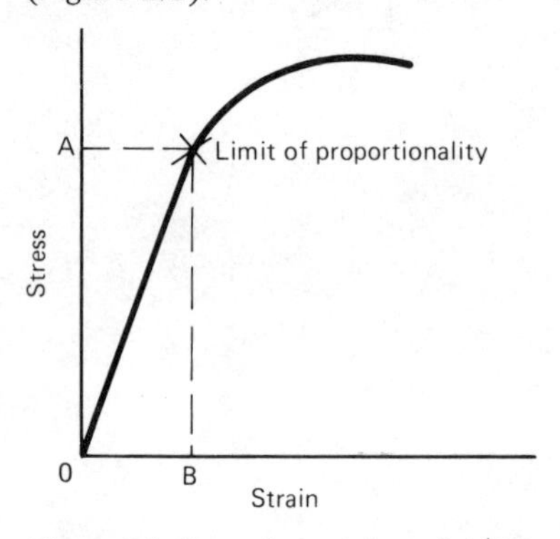

Figure 2.1 Young's modulus = OA/OB

A material is said to be elastic when on removal of an applied pair of forces it returns to its original unstretched length.

Force/extension graphs refer to samples of specific materials with specific dimensions. By plotting stress/strain graphs for given materials the graphs can be used for all sizes of the particular material.

$$\text{Stress} = \frac{\text{force}}{\text{area}}$$

$$\text{Strain} = \frac{\text{extension}}{\text{original length}}$$

For that part of a stress/strain graph where the strain is directly proportional to the stress the ratio of stress to strain, i.e. the slope of the graph, is constant and is called the tensile modulus or Young's modulus E (*Figure 2.1*)

$$\text{Young's modulus} = \frac{\text{stress}}{\text{strain}}$$

The maximum load that can be applied to a sample divided by the original cross-sectional area is called the tensile strength.

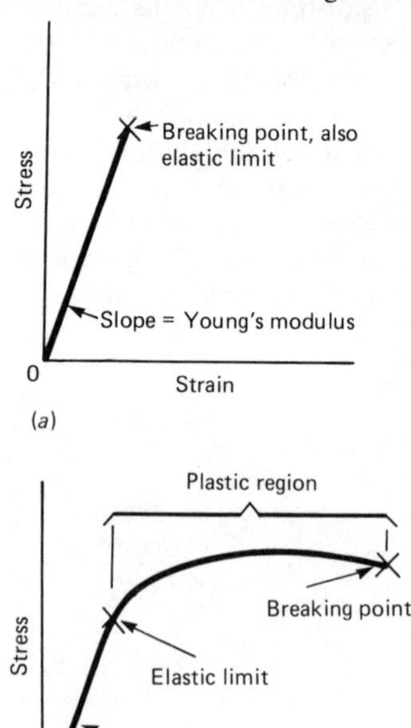

Figure 2.2 (a) Brittle material; (b) ductile material. For stresses at or below the elastic limit the material returns to its original dimensions when the stress is released

If a china cup is broken all the pieces can be stuck back together to give the original shape. For such a material, called a brittle material, the strain is directly proportional to the stress up to the breaking point. The material is elastic up to the breaking point in that no permanent deformation occurs. If a piece of mild steel, say the door panel of a car, is hit it is more likely to dent than break. It shows a permanent deformation, i.e. the material does not return to its original dimensions when the force is no longer applied. The material is said to be ductile and to have behaved plastically (*Figure 2.2*).

Units: stress – N/m^2 or Pa (Pascal, 1 Pa = 1 N/m^2), strain – no units (a ratio), Young's modulus – N/m^2 or Pa.

Examples

2.1 A force of 500 N is found to stretch a strip of material by 1.2 mm. What will be the extension for a force of 250 N if Hooke's law is obeyed?

Ans As the extension is directly proportional to the force, halving the force halves the extension. Hence the extension for a force of 250 N is 0.6 mm.

2.2 What is the stress produced in a wire of cross-sectional area 2.0×10^{-6} m^2 when acted on by a force of 12 N?

Ans

$$\text{Stress} = \frac{\text{force}}{\text{area}} = \frac{12}{2.0 \times 10^{-6}} = 6.0 \times 10^6 \text{ N/m}^2$$

Note: 1 N/m^2 = 1 Pa (Pascal); hence the answer could be written as 6.0×10^6 Pa or 6.0 MPa.

2.3 A wire of diameter 0.70 mm and length 1.0 m gives an extension of 0.80 mm when pulled by a force of 24 N. What is the Young's modulus for the wire if the limit of proportionality has not been exceeded?

Ans

$$\text{Stress} = \frac{\text{force}}{\text{area}} = \frac{24}{\pi[(0.70 \times 10^{-3})/2]^2}$$

$$\text{Strain} = \frac{\text{extension}}{\text{length}} = (0.80 \times 10^{-3})/1.0$$

$$\text{Young's modulus} = \frac{\text{stress}}{\text{strain}} = 7.8 \times 10^{10} \text{ N/m}^2$$

Note: 10^9 N/m^2 = 10^9 Pa = 1 GN/m^2 = 1 GPa (the G stands for the prefix giga); hence the answer could be expressed as 7.8×10^{10} Pa or 78 GN/m^2 or 78 GPa.

Further problems

Take g to be 9.8 m/s^2.

2.4 What is the strain that will be produced in a bar with a Young's modulus of 1.2×10^{11} N/m^2 when it is subjected to a tensile stress of 1.8×10^6 N/m^2?

2.5 The following results were obtained for the load and extension when a tensile testing machine was used to stretch a test piece having a diameter of 7.98 mm. The extensions were measured for an initial gauge length of 50 mm. Plot the force/extension graph and determine a value for Young's modulus.

Load/kN	0	20	40	60	80
Extension/mm	0	0.024	0.048	0.073	0.096

2.6 Your leg bones have a Young's modulus of about 1×10^{10} N/m^2. If the cross-sectional area of the leg bone is 9.50 cm^2, what will be the percentage by which the bone is compressed if it experiences a force of 800 N during walking?

2.7 What is the Young's modulus for brass if a brass wire of length 1.2 m and cross-sectional area 0.10 mm^2 is stretched by 0.45 mm when a load of 0.42 kg is attached to the lower end of the vertically suspended wire? Assume that the wire obeys Hooke's law.

2.8 What is the extension that will occur for a vertical steel wire 3.50 m long and of diameter 0.80 mm when a load of 5.0 kg is attached to its lower end (Young's modulus for steel is 2.1×10^{11} N/m^2)?

2.9 A steel wire 1.2 m long and a copper wire 0.80 m long, each of diameter 2.0 mm, are joined end to end and stretched with a tension of 500 N. What is the elongation of each wire (Young's modulus for steel is 2.1×10^{11} N/m^2 and for copper it is 1.3×10^{11} N/m^2)?

2.10 (a) Explain the difference between a ductile and a brittle material. (b) Which of the following materials would you consider to be brittle and which ductile? Give reasons for your answers.

(i) A plastic teaspoon at room temperature
(ii) A plastic teaspoon at 80°C.
(iii) A metal coat-hanger.
(iv) A glass beaker.

The effect of temperature on solid materials

Notes

In general, the amount by which a piece of material expands is directly proportional to the temperature change θ which causes the expansion and to the initial length of the sample.

Change in length (for temperature change θ) = coefficient of linear expansion × initial length × change in temperature, i.e.

$$L_\theta - L_0 = \alpha L_0 \theta$$

where L_θ is the length after a temperature change θ, L_0 is the initial length and α is the coefficient of linear expansion.

When thermal expansion or contraction is prevented, perhaps by the material being held between fixed-end supports, considerable forces can be produced. For an initial length of L_0, a rise in temperature of θ would mean an expansion of $\alpha L_0 \theta$. If this is prevented it is as if a rod of length $(L_0 + \alpha L_0 \theta)$ has been compressed to L_0.

$$\text{Compressive strain} = \frac{\alpha L_0 \theta}{L_0 + \alpha L_0 \theta}$$

This approximates to

$$\text{Strain} = \alpha\theta$$

If the limit of proportionality is not exceeded,

$$\text{Compressive stress} = E\alpha\theta$$

where E is the Young's modulus. Hence, the force exerted is given by

$$\text{Force} = AE\alpha\theta$$

where A is the cross-sectional area of the material.

An increase in the temperature of a solid produces not only a change in length, but also changes in area A and in volume V.

Change in area = coefficient of area expansion × original area × change in temperature

Change in volume = coefficient of volume expansion × original volume × change in temperature

The coefficient of area expansion is twice the coefficient of linear expansion and the coefficient of volume expansion is three times the coefficient of linear expansion.

A hollow body, e.g. a bottle, expands by the same amount as a solid body having the same external dimensions.

Units: coefficients of expansion – °C^{-1} or K^{-1}, temperature – °C or K.

Examples

2.11 A steel rod has a length of 20.000 0 cm at 20.0 °C. What will be its length at 40.0 °C if the coefficient of linear expansion is 11×10^{-6} °C^{-1}?

Ans

$$\begin{aligned} L_\theta &= L_0(1 + \alpha\theta) \\ &= 20.000\,0\,(1 + 11 \times 10^{-6} \times 20.0) \\ &= 20.004\,4 \text{ cm} \end{aligned}$$

2.12 An aluminium plate contains a circular hole of diameter d 3.000 cm at 18.0 °C. What will be its diameter when the temperature is 78.0 °C (coefficient of linear expansion of aluminium is 23×10^{-6} °C^{-1})?

Ans

$$A_\theta = A_0(1 + 2\alpha\theta)$$

$$\pi \frac{d_\theta^2}{4} = \pi \frac{d_0^2}{4}(1 + 2\alpha\theta)$$

$d_\theta^2 = 3.000^2(1 + 2 \times 23 \times 10^{-6} \times 60.0)$

Diameter, $d_\theta = 3.004$ cm

2.13 What is the change in volume of an aluminium sphere of radius 4.000 cm when it is heated through 50 °C (coefficient of linear expansion of aluminium is 23×10^{-6} °C^{-1})?

Ans $V_\theta - V_0 = 3\alpha V_0 \theta$

$= 3 \times 23 \times 10^{-6} \times \frac{4}{3}\pi \times 4.000^3 \times 50$

Change in volume = 0.925 cm^3

2.14 A bar of mild steel is constrained between two rigid supports so that it cannot expand. What is the stress developed in the bar as a result of a temperature rise of 10 °C if the coefficient of linear expansion is 11×10^{-6} °C^{-1} and the Young's modulus is 2.1×10^{11} N/m^2?

Ans Stress $= E\alpha\theta$

$= 2.1 \times 10^{11} \times 11 \times 10^{-6} \times 10$

$= 2.3 \times 10^7$ N/m^2

The stress is compressive.

Further problems

Take the coefficients of linear expansion of brass, steel and aluminium to be 19×10^{-6} °C^{-1}, 11×10^{-6} °C^{-1} and 23×10^{-6} °C^{-1}, respectively, and the Young's moduli for brass and steel to be 1.2×10^{11} N/m^2 and 2.2×10^{11} N/m^2, respectively.

2.15 By how much will a bar of brass of length 400 mm expand when the temperature of the bar increases by 40 °C?

2.16 What forces are needed to stop a brass rod 20 mm in diameter from expanding when the temperatures increases by 30 °C?

2.17 A steel bar is constrained between two rigid supports. What will be the stress developed in the bar as a result of a temperature drop of 12 °C?

2.18 A steel tape is used to measure a distance. If the measurement is made when the temperature is 30 °C the reading obtained is 202.150 m. The tape had, however, been calibrated at 20 °C. What, therefore, was the true distance?

2.19 Derive an equation showing how the density of a block of material changes with temperature.

2.20 A circular hole in a steel plate has a diameter of 350 mm at 10 °C. By how much will the diameter d change when the temperature increases to 30 °C?

2.21 An aluminium sphere has a diameter of 400 mm at 20 °C. What will be the percentage increase in volume when the sphere is heated to 100 °C?

2.22 At 20 °C a steel sphere has a diameter of 30.00 mm. At the same temperature a brass ring has an internal diameter of 29.90 mm. At what temperature will the ball just pass through the ring?

Static equilibrium

Notes

An object is said to be in static equilibrium when there is no movement or tendency to movement in any direction. The conditions for this to occur are:

(i) There must be no resultant force on the object in any direction.
(ii) The sum of the anticlockwise moments about any axis must equal the sum of the clockwise moments about the same axis.

The first condition can be written in terms of the components of the forces in two mutually perpendicular directions as:

(i) The total of the upwards components of the forces must equal the sum of the downwards components.
(ii) The total of the rightwards horizontal components must equal the sum of the leftwards horizontal components.

The moment of a force is defined as the product of the force and the radius of its potential rotation about the axis concerned.

Examples

Take g to be 9.8 m/s^2.

2.23 A stone of mass 2.0 kg rests on the ground. What are the forces acting on the stone?

Ans As the stone is in equilibrium there must be no resultant force acting in any direction on the stone. The gravitational force acting on the stone is mg, i.e.

$$2.0 \times 9.8 = 19.6 \text{ N}$$

This force is vertical and downwards. There must be a vertical and upwards force of 19.6 N in order to give no resultant force. This upwards force arises from the interaction between the stone and the ground. It is often known as the reaction force.

2.24 A see-saw, of negligible mass, is balanced with a child of mass 40 kg at one end, a distance of 1.6 m from the pivot, and another child of mass 35 kg sitting on the opposite side. At what distance from the pivot should this child be sitting if the see-saw is in equilibrium?

Ans Taking moments about the axis through the pivot gives

Moment for the 40 kg child
$= 40 \times 9.8 \times 1.6$

and

Moment for the 35 kg child
$= 35 \times 9.8 \times d$

where d is the distance of the child from the pivot axis. As there is equilibrium the moment of the 40 kg child must equal the moment of the 35 kg child. Hence

$$35 \times 9.8 \times d = 40 \times 9.8 \times 1.6$$
$$d = 1.8 \text{ m}$$

Further problems

Take g to be 9.8 m/s^2

2.25 A metre rule is balanced on a knife edge at the 500 mm mark. When an object of mass 20 g is placed over the 150 mm mark, where should a 40 g object be placed so that the metre rule is in equilibrium?

2.26 What is the moment (the term torque is often used) about the elbow joint when an object of mass 2 kg is held in the hand of a horizontal outstretched arm, if its distance from the elbow joint is 580 mm?

2.27 A plank of wood is supported in a horizontal position by a vertical rope at each end. The plank of wood is 3.0 m long. If a student of mass 60 kg stands on the plank a distance of 1.0 m from one end, what are the tensions in the supporting ropes due to the presence of the student?

Centres of mass and gravity

Notes

The mass of a body is the sum of all the masses of the small particles that together constitute the body. The gravitational force on a body can be considered in terms of the forces on each of the masses of the small constituent particles. There is, however, one point for a body at which a single gravitational force can be considered to act, regardless of the orientation of the body, which is equivalent to the effects of all the other gravitational forces. The point through which the lines of action of this force passes, in the different orientations, is called the centre of gravity of the body. The effect is as though all the mass of the object was concentrated at that point. Hence, the term centre of mass is sometimes used for the point.

For symmetrical homogeneous objects the centre of mass and the centre of gravity are the geometrical centre. Thus, for a cube the centre of mass is the centre of the cube. All the mass of the cube can be considered to be acting at its centre.

Examples

Take g to be 9.8 m/s^2.

2.28 A uniform rod of length 300 mm has a mass of 1.2 kg. From one end of the rod a mass of 2.0 kg is suspended and from the other end a mass of 3.0 kg. About what axis can the rod be pivoted and be in equilibrium?

Ans See *Figure 2.3.* The mass of the rod is considered to be acting at its centre, i.e. 2 150 mm from the ends. If, at equilibrium,

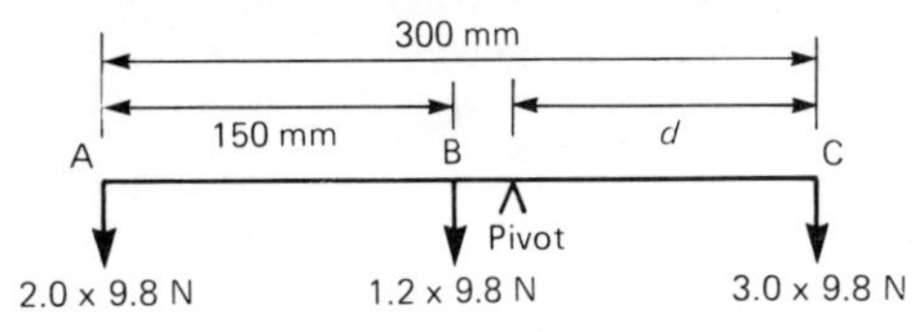

Figure 2.3

the pivot is a distance d from the end at which the 3.0 kg mass is situated then taking moments about the axis through the pivot gives

$$\text{Clockwise moment} = 3.0 \times 9.8 \times d$$
$$\text{Anticlockwise moment} = 2.0 \times 9.8 \times (300-d) + 1.2 \times 9.8 \times (150-d)$$

At equilibrium the clockwise moment equals the anticlockwise moment, so

$$3.0 \times 9.8 \times d = 2.0 \times 9.8 \times (300-d) + 1.2 \times 9.8 \times (150-d)$$
$$d = 126 \text{ mm}$$

2.29 *Figure 2.4* shows an object being held in a hand. If the gravitational force acting on

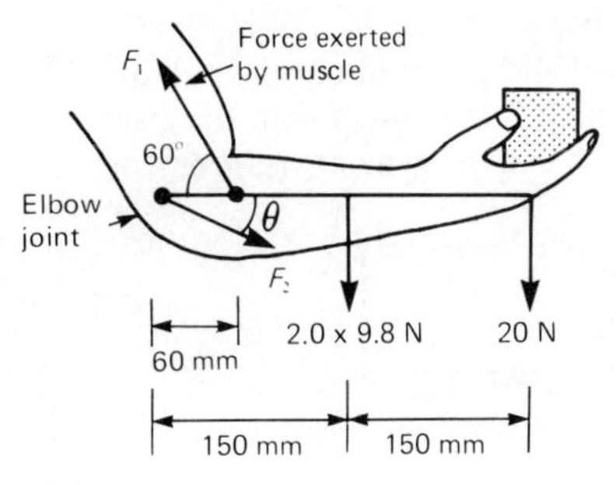

Figure 2.4

the object is 20 N and the forearm is at rest, what is the size of the force F_1 exerted by the muscle? What is the force F_2 exerted at the elbow joint? The forearm and hand have a mass of 2.0 kg with a centre of mass 150 mm from the elbow joint.

Ans Taking moments about the axis through the elbow joint gives

$$\text{Clockwise moment} = 2.0 \times 9.8 \times 150 + 20 \times 300$$
$$\text{Anticlockwise moment} = F_1 \times 60 \sin 60°$$

The term 60 sin 60° is the perpendicular distance between the line of action of the F_1 force and the elbow joint; it is the radius of its potential rotation. As the system is in equilibrium

$$F_1 \times 60 \sin 60° = 2.0 \times 9.8 \times 150 + 20 \times 300$$
$$F_1 = 172 \text{ N}$$

For equilibrium there must be no resultant force in any direction. Thus, for the horizontal direction we must have

$$F_2 \cos \theta = F_1 \cos 60° = 172 \cos 60°$$

In the vertical direction

$$F_2 \sin \theta + 2.0 \times 9.8 + 20 = F_1 \sin 60° = 172 \sin 60°$$

Hence F_2 is 139 N and θ is 52°.

2.30 A uniform ladder of length 3.6 m and mass 18 kg rests with its upper end against a smooth vertical wall and its lower end on horizontal, rough ground. What must be the least coefficient of friction between the ground and the base of the ladder if the ladder is not to slip when it is inclined at 30° to the vertical?

Ans *Figure 2.5* shows the basic arrangement. As the ladder is in equilibrium there must

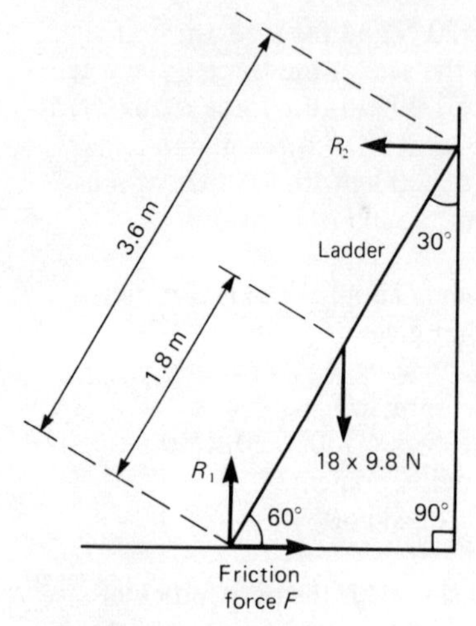

Figure 2.5

be no resultant force in any direction. Thus, for the vertical direction we must have

$$R_1 = 18 \times 9.8 \text{ N}$$

For the horizontal direction

$$R_2 = F$$

But

$$F = \mu R_1$$

Hence

$$R_2 = \mu \times 18 \times 9.8 \text{ N}$$

Taking moments about the axis of contact between the ladder and the ground gives

Clockwise moment = $18 \times 9.8 \times 1.8 \cos 60°$
Anticlockwise moment = $R_2 \times 3.6 \cos 30°$

Hence

$$R_2 \times 3.6 \cos 30° = 18 \times 9.8 \times 1.8 \cos 60°$$
$$\mu \times 18 \times 9.8 \times 3.6 \cos 30° = 18 \times 9.8 \times 1.8 \cos 60°$$

Thus

$$\mu = 0.29$$

Further problems

Take g to be 9.8 m/s^2.

2.31 A horizontal trap door is 1.0 m square and has a mass of 10 kg. The centre of gravity of the door is at its centre. What forces must the hinges and the catch sustain when the trap door is closed? The hinges and the catch are at the opposite edges of the door.

2.32 A sign of mass 4.0 kg is hung from the end of a horizontal, uniform bar of mass 1.2 kg. *Figure 2.6* shows the general arrangement and how the bar is held in the horizontal position. What are the forces at the hinge and the tension in the wire?

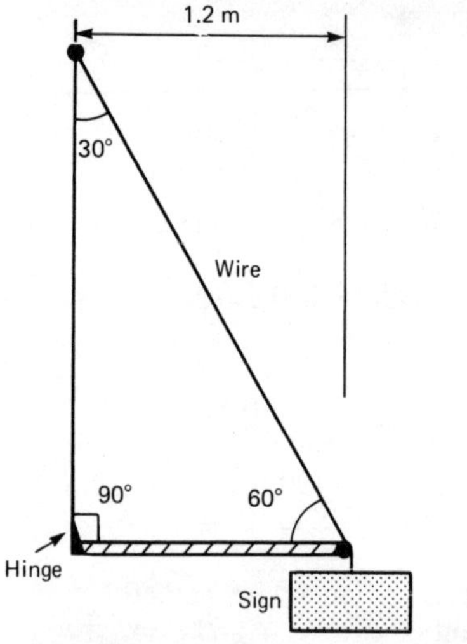

Figure 2.6

2.33 A woman of mass 45 kg walks across a simple plank bridge over a stream. The bridge has a length of 2.4 m and a mass of 16 kg. What are the forces exerted at each end of the bridge by its supports when the woman is (a) 0.5 m, (b) 1.0 m from one end?

2.34 When you use your foot in walking the heel is raised and pivots about the ball of the foot which is in contact with the ground. *Figure 2.7* shows the general arrangement

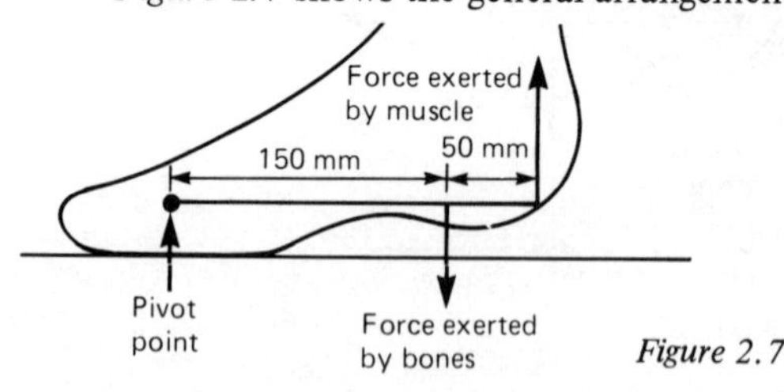

Figure 2.7

and typical dimensions. What are the forces exerted by the muscle and by the lower leg bones if half the weight of the person is taken on the ball of the foot, the person having a mass of 70 kg?

2.35 What is the force that has to be exerted by the biceps muscle when an object of mass 5.0 kg is held horizontally in the outstretched hand? *Figure 2.4* gives the relevant dimensions and mass for the hand–elbow system.

2.36 A man of mass 75 kg is, during push-up exercises, at some instant of time stationary with only his toes and his hands in contact with the ground (*Figure 2.8*). What are the

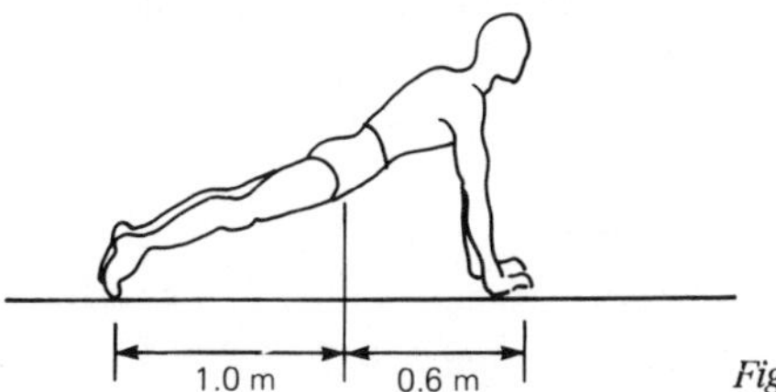

Figure 2.8

forces exerted by each hand and each foot on the ground if his centre of mass is 1.0 m horizontally from his toes and 0.60 m from his hands?

2.37 A uniform ladder 5.0 m long and of mass 20 kg rests with its upper end against a smooth, vertical wall and its lower end on horizontal, rough ground. If the lower end of the ladder is 3.0 m from the base of the vertical wall and the coefficient of static friction between the foot of the ladder and the ground is 0.40 what is the frictional force between the ground and the foot of the ladder when (a) there is nobody on the ladder and (b) when a person of mass 70 kg is half way up the ladder? In either case, would the ladder slip?

2.38 A plank of wood of length 6.0 m and mass 20 kg is stood with its upper end against a smooth, vertical wall and its lower end on horizontal, rough ground. What is the maximum angle to the vertical that the plank can be inclined if it is not to slip (coefficient of static friction between the base of the plank and the ground is 0.50)?

Fluid pressure

Notes

The force applied at right angles to an area divided by that area is called the pressure.

$$\text{Pressure} = \frac{\text{force}}{\text{area}}$$

Fluids can exert pressures. Liquids and gases are fluids. Thus, if there is a height h of a fluid above a horizontal section and the fluid at this section has a cross-sectional area A then the volume of fluid above the section is Ah. If the fluid has a density ρ then this volume has a mass $Ah\rho$. This mass exerts a gravitational force of $Ah\rho g$ on the horizontal section. As this has an area A the pressure is $Ah\rho g/A$. Therefore,

$$\text{Pressure} = h\rho g$$

The pressure is independent of the cross-sectional area of the fluid; indeed the cross-sectional area can vary without changing the pressure.

The Earth's atmosphere exerts a pressure at the surface of the Earth of about 10^5 N/m^2. The unit pascal (Pa) is used for pressure, and 1 Pa = 1 N/m^2

The pressure at the base of a column of liquid of height h is given by

$$\text{Pressure} = h\rho g + A$$

where ρ is the density of the liquid and A the atmospheric pressure acting on the upper surface of the liquid.

The following are some of the basic principles that apply to fluids at rest.

(i) The pressure is the same at all points in the fluid at the same horizontal level.
(ii) The pressure on a surface in the fluid is always at right angles to the surface.
(iii) An external pressure applied to an enclosed fluid at rest is transmitted undiminished to every portion of the fluid and to the walls of the container. This is known as Pascal's principle.

An object immersed in a fluid is acted on by pressures in that fluid. The upper surface of the object is, however, at a different depth to the lower surface. There is thus a pressure difference between the upper and lower surfaces due to this difference in depth, the pressure at the lower surface being greater than that at the upper surface. The result of this is that there is an upthrust acting on the object. The object is said to show buoyancy. The upthrust is equal to the weight of fluid displaced by the object, which is known as Archimedes' principle.

The hydrometer is an instrument used for the measurement of the density of a liquid. When the hydrometer (*Figure 2.9*) is placed in a liquid it

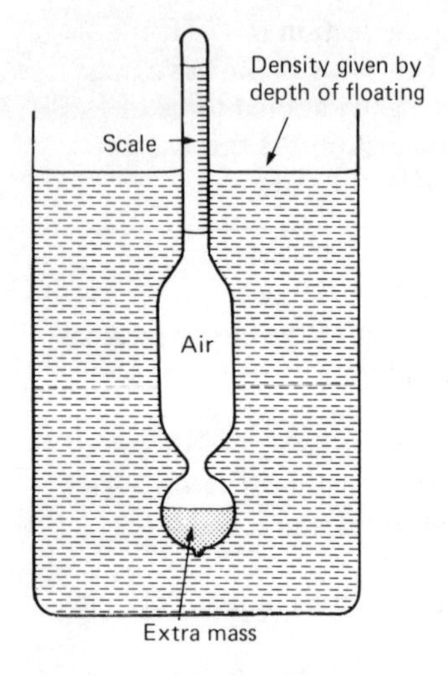

Figure 2.9 A hydrometer

sinks to a depth which depends on the density of the liquid concerned and the weight of the hydrometer concerned. The stem of the hydrometer has a scale marked in density, or relative density.

When the hydrometer is floating there is no resultant force acting on it, the upthrust force being opposite and equal to the weight of the hydrometer. The upthrust is equal to the weight of fluid displaced.

Units: pressure – N/m^2 or Pa (pascal, 1 Pa = 1 N/m^2).

Examples

Take g to be 9.8 m/s^2 and the densities of mercury and water to be 13.5×10^3 kg/m^3 and 1000 kg/m^3, respectively.

2.39 What is the pressure at the base of a column of mercury 760 mm high due solely to the mercury?

Ans

$$\text{Pressure} = h\rho g = 0.760 \times 13.5 \times 10^3 \times 9.81 = 1.01 \times 10^5 \text{ N/m}^2 = 1.01 \times 10^5 \text{ Pa}$$

2.40 One limb of a U-tube manometer is connected to a gas supply, the other limb being open to the atmosphere. The effect of connecting the gas supply is to depress the level of the water in that limb to 160 mm below the level in the limb open to the atmosphere. By how much does the gas pressure differ from the atmospheric pressure and is the gas pressure greater or less than the atmospheric pressure?

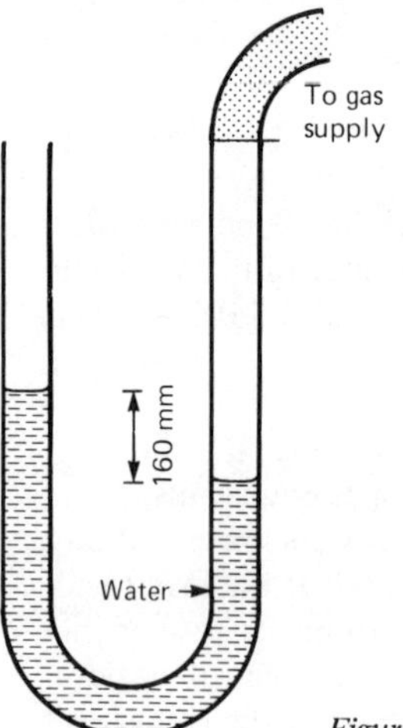

Figure 2.10 A water manometer

Ans *Figure 2.10* shows the form of the U-tube manometer. The gas pressure is greater than that of the atmospheric pressure.

$$\text{Difference in pressure} = h\rho g = 0.160 \times 1000 \times 9.81 = 1.57 \times 10^3 \text{ N/m}^2 \text{ or Pa}$$

2.41 *Figure 2.11* shows a simple barometer, an instrument for the measurement of the atmospheric pressure. (a) What is the pressure at the base of the 760 mm high column of mercury when there is a vacuum above

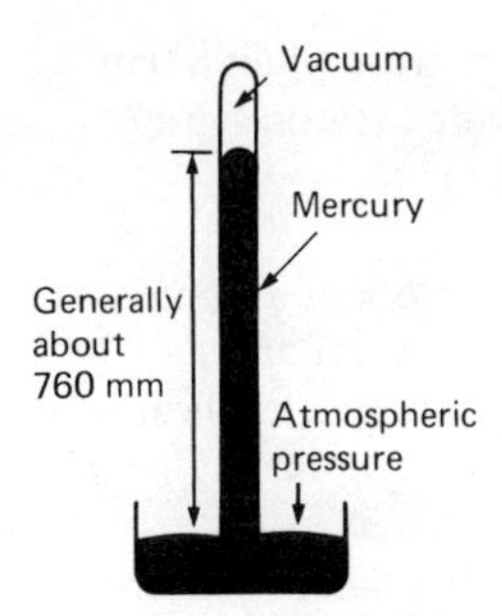

Figure 2.11 A simple barometer

the upper surface of the mercury in the tube? (b) In setting up the barometer air gets into the space above the upper mercury level in the tube and the mercury level falls to 750 mm. What is the pressure of the air above the upper mercury surface in the tube?

Ans (a) As there is a vacuum above the mercury surface the pressure at the base of the mercury column is that due solely to the mercury. Hence

$$\text{Pressure} = h\rho g$$
$$= 0.760 \times 13.5 \times 10^3 \times 9.81$$
$$= 1.01 \times 10^5 \text{ N/m}^2 \text{ or Pa}$$

(b) The atmospheric pressure is that indicated by the barometer when there is no air above the mercury, i.e. 1.01×10^5 N/m^2. This supports a column of mercury 750 mm high with the air pressure acting on the upper surface. Hence, the air pressure must be equal to the atmospheric pressure minus the pressure due to a column of mercury of height 750 mm. The air pressure is thus equivalent to that given by a column of mercury of height 760 – 750 mm.

$$\text{Air pressure} = 0.010 \times 13.5 \times 10^3 \times 9.81$$
$$= 1.32 \times 10^3 \text{ N/m}^2 \text{ or Pa}$$

2.42 A U-tube contains mercury of density 13.5×10^3 kg/m^3. When water is poured into one of the U-tube limbs it is found that there is a column of water of length 100 mm. If the water has a density of 1000 kg/m^3, how much has the level of the mercury in the water-free limb risen above its initial level?

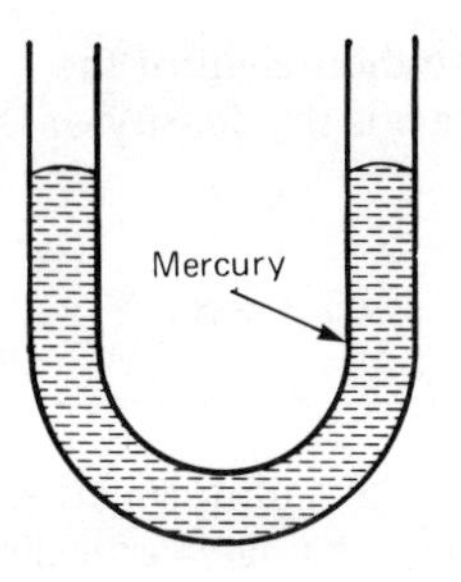

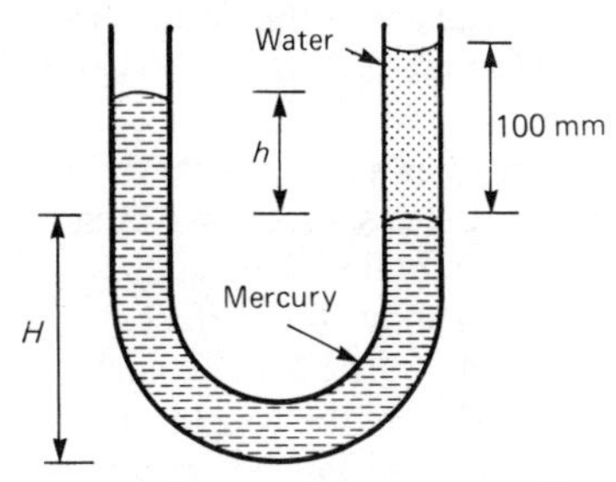

Figure 2.12

Ans *Figure 2.12* shows the initial and final situations. At equilibrium the pressures at each base of the two limbs of the U-tube must be the same. Thus

$$H\rho g + h\rho g = H\rho g + 0.100 \times 1000 \times 9.81$$

where ρ is the density of the mercury. Hence

$$h = \frac{0.100 \times 1000 \times 9.81}{13.5 \times 10^3 \times 9.81}$$
$$= 7.41 \times 10^{-3} \text{ m}$$
$$= 7.41 \text{ mm}$$

Initially, the height of the mercury in the U-tube limb was $H + ½h$. The mercury level in the water-free limb has thus moved by $h/2$, i.e. 3.7 mm.

2.43 A block of wood floats in water with two-thirds of its volume immersed. If the density of water is 1000 kg/m^3, what is the density of the wood?

Ans If the volume of the block of wood is V then the volume immersed is $\frac{2}{3}V$. Hence the volume of water displaced is $\frac{2}{3}V$. The weight of water displaced is thus $\frac{2}{3}V \times 1000 \times 9.8$ N.

This must be equal to the weight of the block, i.e. $V\rho g$ where ρ is the density of the wood. Hence,

$$V\rho \times 9.8 = \tfrac{2}{3}V \times 1000 \times 9.8$$
$$\rho = 670 \text{ kg/m}^3$$

2.44 A weighted test tube with a uniform cross-sectional area of 100 mm^2 floats vertically in water with half of its 100 mm length above the water surface. What length of the test tube will protrude above the liquid surface if it is floated in oil of density 800 kg/m^3?

Ans The test tube is acting as a crude form of hydrometer. If A is the cross-sectional area of the test tube, V its total volume and L the length protruding above the surface, then when it floats $(V - AL)$ is the volume of liquid displaced. Hence, the weight of liquid displaced is $(V - AL)\rho g$, where ρ is the density of the liquid. This upthrust of $(V - AL)\rho g$ must be equal to mg, where m is the total mass of the weighted test tube.

$$(V - AL)\rho g = mg$$

When floating in liquid of density ρ' the length L' is above the surface.

$$(V - AL')\rho' g = mg$$

Hence

$$(V - AL)\rho g = (V - AL')\rho' g$$

Since

$$V = 100 \times 100 \text{ mm}^3$$
$$(100 \times 100 - 100 \times 50)1000 = (100 \times 100 - 100L')800$$

Hence

$$L' = 37.5 \text{ mm}$$

Further problems

Take g to 9.8 m/s^2 and the density of water to be 1000 kg/m^3.

2.45 What is the pressure on a diver when 8.0 m below the surface of a lake (atmospheric pressure is 1.0×10^5 N/m^2)?

2.46 What is the difference in pressure between the blood in a giraffe at its heart and its head if the head is 2.50 m above the heart (density of the blood is 1.05×10^3 kg/m^3)?

2.47 A chisel when struck with a hammer exerts a force of 100 N on a brick. The chisel edge in contact with the brick has an area of 0.75 cm^2. What is the pressure acting on the brick due to the chisel?

2.48 The blood pressure in a vein is 1.6 kN/m^2. What force must be applied to the plunger of a hypodermic needle to inject a fluid into the blood stream if the plunger of the hypodermic has a cross-sectional area of 250 mm^2 and the needle a cross-sectional area of 8.0×10^{-5} mm^2?

2.49 What is the force on the base of a tank 2.0 m long and 1.2 m wide when it is filled with water to a depth of 0.90 m?

2.50 A block of material is found to weigh 56 g in air and 33 g when totally immersed in water. What is the relative density of the material?

2.51 What is the net force acting on a spherical balloon of diameter 10 m when filled with hydrogen of density 0.090 kg/m^3 (air has a density of 1.2 kg/m^3 and the balloon fabric has negligible mass)?

2.52 Why do hot air balloons rise?

2.53 A slab of ice of uniform thickness 250 mm floats in water of density 1000 kg/m^3. The ice has a density of 917 kg/m^3. What should be the minimum area of ice which will just support a man of mass 75 kg without him getting his feet wet?

2.54 A tin can of diameter 70 mm and length 140 mm contains some sand. When placed in water the can floats in a vertical position with half of its height above the water surface. If the density of the water is

1000 kg/m^3, what is the weight of the can and contents?

Expansion of liquids

Notes

In general liquids expand when heated. The change in volume is proportional to the change in temperature and the initial volume.

Change in volume = coefficient of real expansion × change in temperature × initial volume

$$V_\theta - V_0 = \gamma_{real} \times \theta \times V_0$$

where V_θ is the volume after a temperature change of θ, the initial volume being V_0. The term 'real' is used with the coefficient because it refers to the liquid alone.

Liquids are almost invariably held in containers and, thus, when the temperature changes both the liquid and the solid container change in volume. If no correction is made for the container expansion the term 'coefficient of apparent-expansion' is used.

$$\gamma_{real} = \gamma_{app} + \gamma_{container}$$

$\gamma_{container}$ is the cubical coefficient of expansion of the container.

When the temperature of a liquid increases it expands and so the volume occupied by a given mass of liquid increases. This means that the density decreases.

$$V_\theta - V_0 = \gamma_{real} \theta V_0$$

Dividing through by V_0 and rearranging gives

$$V_\theta / V_0 = \gamma_{real} \theta + 1$$

But the mass of the liquid = $V_0 \rho_0 = V_\theta \rho_\theta$, where ρ_0 is the initial density and ρ_θ the density after the temperature change θ. Thus

$$\rho_0 / \rho_\theta = \gamma_{real} \theta + 1$$

$$\rho_0 = \rho_\theta \, (1 + \gamma_{real} \theta)$$

Units: coefficients of expansion – $°C^{-1}$ or K^{-1}, temperature – °C or K.

Examples

2.55 A vertical glass tube contains a column of water. At 12 °C the height of the water column is 400 mm. What will be the height of the water column at 32 °C (coefficient of linear expansion of glass is $9 \times 10^{-6}\ °C^{-1}$ and coefficient of real expansion of water is $210 \times 10^{-6}\ °C^{-1}$)?

Ans $\gamma_{real} = \gamma_{app} + \gamma_{container}$

$\gamma_{container}$ is three times the linear coefficient, i.e. $27 \times 10^{-6}\ °C^{-1}$. Thus, the coefficient of apparent expansion is $183 \times 10^{-6}\ °C^{-1}$. The volume of water at 12 °C is $0.400 \times A$, where A is the cross-sectional area of the tube at 12 °C. Hence

$$0.400 \times A - L \times A = 183 \times 10^{-6} \times 0.400 \times A \times (32-12)$$

L is the length of the water column at 32 °C. The cross-sectional area of the tube has been assumed not to change with temperature; that is why the coefficient of apparent expansion is being used. The result is

$$L = 0.398 \text{ m}$$

2.56 The density of mercury is $13.595 \times 10^3 \text{ kg/m}^3$ at 0 °C. What will be the density of mercury at 25 °C if the coefficient of real expansion for mercury is $180 \times 10^{-6}\ °C^{-1}$?

Ans $\rho_0 = \rho_\theta(1 + \gamma\theta)$

$$\rho_\theta = \frac{13.595 \times 10^3}{(1 + 180 \times 10^{-6} \times 25)} = 13.534 \times 10^3 \text{ kg/m}^3$$

Further problems

Take the coefficient of linear expansion of glass to be $3 \times 10^{-6}\ °C^{-1}$ and the coefficients of real expansion of mercury and alcohol to be $180 \times 10^{-6}\ °C^{-1}$ and $1080 \times 10^{-6}\ °C^{-1}$, respectively.

2.57 A vertical glass tube of length 1.000 m and internal diameter 10 mm contains mercury to a height of 990 mm at 0 °C. To what temperature should the tube be heated if the mercury is to begin to overflow (neglect any consideration of surface tension effects)?

2.58 What would have been the answer to the previous question if, instead of mercury, the liquid used had been ethyl alcohol with a coefficient of real expansion of 1080×10^{-6} °C^{-1}?

2.59 What is the percentage change in volume of the mercury in a mercury-in-glass thermometer when the temperature changes from 0 °C to 100 °C?

2.60 What is the percentage change in the density of ethyl alcohol when the temperature changes from 0 °C to 20 °C?

The ideal gas laws

Notes

Boyle's law:
For a fixed mass of gas at a constant temperature the product of the pressure and the volume is a constant, i.e.

$$pV = \text{a constant}$$

This can be expressed as the volume being inversely proportional to the pressure, i.e.

$$V \propto 1/p$$

Charles' law:
For a fixed mass of gas at a constant pressure the volume divided by the temperature, when expressed on the Kelvin scale, is a constant, i.e.

$$V/T = \text{a constant}$$

This can be expressed as the volume being directly proportional to the temperature, when expressed on the Kelvin scale, i.e.

$$V \propto T$$

Pressure law:
For a fixed mass of gas at constant volume the pressure divided by the temperature, when expressed on the Kelvin scale, is a constant, i.e.

$$p/T = \text{a constant}$$

This can be expressed as the pressure being directly proportional to the temperature, when expressed on the Kelvin scale, i.e.

$$p \propto T$$

The above relationships define an ideal gas. The relationships were, at one time, used as a means of defining a temperature scale, known as the ideal gas temperature scale or Absolute temperature scale. The Kelvin scale of temperature is, however, now used. Although it is defined in a different way the temperatures on the Absolute and Kelvin scale are essentially identical and are obtained by adding 273 to the temperatures on the Celsius scale (for an accuracy to only three figures).

The above relationships can be combined to give the equation

$$pV/T = \text{a constant}$$

The value of the constant depends on the mass of gas and the nature of the gas concerned. It can be written as

$$pV/T = mR$$

where m is the mass of the gas and R is a constant, known as the specific or characteristic gas constant, which depends on the gas concerned. Another way of writing this relationship is

$$pV/T = mR_0/M$$

where M is the molar mass of the gas concerned and R_0 is the universal gas constant and is the same for all ideal gases.

It follows from the above equation that one mole of an ideal gas at a temperature of 0 °C and a pressure of 1.01×10^5 N/m^2 (normal atmospheric pressure) occupies a volume of 2.24×10^{-2} m^3.

Units: pressure – N/m^2 or Pa, volume – m^3, temperature – K, mass – kg.

Examples

2.61 2.0 m^3 of air are stored at a pressure of 8×10^5 N/m^2. What would be the volume of the air at a pressure of 1×10^5 N/m^2 if the temperature is unchanged?

Ans pV = a constant
$1 \times 10^5 \times V = 8 \times 10^5 \times 2.0$
$V = 16.0$ m^3

2.62 A gas is stored under constant pressure at 0 °C. To what temperature should it be heated if the gas volume is to double?

Ans V/T = a constant
$V/273 = 2V/T$
$T = 546$ K $= 273$ °C

2.63 A gas is stored in a cylinder at a pressure of 6.0×10^5 N/m^2 and at a temperature of 15 °C. What will be the pressure of the gas at 35 °C if the volume of the gas does not change?

Ans p/T = a constant
$p/308 = (6.00 \times 10^5)/288$
$p = 6.4 \times 10^5$ N/m^2

2.64 A barometer tube 100 cm long contains some air above the mercury surface. When the tube is vertical the mercury level is at a height of 70 cm. When the tube is inclined at 30° to the vertical the length of the mercury column, measured along the tube, is 78 cm. What is the atmospheric pressure?

Ans When the tube is vertical the volume occupied by the air, above the mercury surface, is $(100 - 70)A \times 10^{-2}$, where A is the cross-sectional area of the tube. The pressure of the air is $(h - 70) \times 10^{-2}\ \rho g$, where h is the height that the barometer would have stood at if no air had been above the mercury and ρ is the density of the mercury.

When the tube is inclined the volume occupied by the air is $(100 - 78)A \times 10^{-2}$ and the air pressure in the tube is $(h - 78 \cos 30°) \times 10^{-2}\ \rho g$. Applying Boyle's law gives

$$(100 - 70)A \times 10^{-2} \times (h - 70) \times 10^{-2}\ \rho g = (100 - 78)A \times 10^{-2} \times (h - 78 \cos 30°) \times 10^{-2}\ \rho g$$

Hence

$$h = 76.6 \text{ cm}$$

This is the height that the mercury would stand at in a vertical barometer with no air above the mercury surface. This is an atmospheric pressure of $76.6 \times 10^{-2} \times 13.5 \times 10^3 \times 9.81$ N/m^2, where the density of mercury has been taken as 13.5×10^3 kg/m^3. The atmospheric pressure is thus 1.01×10^5 N/m^2.

2.65 One mole of a gas at a temperature of 0 °C and a pressure of 1×10^5 N/m^2 has a volume of 22.4×10^{-3} m^3. Calculate the value of the universal gas constant if the gas is an ideal gas.

Ans $pV/T = mR_0/M$

For one mole $m = M$, thus

$$R_0 = \frac{1 \times 10^5 \times 22.4 \times 10^{-3}}{273}$$

$$= 8.21 \text{ N m mol}^{-1} \text{ K}^{-1}$$

2.66 What is the volume occupied by 2.0 kg of air at a pressure of 2.0×10^5 N/m^2 and a temperature of 20 °C if the characteristic gas constant for air is 287 N m kg^{-1} K^{-1}?

Ans $pV/T = mR$
$(2.0 \times 10^5 \times V)/293 = 2.0 \times 287$
$V = 0.841$ m^3

Further problems

Take g to be 9.8 m/s^2 and the density of mercury to be 13.5×10^3 kg/m^3.

2.67 A gas has a volume of 0.20 m^3 when at a pressure of 1.2×10^5 N/m^2. What pressure is needed to compress the gas into a volume of 0.12 m^3 if the temperature remains constant and the gas can be assumed to be an ideal gas?

2.68 The gas in a cylinder is under a pressure of 5.0×10^5 N/m^2 at a temperature of 18 °C. What will be the pressure if the temperature of the gas rises to 35 °C (expansion of the cylinder can be assumed to be negligible)?

2.69 What is the mass of oxygen in an oxygen tank at 20 °C if the tank has a volume of 0.20 m^3 and the oxygen pressure in the tank is 4.0×10^5 N/m^2 (mass of one mole of oxygen is 16 g and the universal gas constant is 8.31 N m mol^{-1} K^{-1}).

2.70 A capillary tube with a uniform internal cross-section contains a thread of mercury of length 140 mm. When the tube is vertical this thread of mercury encloses an air column of length 220 mm between it and the closed end of the tube, the other, lower end of the tube being open to the atmosphere. When the tube is placed vertically with the open end upwards the length of the air column decreases to 150 mm. What is the atmospheric pressure A?

2.71 An air bubble has a diameter of 30 mm at the bottom of a lake where the pressure is 2.5×10^5 N/m^2 and the temperature 7 °C. What will be the diameter of the air bubble at the surface where the pressure is 1.0×10^5 N/m^2 and the temperature 20 °C?

2.72 A simple barometer consists of a tube with an internal cross-sectional area of 100 mm^2 and a length of 900 mm. The mercury in the tube stands at a height of 755 mm, the temperature being 20 °C. Air is introduced into the space above the mercury and the mercury level drops to 700 mm. What is the mass of the air introduced (the mean mass of one mole of air is 29 g and the universal gas constant is 8.31 N m mol^{-1} K^{-1})?

Real gases

Notes

Figure 2.13 shows the pressure/volume relationships for carbon dioxide. This type of result was first obtained by T. Andrew. The graph lines are known as isothermals in that for each line the temperature is constant. The temperature of 31.4 °C is known as the critical temperature. For temperatures below this value carbon dioxide can be liquefied by the application of a suitable pressure, above this temperature liquefaction is not possible. For pressure/volume conditions below the critical temperature and within the region bounded by the dashed line, the carbon dioxide exists as a mixture of liquid and vapour. For this region the volume changes with no change in pressure. For temperatures below the critical temperature and pressure/volume conditions which give values to the left of the dashed line carbon dioxide exists in the liquid state; for values to the right of the dashed line carbon dioxide exists as a vapour. At temperatures greater than the critical temperature carbon dioxide exists as a gas.

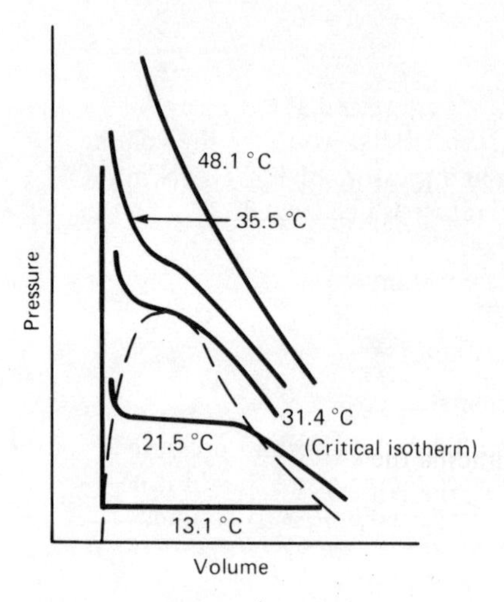

Figure 2.13 Pressure/volume relationships for carbon dioxide

Examples

2.73 The following data are for water.

Critical temperature 647.2 K
Critical pressure 217.7×10^5 N/m^2
Critical volume 45.0×10^{-6} m^3/mole

(a) What is the maximum temperature at which water can be changed from steam into liquid? (b) What pressure is needed if water at the critical temperature is to be changed from steam to liquid? (c) Would you expect water to follow the ideal gas equations when at normal atmospheric pressure and room temperature?

Ans (a) 647.2 K; at temperatures greater than this the liquid cannot be obtained, however great the pressure
(b) 217.7×10^5 N/m^2
(c) No

Further problems

2.74 For ammonia the critical temperature is 132.5 °C, the critical volume being 72.4×10^{-6} m^3 and the critical pressure 111.5×10^5 N/m^2. (a) What is the maximum temperature at which ammonia can be liquefied? (b) What pressure is needed to liquefy ammonia at the critical temperature?

2.75 Use *Figure 2.13* for this question.
(a) Describe how the state of carbon dioxide changes when a sample at room temperature and low pressure is subject to a steadily increasing pressure. (b) How would your answer to (a) differ if the carbon dioxide was at a temperature of 35 °C?

2.76 Use *Figure 2.14* for this question. What is the physical state of water under the following conditions?
(a) Pressure 200×10^5 N/m^2 and temperature 360 °C

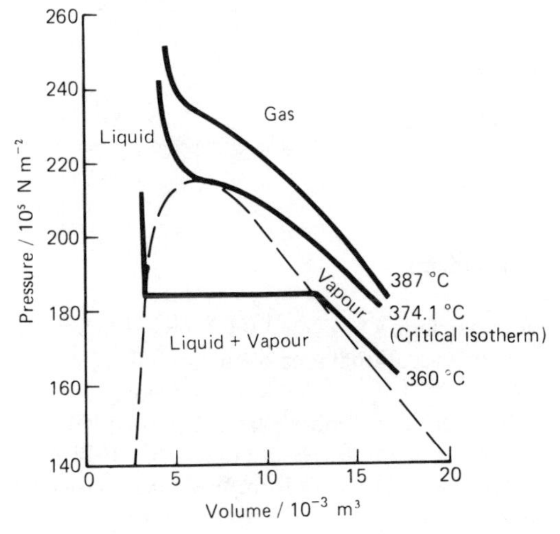

Figure 2.14 Pressure/volume relationships for water

(b) Pressure 240×10^5 N/m^2 and temperature 390 °C
(c) Pressure 200×10^5 N/m^2 and temperature 370 °C

Fluid flow

Notes

The path followed by an element of moving fluid is called a line of flow. With steady flow the velocity at any one point in the fluid is constant, not varying with time. A streamline is defined as a line the tangent to which, at any point, gives the direction of fluid motion at that point. There is thus no velocity component at right angles to a streamline. In steady flow, the streamlines coincide with the lines of flow. For the flow of a fluid down a tube formed by a set of streamlines, the mass of fluid leaving the tube in unit time equals the mass entering in the same time. It is assumed that the fluid is incompressible. Thus, if A_1 is the cross-sectional area at the inlet to the tube and v_1 the velocity at the point of inlet, the volume entering in time t is A_1v_1t. If the fluid has a density ρ then the mass entering is $A_1v_1t\rho$. The mass leaving the tube is $A_2v_2t\rho$, where A_2 is the cross-sectional area of the outlet and v_2 the velocity at that point. Thus

$$A_1v_1t\rho = A_2v_2t\rho$$
$$A_1v_1 = A_2v_2$$

This relationship is known as the equation of continuity.

If the cross-sectional area of the tube through which the fluid is flowing is reduced then the velocity must increase. When the velocity increases there is an acceleration. An acceleration requires a force. This is provided by a pressure difference in the fluid. Thus, when the velocity increases there is a pressure drop.

A fundamental equation in fluid mechanics is Bernouilli's equation. This can be written as

$$p_1 + \rho\frac{v_1^2}{2} + \rho gh_1 = p_2 + \rho\frac{v_2^2}{2} + \rho gh_2$$

where p_1 is the pressure at height h_1 in the fluid which moves with velocity v_1. Similarly, p_2 is the pressure at a height h_2 and velocity v_2. This equation applies when the motion is steady, e.g. no turbulence, and the fluid is incompressible.

Examples

2.77 A garden hose has an internal diameter of 15 mm. If a nozzle fixed to the open end of the hosepipe has an internal diameter of 2.0 mm what will be the velocity with which the water leaves the nozzle when it has a velocity of 1.0 m/s in the hosepipe?

Ans

$$A_1 v_1 = A_2 v_2$$

$$\pi \frac{d_1^2}{4} v_1 = \pi \frac{d_2^2}{4} v_2$$

Hence

$$v_2 = (15^2 \times 1.0)/2.0^2$$
$$= 56 \text{ m/s (to two significant figures)}$$

2.78 A tank of water has a small hole in its side at a point 2.0 m below the water surface in the tank. The hole has an area of 3.0 mm^2. If the tank is open to the atmosphere with what velocity will the water emerge from the hole? What will be the volume of water escaping per second?

Ans Applying Bernoulli's equation gives

$$p_1 + (\rho v_1^2)/2 + \rho g h = p_2 + (\rho v_2^2)/2$$

where h is the difference in height between the exit hole and the water surface, p_1 is the pressure at the open surface in the tank, i.e. atmospheric pressure, and p_2 is the pressure at the exit hole. The difference in height between these two points is so small that there will be virtually no difference in the atmospheric pressure between the two points and thus p_1 and p_2 can be considered to be identical. The velocity of the water in the tank, i.e. v_1, will be considerably smaller than that of the water emerging from the hole, i.e. v_2, and can be neglected. Thus

$$\rho g h = (\rho v_2^2)/2$$

Hence

$$v_2^2 = 2gh$$

This is the same velocity as that acquired by any body falling through a height h. The relationship is known as Torricelli's theorem. Hence

$$v_2^2 = 2 \times 9.8 \times 2.0$$
$$v_2 = 6.3 \text{ m/s}$$

The volume rate of flow is Av, i.e. 1.89×10^{-5} m^3/s

2.79 Water flows through a horizontal pipe which gradually tapers from a cross-sectional area of 50 mm^2 to 20 mm^2. If the pressure difference between these two different cross-sections of pipe is 20 kN/m^2, what is the volume of water passing through the pipe per second (density of water is 1000 kg/m^3)?

Ans Applying the equation of continuity gives

$$A_1 v_1 = A_2 v_2$$
$$50 \times 10^{-6} v_1 = 20 \times 10^{-6} v_2$$
$$5v_1 = 2v_2$$

Applying Bernouilli's equation gives

$$p_1 + (\rho v_1^2)/2 + \rho g h = p_2 + (\rho v_2^2)/2 + \rho g h$$

$$v_2^2 - v_1^2 = \frac{2}{\rho}(p_1 - p_2)$$

$$25v^2/4 - v_1^2 = (2 \times 20 \times 10^3)/1000$$

Hence

$$v_1 = 2.8 \text{ m/s}$$
$$v_2 = 7.0 \text{ m/s}$$

$$\text{Volume rate of flow} = Av_2 = 20 \times 10^{-6} \times 2.76$$
$$= 5.5 \times 10^{-5} \text{ m}^3/\text{s}$$

Further problems

2.80 A constriction in a horizontal pipe reduces the cross-sectional area of the pipe from 10 cm^2 to 1.0 cm^2. What is the velocity of a fluid in the full bore part of the tube if there is a drop of pressure of 0.5×10^7 N/m^2 when the fluid passes from the full to narrow bore of the pipe (the fluid has a density of 1000 kg/m^3)?

2.81 What would be the speed of outflow of water from a hole in the base of a full bucket of water? Guess any data you need.

2.82 The shape of an aircraft wing is designed to give a higher speed of air flow over the top of the wing than below it. What is the pressure difference between positions immediately below and immediately above the wing if the air speeds are 100 m/s below and 110 m/s above (density of the air is 1.3 kg/m^3)?

2.83 An open water tank discharges water to the atmosphere through a pipe that falls a distance of 3.0 m from the water level in the tank. The pipe has a uniform cross-sectional area of 80 mm^2. What is (a) the velocity with which the water leaves the pipe, (b) the volume of water discarded per second (density of water is 1000 kg/m^3)?

2.84 A horizontal pipe of diameter 40 mm has a constriction which reduces the pipe diameter to 5.0 mm. If water flows along the pipe with a velocity of 0.60 m/s in the 40 mm diameter part, what will be (a) the velocity of the water at the constriction, (b) the volumetric rate of flow through the pipe?

Viscosity

Notes

With steady fluid flow past a surface the streamlines are parallel to the surface and the flow is said to be laminar (a laminar is a thin layer). The fluid in contact with the surface is found to have the same velocity as the surface, thus if the surface is at rest the fluid layer immediately adjacent to the surface is at rest. The velocities of other layers of fluid increases uniformly with distance away from the surface, i.e. there is a constant velocity gradient. The fluid flow past the surface can thus be considered to be like layers of fluid sliding over each other.

The presence of the surface exerts a drag on the fluid. This force, which opposes relative motion between fluid and surface, is called a viscous force and the fluid is said to show viscosity.

$$\frac{\text{Viscous force}}{\text{Area of surface}} = \text{coefficient of viscosity} \times \text{velocity gradient}$$

or

$$\frac{F}{A} = \eta \frac{dv}{dy}$$

The volume of fluid flowing through a pipe per second $\dot{V}$ is related to the pressure difference p between the ends of the pipe by

$$\dot{V} = \frac{\pi p r^4}{8L\eta}$$

where r is the radius of the pipe, L the length of the pipe and η the coefficient of viscosity of the fluid. This equation is known as **Poiseuille's equation.**

For a sphere falling in a fluid, at such a speed that the flow is laminar, the viscous drag acting on the sphere is $6\pi\eta r v$, where r is the radius of the sphere and v its velocity. This is known as **Stokes' law.** When a sphere starts to fall in a fluid it accelerates and as its velocity increases so does the viscous drag. Eventually, there is no net force acting on the sphere and it reaches a constant velocity, known as the terminal velocity. When this occurs

Gravitational force
= viscous drag + buoyancy force

$$mg = 6\pi\eta r v + V\sigma g$$

where m is the mass of the sphere, σ the fluid density and V the volume of the sphere.

$$m = \frac{4}{3}\pi r^3 \rho g$$

and

$$V = \frac{4}{3}\pi r^3$$

where ρ is the density of the sphere, assumed to be homogeneous.

Units: coefficient of viscosity – N s m^{-2}, force – N, density – kg m^{-3}, velocity – m s^{-1}, area – m^2, distance – m.

Examples

Take g to be 9.8 m/s^2.

2.85 Water flows with laminar flow over a horizontal flat plate 0.20 m by 0.10 m. If the velocity gradient of the water immediately adjacent to the plate is 50 m s^{-1} m^{-1} and the coefficient of viscosity is 1.0×10^{-3} N s m^{-2}, what is the viscous drag experienced by the plate?

Ans $$\frac{F}{A} = \eta \frac{dv}{dy}$$

$$\frac{F}{0.20 \times 0.10} = 1.0 \times 10^{-3} \times 50$$

$$F = 0.001 \text{ N}$$

2.86 What is the rate of flow of water through a tube of radius 3.0 mm and length 200 mm if the pressure difference between the two ends of the tube is 3.0×10^3 N/m^2 and the coefficient of viscosity for water is 0.001 N s m^{-2}?

Ans Poiseuille's equation gives

$$\dot{V} = \frac{\pi p r^4}{8 L \eta}$$

$$= \frac{\pi \times 3.0 \times 10^3 \times (3.0 \times 10^{-3})^4}{8 \times 200 \times 10^{-3} \times 0.001}$$

$$= 4.8 \times 10^{-4} \text{ m}^3/\text{s}$$

2.87 What is the terminal velocity attained by a steel (density 8000 kg/m^3) ball bearing of radius 1.0 mm falling in oil of coefficient of viscosity 1.0 N s m^{-2} and density 900 kg/m^3?

Ans Applying Stokes' law for equilibrium conditions,

$$\frac{4}{3}\pi r^3 \rho g = 6\pi\eta r v + \frac{4}{3}\pi r^3 \sigma g$$

$$v = \frac{2r^2(\rho - \sigma)g}{9\eta}$$

$$= \frac{2 \times 1.0 \times 10^{-3} \times (8000 - 900) \times 9.8}{9 \times 1.0}$$

$$= 15 \text{ m/s}$$

Further problems

Take g to be 9.8 m/s^2.

2.88 An oil drop of density 900 kg/m^3 has a terminal velocity of 0.15 m/s when falling in air. If the air has a density of 1.2 kg/m^3 and a coefficient of viscosity of 18×10^{-6} N s m^{-2}, what is the radius of the oil drop if it is assumed to be spherical during its fall?

2.89 What is the terminal velocity of a steel sphere (density 8500 kg/m^3) of radius 1.0 mm when falling in glycerine having a coefficient of viscosity of 1.5 N s m^{-2} and a density of 1300 kg/m^3?

2.90 What will be the terminal velocity of an air bubble rising in water (coefficient of viscosity 1.0×10^{-3} N s m^{-2}) if it has a diameter of 1.0 mm (air has a density of 1.2 kg/m^3 and water one of 1000 kg/m^3)?

2.91 What is the rate of flow of water through a tube of diameter 2.0 mm and length 0.50 m if the difference in pressure between the two ends of the tube is 5.0×10^4 N/m^2 (coefficient of viscosity of water is 1.0×10^{-3} N s m^{-2})?

The mole

Notes

The quantity of matter that contains the same number of particles as there are atoms in 12 g of the carbon-12 isotope is called a mole. The number of particles in a mole is called Avogadro's constant and has a value of about 6.02×10^{23}.

Examples

2.92 What is the mass in grams of two moles of carbon dioxide (CO_2) (molar mass of carbon is 12 g and that of atomic oxygen is 16 g)?

Ans $12 + 2 \times 16 = 44$ g

2.93 Estimate the number of atoms in 1 g of water (H_2O). Hydrogen has a molar mass of 1 g, atomic oxygen one of 16 g and Avogadro's constant is 6×10^{23}.

Ans One mole of water has a mass of $2 \times 1 + 16 = 18$ g. Hence 1 g of water is 1/18th of a mole and thus has

$$6 \times 10^{23}/18 = 3 \times 10^{22}$$

particles

Further problems

Take the molar masses of atomic oxygen and atomic nitrogen to be 16 and 14, respectively, and Avogadro's constant to be 6×10^{23}.

2.94 Estimate the number of molecules in 1.0 kg of oxygen.

2.95 How many oxygen molecules are there in a jar containing 5 g of oxygen?

2.96 If the jar in the previous question had an internal volume of 1.5×10^{-3} m^3, what is the number of oxygen molecules per unit volume?

2.97 Nitrogen has a density, at 0 °C and atmospheric pressure, of 1.25 kg/m^3. What is the number of nitrogen molecules in one cubic metre under these conditions?

The kinetic theory of gases

Notes

The kinetic theory considers a gas to be composed of molecules in constant motion and that it is the collisions of these molecules with the walls of the gas container which are responsible for the pressure. The following assumptions enable a gas equation similar to that of ideal gases to be derived.

(i) The gas consists of identical particles.
(ii) These particles, usually molecules, are in random motion.
(iii) The total number of molecules in any sample of gas is large.
(iv) The volume of the molecules is small in comparison with the total volume occupied by a gas.
(v) No forces act on the molecules except during collisions.
(vi) Collisions between molecules and the walls of a container are perfectly elastic.
(vii) Intermolecular collisions are rare.
(viii) The molecules obey Newton's laws of motion.

Consider a single molecule moving backwards and forwards between two parallel walls of a container a distance L apart. If the speed of the molecule between the walls is v then in a time t it would cover a distance of vt and so make $vt/2L$ collisions with one end wall. If the molecule has a mass m then the change in momentum on colliding with the wall is $2mv$ per collision and so in time t the total change in momentum is $2mv \times vt/2L$. There are, however, many molecules in the container and they will be moving in all directions. The velocities of all the molecules can be resolved into three mutually perpendicular directions and thus of the total number n we can consider $n/3$ to be moving in the direction we considered for the single molecule. Thus, the total change in momentum in time t is $(n/3) \times 2mv \times vt/2L$, i.e. $\frac{1}{3}nmv^2 t/L$. Force is rate of change of momentum and so the force on the container wall is $\frac{1}{3}nmv^2/L$. Pressure is force per unit area and so equals $\frac{1}{3}nmv^2/LA$ where A is the area of the wall. But LA is the volume V of the container. Thus

$$pV = \tfrac{1}{3}nmv^2$$

The molecules do not all have the same speed, thus in the equation we have to use the average value of all the v^2 terms, i.e. $\overline{v^2}$. The square root of $\overline{v^2}$ is referred to as the root mean square velocity.

If we consider 1 mole of gas then

$$pV = R_0 T$$

the ideal gas equation. The kinetic model gives

$$pV = \tfrac{1}{3}Nm\overline{v^2}$$

where N is the number of molecules in one mole, i.e. Avogadro's constant. This would seem to suggest that if we consider the kinetic energy of the molecules to be proportional to the temperature, i.e. $\frac{1}{2}m\overline{v^2}$, we have

$$pV \propto T$$

Examples

2.98 What is the root mean square speed of the molecules in air if air has a density of 1.2 kg/m³ at a pressure of 1×10^5 N/m²?

Ans $pV = \frac{1}{3} nm\overline{v^2}$

But the density $\rho = \dfrac{nm}{V}$, hence

$$\overline{v^2} = \frac{3p}{\rho} = \frac{3 \times 10^5}{1.2}$$

Hence the root mean square speed is 500 m/s

Further problems

2.99 What is Brownian motion? How can it be explained in terms of simple kinetic theory of gases?

2.100 If the Kelvin temperature of a gas in a container is doubled, how does the root mean square speed of the molecules change?

2.101 The temperature of a container of gas is raised from 15 °C to 30 °C. How does the root mean square speed change?

2.102 What is the root mean square speed of hydrogen molecules at a density of 0.09 kg/m³ under a pressure of 1×10^5 N/m²?

2.103 What is pressure due to in terms of the kinetic theory?

2.104 How would the pressure of a gas change if the volume and temperature remained constant, but the number of molecules in the gas container were doubled?

2.105 How do the root mean square velocities of hydrogen and oxygen compare at the same temperature (molar mass of atomic hydrogen is 1 g and that of atomic oxygen is 16 g)?

The atomic model of solids

Notes

A crystalline substance is one in which there are particles in the solid arranged in an orderly manner. Metals are crystalline materials, generally composed of a large number of small crystals rather than a single large crystal. There are, however, invariably faults in the orderly packing, such faults being known as dislocations. These dislocations are a vital component in the explanation of the behaviour of metals.

Further problems

2.106 Why is the presence of dislocations in metals necessary if the behaviour of metals when subject to stress is to be explained?

2.107 What evidence is there for solid metals being crystalline?

2.108 Glass is said to be non-crystalline. What does this imply for the arrangement of atoms in glass?

2.109 Rubber bands can be stretched a considerable amount before they break. Explain this in terms of the atomic arrangement in rubber.

Examination questions

2.110 Two copper wires A and B, of the same, known areas of cross-section, are subjected to measured stretching forces and the corresponding extensions are measured. The results, on a force/extension graph, are shown in *Figure 2.15*.

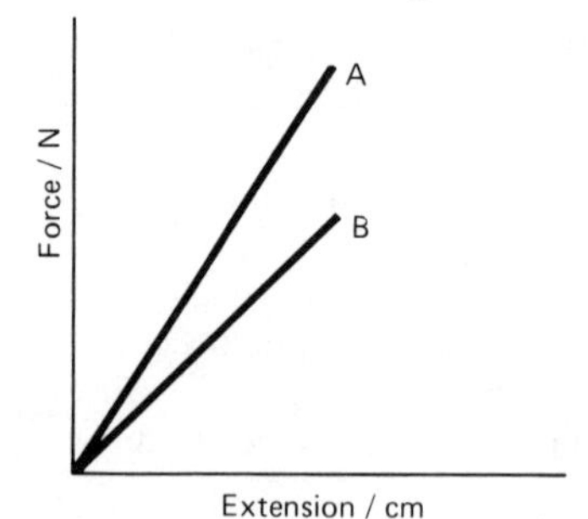

Figure 2.15

Explain what deduction you could make about the difference between the two wires.

Define the quantities which you would plot to get the same graph for both wires. How would you use this second graph to evaluate an important physical constant of copper?

(University of London)

2.111 (a) A heavy, rigid bar is supported horizontally from a fixed support by two vertical wires, A and B, of the same initial length and which experience the same extension. If the ratio of the diameter of A to that of B is 2 and the ratio of Young's modulus of A to that of B is 2, calculate the ratio of the tension in A to that in B.
(b) If the distance between the wires is D, calculate the distance x of wire A from the centre of gravity of the bar.

(JMB)

2.112 It is modern practice for railways to use long, welded steel rails. If these are laid without any stress in the rails on a day when the temperature is 15 °C, estimate the internal force when the temperature of the rails is 35 °C. You may assume that the rails are so held that there is no longitudinal movement and that the cross-sectional area of the rails is 7.0×10^{-3} m^2.

The Young modulus for steel = 2.0×10^{11} N m^{-2} and the linear expansivity = 1.0×10^{-5} K^{-1}.

(AEB)

2.113 A mercury barometer, with a scale attached, has a little air above the mercury. The top of the tube is 1.00 m above the level of the mercury in the reservoir. When the tube is vertical the height of the mercury column is 700 mm. When the tube is inclined at 60° to the vertical the reading of the mercury level on the scale is 950 mm. To what height would the mercury have risen in the vertical tube had there not been any air in it?

(University of London)

2.114 (a) Describe how you would determine the density of air at the temperature and pressure of the laboratory. Mention the chief source of error in your method, and indicate the accuracy that you would expect for your result.
(b) Show that for an ideal gas of relative molecular mass M_r the density ρ at pressure p and temperature T is given by the formula

$$\rho = a\frac{pM_r}{R_0T}$$

where the constant $a = 10^{-3}$ kg mol^{-1}.
(c) The value of M_r for uranium hexafluoride is 311 for the ^{235}U isotope and 314 for the ^{238}U isotope. Both exist as vapour at pressure 1.01×10^5 Pa and temperature 56 °C.

For these conditions, calculate the densities of pure samples of the two hexafluorides, and the density of a mixture of the two that contains 20 per cent by volume of the ^{235}U compound.

Take the molar gas constant R_0 to be 8.31 J mol^{-1} K^{-1}.

(Oxford Local Examinations)

2.115 (a) Explain the terms lines of flow and streamlines when applied to fluid flow and deduce the relationship between them in laminar flow.
(b) State Bernouilli's equation, define the physical quantities which appear in it and the conditions required for its validity.
(c) The depth of water in a tank of large cross-sectional area is maintained at 20 cm and water emerges in a continuous stream out of a hole 5 mm in diameter in the base. Calculate

(i) the speed of efflux of water from the hole,
(ii) the rate of mass flow of water from the hole.

Density of water = 1.00×10^3 kg m^{-3}.

(JMB)

2.116 (a) State four of the basic assumptions made in developing the simple kinetic theory for an ideal gas.
(b) The theory derives the formula

$$p = \frac{1}{3}\rho\overline{c^2}$$

where p is the pressure of the gas, ρ is the density of the gas and $\overline{c^2}$ is the mean square speed of the molecules. Explain more fully what is meant by $\overline{c^2}$ and explain its significance in relation to the temperature of a gas.
(c) Describe briefly the experiments which Andrews performed on carbon dioxide. (A detailed description of the apparatus is not required.)

(i) Draw graphs to show the pressure/volume relationship which Andrews obtained for various temperatures. Indicate on your diagram the various states of the carbon dioxide.
(ii) Use your graph to explain the meaning of critical temperature. What is its significance in connection with the liquefaction of gases?

(AEB)

Answers

2.4 Ans Stress/strain $= E$, hence strain $= 1.5 \times 10^{-5}$

2.5 Ans Stress/strain graph gives $E = 8.30 \times 10^{11}$ N/m^2

2.6 Ans Stress/strain $= E$, hence strain $= 8.4 \times 10^{-3}$ %

2.7 Ans Stress/strain $= E$, where stress = force/area and strain = extension/length, hence $E = 110 \times 10^9$ N/m^2

2.8 Ans Stress/strain $= E$, hence as strain = extension/length the extension is 1.6×10^{-3} m

2.9 Ans Each wire can be considered to be separately under the tension of 500 N, hence the elongation for steel is 9.1×10^{-4} m and that for copper is 9.8×10^{-4} m

2.10 Ans (a) See the notes
(b) (i) and (iv) are brittle; (ii) and (iii) are ductile

2.15 Ans Expansion $= L_0\alpha\theta = 0.30$ mm

2.16 Ans Force $= AE\alpha\theta = 2.1 \times 10^4$ N

2.17 Ans Stress $= E\alpha\theta = 2.9 \times 10^7$ N/m^2

2.18 Ans True distance $= 202.150(1 + \alpha\theta) = 202.172$ m

2.19 Ans $V_\theta = V_0(1 + 3\alpha\theta)$, $\rho_\theta = m/V_\theta$, $\rho_0 = m/V_0$, hence $\rho_\theta = \rho_0/(1 + 3\alpha\theta)$

2.20 Ans $d_\theta^2 = d_0^2(1 + 2\alpha\theta)$, hence change in diameter is 0.077 mm

2.21 Ans Volume increase $= 3\alpha\theta \times 100 = 0.55\%$

2.22 Ans $d_\theta(1 + \alpha\theta)$ will have the same value for both sphere and ring when the temperature is 440 °C

2.25 Ans Take moments about the pivot, hence the 40 g object should be placed at 675 mm

2.26 Ans Moment $= 2 \times 9.8 \times 580 \times 10^{-3} = 11$ N m

2.27 Ans Neglect any consideration of the mass of the plank as only the extra forces due to the student's presence are required. $T_1 + T_2 = 60 \times 9.8$ and taking moments about one end $60 \times 9.8 \times 1.0 = 3.0T_2$, hence $T_1 = 392$ N and $T_2 = 196$ N

2.31 Ans Each must sustain $\frac{1}{2} \times 10 \times 9.8 = 49$ N

2.32 Ans Taking moments about the hinge gives $T \times 1.2 \cos 30° = 4.0 \times 9.8 \times 1.2$, hence tension $T = 45.3$ N. Equating forces in the horizontal direction gives $R \cos\phi = T \cos 60°$, and in the vertical direction gives $R \sin\phi + T \sin 60° = 4.0 \times 9.8$, hence the hinge reaction $R = 22.5$ N and $\phi = 0°$, i.e. the reaction is horizontal

2.33 Ans (a) $R_1 + R_2 = (16 + 45)9.8$ N. Taking moments about the nearest end to the woman gives $45 \times 9.8 \times 0.5 + 16 \times 9.8 \times 1.2 = R_2 \times 2.4$, hence $R_2 = 170$ N and R_1 428 N
(b) At the 1.0 m distance $R_2 = 262$ N and $R_1 = 336$ N

2.34 Ans Taking moments about the muscle, $F \times 50 \times 10^{-3} = 35 \times 9.8 \times 200 \times 10^{-3}$, hence the force exerted by the bones $F = 1400$ N. Force exerted by the muscle $= 1400 - 35 \times 9.8 = 1100$ N

2.35 Ans See **2.29**. Taking moments about the elbow joint gives $2 \times 9.8 \times 150 \times 10^{-3} + 5.0 \times 9.8 \times 300 \times 10^{-3} = F_1 \times 60 \times 10^{-3} \sin 60°$, hence $F_1 = 340$ N

2.36 Ans Taking moments about the toes, $75 \times 9.8 \times 1.0 = F \times 1.6$, hence $F = 459$ N. Thus the force acting on each hand is about 230 N. Force exerted by the toes $= 75 \times 9.8 - 459 = 276$ N, hence the force exerted by the toes on each foot equals about 138 N

2.37 Ans See **2.30.**
(a) $R_1 = 20 \times 9.8 = 196$ N;
$F = \mu R_1 = 0.4 \times 196 = 78.4$ N
This is the maximum force if slipping is not to occur. The actual frictional force is given by taking moments about the base of the ladder, $R_2 \times 4 = 20 \times 9.8 \times 1.5$, hence $R_2 = 73.5$ N. This equals the frictional force, thus slipping will not occur.
(b) $R_1 = 90 \times 9.8 = 882$ N;
$F = \mu R_1 = 353$ N. taking moments about the base of the ladder gives $R_2 \times 4 = 20 \times 9.8 \times 1.5 + 70 \times 9.8 \times 1.5$, hence $R_2 = 331$ N. This equals the frictional force and slipping will not occur

2.38 Ans Frictional force $= 0.5 \times 20 \times 9.8$ N
This equals the reaction at the wall. Taking moments about the base of the plank gives $0.5 \times 20 \times 9.8 \times 6 \cos\theta = 20 \times 9.8 \times 3 \sin\theta$, hence $\theta = 45°$

2.45 Ans $p = h\rho g + A = 1.78 \times 10^5$ N/m^2

2.46 Ans $p = h\rho g = 2.57 \times 10^4$ N/m^2

2.47 Ans $p = F/A = 1.3 \times 10^6$ N/m^2

2.48 Ans $1.6 \times 10^3 = F/250 \times 10^{-6}$, hence $F = 0.40$ N

2.49 Ans $p = F/A = h\rho g$, hence $F = 2.1 \times 10^4$ N

2.50 Ans $33 \times 10^{-3} g = 56 \times 10^{-3} g - V \times \rho_w g$, hence $V = 23 \times 10^{-3}/\rho_w$. But $\rho = 56 \times 10^{-3}/V$, hence $\rho/\rho_w = 56/23 = 2.4$ (ρ is the density of the material and ρ_w that of water)

2.51 Ans Net force = upthrust − weight = 5.7 kN

2.52 Ans Hot air has a lower density than cold air and so, since the balloon contains hot air in a cold air environment, there is a net upwards force acting on the balloon

2.53 Ans Net upwards force = upthrust − weight of ice = 75 g N, hence the area required is 3.61 m^2

2.54 Ans Upthrust = weight, hence weight = 2.64 N

2.57 Ans Apparent coefficient of expansion $= 171 \times 10^{-6}$ °C^{-1}, hence the temperature $t = 59.1$ °C

2.58 Ans 9.4 °C

2.59 Ans $\alpha\theta \times 100 = 1.71\%$

2.60 Ans $\rho_{20} = \rho_0/(1 + \gamma\theta)$; this approximates to $\rho_{20} = \rho_0(1 - \gamma\theta)$, using the binomial theorem. Hence $(\rho_{20} - \rho_0)/\rho_0 = -\gamma\theta$ and so the percentage change is 2.16%

2.67 Ans pV = constant, hence $p = 2.0 \times 10^5$ N/m^2

2.68 Ans p/T = a constant, hence $p = 5.3 \times 10^5$ N/m^2

2.69 Ans $pV = mR_0T/M$, hence $m = 0.53$ kg

2.70 Ans pV = a constant, hence, as initially $p_1 = A - h\rho g$ and finally $p_2 = A + h\rho g$, A is 0.979×10^5 N/m^2

2.71 Ans pV/T = a constant, hence the diameter = 48.5 mm

2.72 Ans $pV = mR_0T/M$, hence $m = pVM/R_0T$ = $55 \times 10^{-3} \times 13.5 \times 10^3 \times 9.8 \times 200 \times 10^{-3} \times 100 \times 10^{-6} \times 29 \times 10^{-3}/(8.31 \times 293) = 1.73 \times 10^{-6}$ kg

2.74 Ans (a) 132.5 °C
(b) 111.5×10^5 N/m^2

2.75 Ans (a) Vapour to vapour plus liquid to liquid
(b) Gas at all pressures

2.76 Ans (a) Liquid
(b) Gas
(c) Vapour

2.80 Ans $p_1 + \frac{1}{2}\rho v_1^2 = p_2 + \frac{1}{2}\rho v_2^2$, also $A_1 v_1 = A_2 v_2$, hence $v_1 = 10$ m/s

2.81 Ans $v^2 = 2gh$ (see **2.78**), hence for a bucket 45 cm high, v is about 3 m/s

2.82 Ans $p_1 + \frac{1}{2}\rho v_1^2 = p_2 + \frac{1}{2}\rho v_2^2$, hence $p_1 - p_2 = 1.37 \times 10^3$ N/m^2

2.83 Ans (a) $v^2 = 2gh$, hence $v = 7.7$ m/s
(b) Volume per second = $Av = 6.2 \times 10^{-4}$ m^3/s

2.84 Ans (a) $A_1 v_1 = A_2 v_2$, hence $v_2 = 38.4$ m/s
(b) Volume per second = Av = 7.5×10^{-4} m^3/s

2.88 Ans See **2.87**. Radius = 3.7×10^{-5} m

2.89 Ans See **2.87**. Velocity = 0.011 m/s

2.90 Ans $\frac{4}{3}\pi r^3 \rho g + 6\pi\eta r v = \frac{4}{3}\pi r^3 \sigma g$, hence $v = 2.18$ m/s

2.91 Ans See **2.86**. $\dot{V} = 3.9 \times 10^{-5}$ m^3/s

2.94 Ans Molecular mass is 32 g, hence the number of molecules is $6 \times 10^{23} \times 1.0/(32 \times 10^{-3}) = 0.19 \times 10^{26}$

2.95 Ans $6 \times 10^{23} \times 5/32 = 0.94 \times 10^{23}$

2.96 Ans $0.94 \times 10^{23}/(1.5 \times 10^{-3})$ = 0.63×10^{26} molecules m^{-3}

2.97 Ans $6 \times 10^{23} \times 1.25/(28 \times 10^{-3})$ = 0.27×10^{26} molecules

2.99 Ans Brownian motion is the random motion of small particles, e.g. ash particles in air. This motion is due to buffeting of the particles by molecular bombardment

2.100 Ans $\overline{v^2} \propto T$, hence speed increases by $2^{1/2}$

2.101 Ans $(303/288)^{1/2}$

2.102 Ans $(3p/\rho)^{1/2}$, see **2.98**;
$(\overline{v^2})^{1/2}$ 1.8 km/s

2.103 Ans Molecular bombardment

2.104 Ans The pressure is directly proportional to the number of molecules and is thus doubled

2.105 Ans $\overline{v^2} \propto 1/m$, hence the hydrogen molecules are four times faster

2.106 Ans Without dislocations metals would all have to have very high strengths. Normally they do not

2.107 Ans Metal crystals; polycrystalline form showing when metal surfaces are etched; X-ray crystallography

2.108 Ans This implies non-orderly packing of the atoms. There may be some short range order, but no order over the long range

2.109 Ans Think of rubber as being like a piece of netting which, in the unstressed state, is crumpled up. When stressed the netting tangle becomes uncrumpled and assumes an orderly form. It is this which gives rise to the very large extensions. Rubber consists of chains of atoms with some cross-linking between the chains

2.110 Ans A has a shorter length than B. Plot the force against strain, i.e. extension/original length and then stress/strain = Young's modulus

2.111 Ans (a) Force $= EA \times$ strain, hence tension in A is 4 times that in B
(b) Take moments about the centre of gravity to give $4Fx = (D - x)F$, hence $x = D/5$

2.112 Ans Force $= AE\alpha\theta = 2.8 \times 10^5$ N

2.113 Ans $pV =$ a constant; the initial pressure of the air is atmospheric pressure minus the pressure due to 700 mm of mercury, the volume being $(1.00 - 0.700)A$, where A is the cross-sectional area of the tube. Finally, the pressure is atmospheric pressure minus that due to 950 cos 60° mm of mercury, the volume being $(1.00 - 0.950)A$. Hence atmospheric pressure is that due to 745 mm of mercury

2.114 Ans (a) You could measure the volume of a large container, then weigh it full of air and when evacuated
(b) $pV = mR_0T/M$, hence, as $\rho = m/V$ we have $\rho = pM/R_0T$. The mass M of a mole is given in kilogrammes and thus is $10^{-3} \times M_r$, where M_r is the relative molecular mass in terms of carbon as 12 g
(c) Densities are: ^{235}U 11.49 kg/m^3, ^{238}U 11.60 kg/m^3 and mixture 11.58 kg/m^3

2.115 Ans (a) and (b) See the notes in this chapter
(c) (i) $v^2 = 2gh$, hence $v = 1.98$ m/s
(ii) 3.88×10^{-2} kg/s

2.116 Ans See the notes in this chapter

3 Energy

Forms of energy

Notes

Formulae can be devised to represent various forms of energy and they are always devised in such a way that energy is conserved in any interaction. Thus, kinetic energy, the energy associated with a moving object by virtue of its motion, is defined as

$$\text{Kinetic energy} = \tfrac{1}{2}mv^2$$

where m is the mass of the moving object and v its velocity.

When an object is acted on by a net force and moves, the product of the displacement s (produced in the direction of the force) and the force F is called work.

$$\text{Work} = Fs$$

Work describes energy as it is being transferred from one object to another.

Potential energy is the energy an object possesses by virtue of its position. Gravitational potential energy is the energy a body has by virtue of its position in the gravitational field.

$$\text{Gravitational potential energy} = mgh$$

where m is the mass of the object, h its height above some datum line and g the acceleration due to gravity. When an object falls and accelerates in the gravitational field it loses potential energy and gains kinetic energy. The potential energy lost equals the gain in kinetic energy, i.e. energy is conserved.

There are other ways in which an object can have potential energy. For example, when an object, attached to the free end of a tethered spring, is moved with, or against, the force exerted by the spring. If the spring obeys Hooke's law then the force F is proportional to the extension x (see *Figure 3.1*). Thus

$$F = kx$$

where k is the force constant. Thus, when the spring is caused to extend from x_0 to x, the force changes from F_0 to F. The average force is thus $\tfrac{1}{2}(F_0 + F)$, hence

$$\text{Average force} = (kx_0 + kx)/2$$

Hence

$$\text{Work} = \left(\frac{kx_0 + kx}{2}\right) \times (x - x_0)$$

$$= \tfrac{1}{2}kx^2 - \tfrac{1}{2}kx_0^2$$

There is, thus, an energy of $\tfrac{1}{2}kx^2$ associated with the position of an object at displacement x.

$$\text{Elastic potential energy} = \tfrac{1}{2}kx^2$$

The area under the force/extension graph between the displacements x_0 and x represents the elastic potential energy stored in the spring when it has been stretched from x_0 to x.

The energy stored in a stretched strip of material is equal to the average force multiplied by the distance through which the material has been extended, assuming that Hooke's law is obeyed. If the force increases from zero to F then the average force is $\tfrac{1}{2}F$. Hence

$$\text{Energy stored} = \tfrac{1}{2}Fx$$

where x is the extension, and so

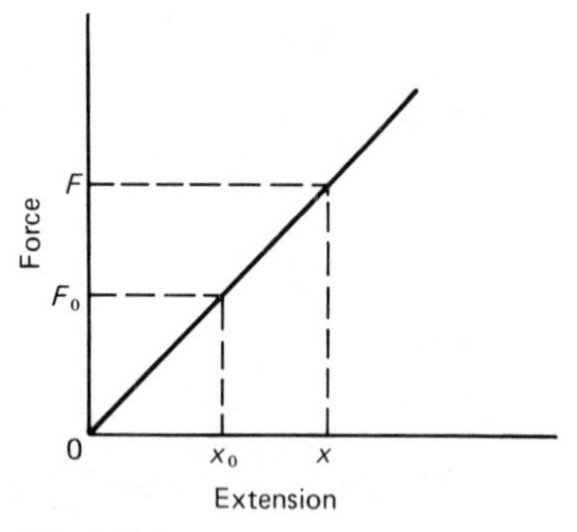

Figure 3.1

$$\frac{\text{Energy stored}}{\text{Volume of material}} = \frac{\tfrac{1}{2}Fx}{\text{volume of material}}$$

$$\frac{\text{Energy stored}}{\text{Volume of material}} = \tfrac{1}{2} \times \text{stress} \times \text{strain}$$

Energy is stored in a compressed gas. For a gas being compressed in a cylinder by a piston (see *Figure 3.2*)

$$\text{Work done} = F\Delta L = pA\Delta L = p\Delta V$$

where ΔV is the change in volume.

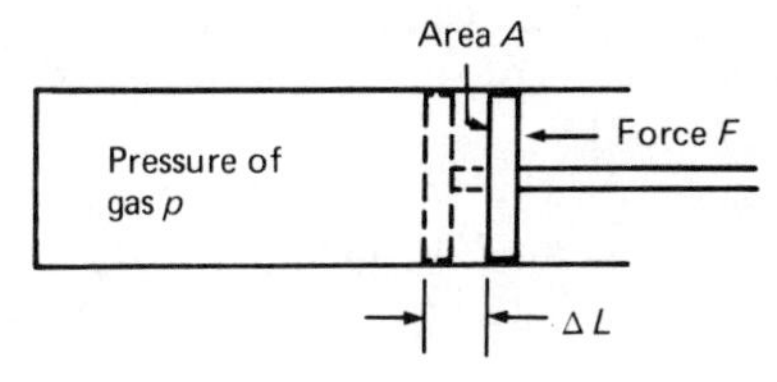

Figure 3.2

The energy associated with a body by virtue of it rotating is given by

$$\text{Rotational kinetic energy} = \tfrac{1}{2}I\omega^2$$

where I is the moment of inertia and ω the angular velocity. The work done in rotating an object by means of a torque T through an angle θ is

$$\text{Work} = T\theta$$

The power, i.e. rate of doing work, when an object rotates at f revolutions per second is

$$\text{Power} = 2\pi fT$$

Units: force – N, distance – m, work – N m or J (joule, 1 J = 1 N m), energy (all forms) – J, moment of inertia – kg m^2, angular velocity – s^{-1}, torque – N m, power – J/s or W (watt, 1 W = 1 J/s).

Examples

Take g to be 9.8 m/s^2.

3.1 A block of mass 1.0 kg is pushed 1.2 m up an inclined plane which is at 30° to the horizontal. If the velocity of the block remains constant, what is the work done?

Ans There are two ways in which the answer can be obtained. One is to calculate the change in potential energy due to the work done.

$$\text{Change in potential energy} = mgh$$

where h is the vertical height change.

$$\text{Change in potential energy} = 1.0 \times 9.8 \times 1.2 \sin 30^\circ = 5.9 \text{ J}$$

This change in potential energy results from the work done and, as no other form of energy is involved, must be equal to the work done.

The other way of obtaining the answer is to calculate directly the work done. The component of the displacement in the direction of the force is 1.2 sin 30°. Hence the work done is

$$\text{Work} = \text{force} \times \text{distance} = 1.0 \times 9.8 \times 1.2 \sin 30^\circ = 5.9 \text{ J}$$

3.2 An object of mass 2.0 kg starts from rest and slides down a frictionless incline which makes an angle of 30° to the horizontal. If it slides 2.5 m down the plane what will be its velocity?

Ans The potential energy lost by the object is mgh, where h is the vertical height through which the object has moved.

$$\text{Potential energy lost} = 2.0 \times 9.8 \times 2.5 \sin 30^\circ \text{ J}$$

This potential energy is transformed into kinetic energy. Hence

$$\text{Kinetic energy} = \tfrac{1}{2}mv^2 = 2.0 \times 9.8 \times 2.5 \sin 30^\circ$$

$$v = 4.9 \text{ m/s}$$

3.3 A bullet of mass 3.0 g moving with a velocity of 400 m/s hits a tree and comes to rest, inside the tree, after penetrating to a depth of 60 mm. What is the average force of retardation during the passage of the bullet within the tree?

Ans Kinetic energy of the bullet on hitting the tree is given by

$$\text{Kinetic energy} = \tfrac{1}{2}mv^2 = \tfrac{1}{2} \times 3.0 \times 10^{-3} \times 400^2 = 240\ \text{J}$$

This energy is used in penetrating to a depth of 60 mm. Hence

$$\text{Average force} \times \text{distance} = 240$$
$$\text{Average force} = 240/(60 \times 10^{-3}) = 4000\ \text{N}$$

3.4 An object of mass 50 g moving with a velocity of 2.0 m/s strikes another object of mass 20 g. If this object was initially at rest, what will be the velocities of the two objects after the collision if (a) the collision is perfectly elastic, (b) the collision is inelastic and the two objects combine at the collision?

Ans For a perfectly elastic collision kinetic energy is conserved as well as the momentum. In an inelastic collision only the momentum is conserved.

(a) Conservation of kinetic energy gives

$$\tfrac{1}{2} \times 50 \times 10^{-3} \times 2.0^2 = \tfrac{1}{2} \times 50 \times 10^{-3}\, v_1^2 + \tfrac{1}{2} \times 20 \times 10^{-3}\, v_2^2$$

where v_1 and v_2 are the velocities after the collision. Conservation of momentum gives

$$50 \times 10^{-3} \times 2.0 = 50 \times 10^{-3}\, v_1 + 20 \times 10^{-3}\, v_2$$

The two simultaneous equations can be solved for v_1 and v_2. The result is that v_1 can have the values of either 2.0 m/s or 6/7 m/s. The values of v_2 are 0 or 20/7 m/s, respectively.

(b) For the inelastic collision, conservation of momentum gives

$$50 \times 10^{-3} \times 2.0 = 70 \times 10^{-3}\, v$$

where v is the velocity of the combined object after the collision. Hence v is 10/7 m/s.

3.5 An object of mass 2.0 kg is dropped from a height of 20 cm onto a spring having a force constant of 1500 N/m. What is the maximum distance the spring will be compressed?

Ans Potential energy lost by the falling object $= 2.0 \times 9.8 \times 20 \times 10^{-2}$ J

This energy is converted into elastic potential energy. Hence, if the spring is assumed to obey Hooke's law,

$$\tfrac{1}{2}kx^2 = 2.0 \times 9.8 \times 20 \times 10^{-2}$$

and, as k is 1500 N/m, the value of x is 7.23×10^{-2} m, i.e. 7.23 cm.

3.6 A sample of steel has a Young's modulus of $220 \times 10^9\ \text{N/m}^2$. What will be the energy stored in unit volume of steel when subjected to a strain of 0.001?

Ans Energy stored per unit volume $= \tfrac{1}{2} \times \text{stress} \times \text{strain}$

As Young's modulus is stress/strain,

$$\text{Energy stored per unit volume} = \tfrac{1}{2} \times \text{Young's modulus} \times (\text{strain})^2 = \tfrac{1}{2} \times 220 \times 10^9 \times (0.001)^2 = 110 \times 10^3\ \text{J/m}^3$$

3.7 What is the energy stored in a flywheel having a moment of inertia of $200\ \text{kg m}^2$ and rotating at 4.0 revolutions per second?

Ans

$$\text{Rotational kinetic energy} = \tfrac{1}{2}I\omega^2 = \tfrac{1}{2}I(2\pi f)^2 = \tfrac{1}{2} \times 200 \times (2\pi \times 4.0)^2 = 63\,000\ \text{J}$$

3.8 What is the output torque of a motor shaft which rotates at 50 revolutions per second and gives an output of 1.2 kW?

Ans $\text{Power} = T \times 2\pi f$

Hence

$$T = \frac{1.2 \times 10^3}{2\pi \times 50} = 3.8\ \text{N m}$$

Further problems

Take g to be $9.8\ \text{m/s}^2$.

3.9 An object of mass 200 g falls through a height of 2.0 m into sand. What is the average resistive force of the sand if the object penetrates the sand to a depth of 250 mm?

3.10 A car goes up an incline with a constant velocity. If the incline is at 5° to the horizontal what is the energy used in the car (of mass 1500 kg) travelling 100 m up such an incline (neglect friction)?

3.11 A car goes up a road inclined at 5° to the horizontal at a constant velocity of 50 km/h. What is the power developed by the car if it has a mass of 1500 kg (neglect friction)?

3.12 What would be the answer to the previous question if the coefficient of friction between the car tyres and the road was 0.4?

3.13 An object of mass 400 g is projected along a horizontal surface with an initial velocity of 3.0 m/s. The object comes to rest in a distance of 2.4 m. What is the average frictional force acting on the object?

3.14 An object of mass 500 g is projected up an inclined plane, at 5° to the horizontal. If it starts at the bottom of the plane with a velocity of 3.0 m/s and slides up the plane a distance of 3.0 m, what is the average frictional force between the object and the plane?

3.15 Sketch graphs showing how the momentum and kinetic energy of two balls change when they collide if (a) the collision is completely elastic, (b) the collision is inelastic.

3.16 A golfer hits a golf ball of mass 20 g and gives it an initial velocity of 50 m/s. If the contact between the golf club and the ball is perfectly elastic and has a duration of 0.015 s, what is (a) the average force exerted by the club on the ball, (b) the initial kinetic energy of the ball and (c) the work done on the ball by the club?

3.17 What is the kinetic energy, relative to its initial energy, of a neutron after it collides with a carbon nucleus which has a mass 12 times greater than that of the neutron? Assume that the collision is head-on between the neutron and the carbon nucleus, which is initially at rest. Also assume that the collision is perfectly elastic.

3.18 A pump lifts 120 m^3 of water per hour from a reservoir to a point 4.0 m above the reservoir water level. It discharges the water at this level with a velocity of 7.0 m/s. What is the power that has to be supplied to the pump if it has an efficiency of 60% (density of water is 1000 kg/m^3)?

3.19 A wagon of mass 15 000 kg travelling at 4.0 m/s overtakes and collides with another wagon of mass 10 000 kg travelling in the same direction at 2.0 m/s. After the collision the two wagons lock together. What will be their velocity and the loss in kinetic energy?

3.20 A pile driver hammer has a mass of 1000 kg and falls through a vertical distance of 3.0 m before colliding with the pile. If the pile has a mass of 400 kg and is driven 0.12 m into the ground, what is (a) the velocity of the pile driver hammer immediately before the impact, (b) the common velocity of the pile and the hammer immediately after the collision if there is no rebound and (c) the loss in kinetic energy at impact?

3.21 A bullet of mass 3.0 g is fired into a block of wood of mass 1.0 kg. The block of wood is supported by vertical strings, like a pendulum. After the bullet has struck and become embedded in the block of wood, the block swings in a vertical arc and rises through a vertical height of 45 mm. What was the speed of the bullet?

3.22 A toy gun uses a compressed spring as the energy source for the firing of small pellets. If the spring has a force constant of 50 N/m and is compressed by 100 mm, how high can the gun fire a pellet of mass 5.0 g vertically?

3.23 What work is needed to compress a spring having a force constant of 100 N/m from zero compression to 20 mm compression? What work is needed to then compress the spring by a further 20 mm?

3.24 What is the strain energy stored in a bar of steel with a rectangular cross-section of 40 mm by 20 mm and a length of 0.8 m when it is subjected to an axial load of 120×10^3 N (Young's modulus for steel is 210×10^9 N/m^2)?

3.25 A steel wire of diameter 2.0 mm and length 2.0 m is subjected to a strain of 2.0×10^{-4}. What is the energy stored in the wire if the steel has an elastic modulus of 210×10^9 N/m^2?

3.26 A flywheel has a moment of inertia of 300 kg m^2 and is rotating at 4.0 revolutions per second. What is the rotational kinetic energy of the flywheel?

3.27 What is the energy stored in a flywheel with a moment of inertia of 5.0 kg m^2 when it is rotating at 10 revolutions per second? What is the energy released by the flywheel when it slows down to 5 revolutions per second?

3.28 What is the power input for a flywheel if it is kept rotating at 8.0 revolutions per second by a torque of 30 N m?

3.29 What is the uniform torque needed to accelerate a flywheel of moment of inertia 6.0 kg m^2 from rest to 50 revolutions per second in 10 s? What is the gain in rotational kinetic energy as a result of the acceleration?

3.30 What should be the moment of inertia of a flywheel if it is to store 10^5 J of energy when rotating at 10 revolutions per second?

Heat

Notes

Heat is a form of energy, being the energy that flows from one object to another because of a temperature difference between them. The heat capacity of an object is defined as the amount of heat energy Q supplied to the body divided by the corresponding temperature rise θ,

$$\text{Heat capacity} = \frac{\text{heat energy supplied}}{\text{change in temperature produced}}$$

The specific heat capacity is the heat capacity per unit mass.

$$\text{Specific heat capacity} = \frac{\text{heat energy supplied}}{\text{mass} \times \text{change in temperature}}$$

The molar heat capacity is the heat capacity per mole of material.

When a current passes through a resistor, the resistor becomes warm. The electrical energy giving the temperature change is given by

$$\text{Electrical energy} = IVt$$

where I is the current, V the potential difference across the resistor and t the time for which the current is passed through the element concerned. The rate of transformation of energy is power, hence

$$\text{Power} = IV$$

A change in temperature can be produced by rubbing two objects against each other. An object is being moved through a distance against a frictional force. The energy transfer is the product of this force and the distance.

The specific heat capacities of solids and liquids can be measured as follows.

(i) The method of mixtures in which two substances at different temperatures are 'mixed' and the resulting mixture temperature determined. Thus, for a block of solid material heated in steam, a known temperature, and then dropped into a lagged calorimeter containing a measured amount of water at a known temperature

Heat lost be the hot block in cooling to the mixture temperature
= heat gained by the water and calorimeter in rising to that mixture temperature (neglecting heat losses to the surroundings)

(ii) Electrical methods in which an electrical heating element is inserted into the substance concerned and a measured amount of electrical energy transformed into a determined temperature change.

Energy supplied by heater = heat gained by the substance + any heat gained by the container of that substance (neglecting heat losses to the surroundings)

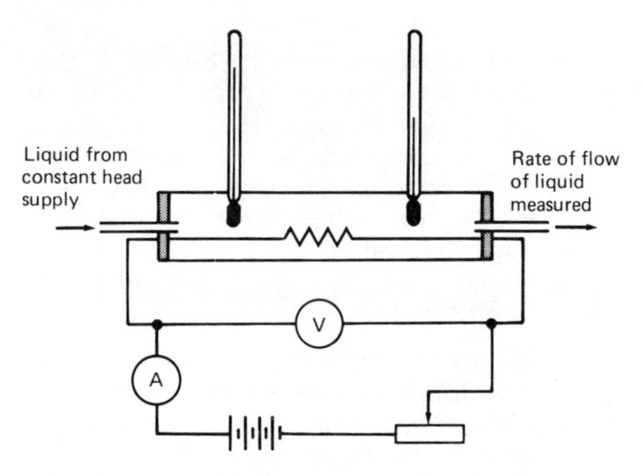

Figure 3.3 A simple form of continuous flow calorimeter

(iii) A continuous flow method (*Figure 3.3*) for liquids or gases which involves the continuous flow of the fluid over a heating element, the electrical energy to which is measured. When steady state conditions exist,

Energy supplied by heater per second = heat gained by fluid per second + heat losses per second from the apparatus (h)

Thus

$$IV = mc(\theta_2 - \theta_1) + h$$

where m is the mass of fluid passing through the apparatus per second, c its specific heat capacity, θ_1 the temperature of the fluid on entering the apparatus and θ_2 that on leaving. If the experiment is repeated for the same temperatures the heat losses are the same.

$$I'V' = m'c(\theta_2 - \theta_1) + h$$

Heat h can be eliminated from these two simultaneous equations and c obtained.

Units: mass – kg, temperature – °C or K, specific heat capacity – J kg^{-1} °C^{-1} or J kg^{-1} K^{-1}, heat energy – J, heat capacity – J °C^{-1} or J K^{-1}, current – A, potential difference – V, electrical energy – J, power – W.

Examples

3.31 An electric kettle has a heat capacity of 400 J °C^{-1} and initially contains 1.0 kg of water at 20 °C. If electrical power is supplied at 2.0 kW, how long will it take the water in the kettle to reach 100 °C if heat losses to the surroundings can be ignored (specific heat capacity of water is 4200 J kg^{-1} °C^{-1})?

Ans Energy needed
$= 400 \times (100 - 20) + 1.0 \times 4200 \times (100 - 20)$
$= 3.6 \times 10^5$ J

This energy is supplied by the electricity. If the power is on for time t, then

$$2.0 \times 10^3 t = 3.6 \times 10^5$$
$$t = 1.8 \times 10^2 \text{ s}$$

3.32 A block of copper of mass 120 g and at a temperature of 100 °C is dropped into a calorimeter containing 250 g of water at 20 °C. If the calorimeter has a heat capacity of 40 J °C^{-1}, the copper a specific heat capacity of 380 J kg^{-1} °C^{-1}, the water a specific heat capacity of 4200 J kg^{-1} °C^{-1} and no heat losses occur, what will be the resulting 'mixture' temperature of the water and the copper?

Ans Heat lost by the copper
$= 120 \times 10^{-3} \times 380 \times (100 - \theta)$

where θ is the mixture temperature.

Heat gained by calorimeter plus contents
$= 40 \times (\theta - 20) + 250 \times 10^{-3} \times 4200 \times (\theta - 20)$

As no heat losses occur the heat lost equals the heat gained. Hence θ is 23.2 °C.

3.33 In a continuous flow calorimeter experiment to determine the specific heat capacity of water the following readings were obtained.

Inlet temperature of water = 18.5 °C
Exit temperature of water = 22.5 °C
Current = 1.35 A
Potential difference = 5.0 V
Rate of flow of water = 0.39 g/s

The experiment was then repeated to give the same temperatures at the inlet and exit points.

Current = 2.2 A
Potential difference = 6.0 V
Rate of flow of water = 0.75 g/s

What is the specific heat capacity of the water?

Ans $IV = mc(\theta_2 - \theta_1) + h$
$I'V' = m'c(\theta_2 - \theta_1) + h$

Subtracting the two equations gives

$$IV - I'V' = (m - m') \times c \times (\theta_2 - \theta_1)$$

Hence

$$c = \frac{1.35 \times 5.0 - 2.2 \times 6.0}{(0.39 - 0.75) \times (22.5 - 18.5) \times 10^{-3}}$$
$$= 4479 \text{ J kg}^{-1} \text{ °C}^{-1}$$

3.34 A bullet travelling at 300 m/s strikes and becomes embedded in the target. What will be the rise in temperature θ of the bullet if 60% of the energy is used to heat the bullet of specific heat capacity 120 J kg^{-1} °C^{-1}?

Ans Kinetic energy of the bullet $= \frac{1}{2}m \times 300^2$

where m is the mass of the bullet.

Heat gained by the bullet $= m \times 120 \times \Delta\theta$

As 60% of the loss in kinetic energy is transformed into the heat for the bullet

$$m \times 120 \times \Delta\theta = \tfrac{1}{2}m \times 300^2 \times 60/100$$

Hence the change in temperature, $\Delta\theta$, is 225 °C.

3.35 One method of measuring the power developed by a motor is to wrap a band brake round the motor shaft or flywheel (*Figure 3.4*). When the shaft rotates the readings of the two spring balances on each end of the band differ. Calculate the power developed by a motor with such an arrangement when the tensions on the two sides, i.e. F_2 and F_1, are 120 N and 370 N, respectively, and the shaft is rotating at 50 revolutions per second. The band is wrapped round the flywheel of radius 0.40 m.

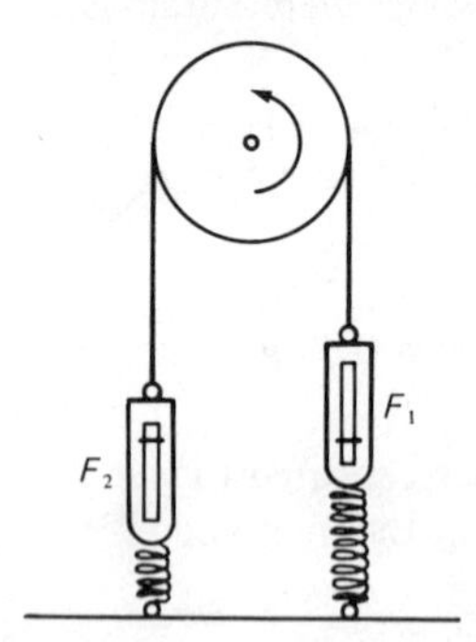

Figure 3.4

Ans Frictional force $= F_1 - F_2$

For one revolution of the shaft the distance moved against the frictional force is $2\pi r$, where r is the radius of the shaft or flywheel around which the band is wrapped. For f revolutions per second the work done in 1 s is

Work done per second $= (F_1 - F_2) \times 2\pi rf$

Hence, as this is the power

Power $= (370 - 120) \times 2\pi \times 0.40 \times 50$
$= 31$ kW (to two significant figures)

This power is developed as heat at the band brake.

Further problems

Take the specific heat capacity of water to be 4200 J kg^{-1} °C^{-1}.

3.36 An aluminium pan has a heat capacity of 180 J °C^{-1} and contains 800 g of water. What heat energy is needed to raise the temperature of the pan and its contents by 60 °C?

3.37 A car of mass 1000 kg and travelling at 4.0 m/s is brought to rest by its brakes. What is the heat energy transferred to its brakes during the braking operation?

3.38 What mass of liquid of specific heat capacity 2100 J kg^{-1} °C^{-1} and at a temperature of 50 °C must be mixed with 150 g of water at 20 °C if the final temperature of the mixture is to be 40 °C? Assume that no heat losses occur from the system and that the effect of any container can be neglected.

3.39 A heating coil rated at 150 W is immersed in 200 g of a liquid of specific heat capacity

$2400\ J\ kg^{-1}\ °C^{-1}$. If the heating element is switched on for 3.0 minutes, what temperature change will occur? Assume that no heat losses occur and that the effect of any container can be neglected.

3.40 An insulated copper can of mass 200 g contains 120 g of water at 18 °C. A block of iron of mass 50 g and at a temperature of 100 °C is placed in the water in the calorimeter. What will be the 'mixture' temperature obtained (specific heat capacity of copper is $380\ J\ kg^{-1}\ °C^{-1}$ and that of iron is $480\ J\ kg^{-1}\ °C^{-1}$)?

3.41 A steel ball-bearing is dropped from a height of 4.0 m to the ground. If 50% of the energy immediately prior to the impact appears as heat in the ball, by how much will the temperature of the ball change (specific heat capacity of the steel is $480\ J\ kg^{-1}\ °C^{-1}$)?

3.42 Describe a method that can be used for the measurement of the specific heat capacity of a liquid, explaining carefully how errors can be minimized. The liquid is available in large quantities.

3.43 Explain how the continuous flow calorimeter is used for the measurement of the specific heat capacity of a fluid. Explain how heat losses are taken into account in obtaining a result and why the heat capacity of the calorimeter plays no part in the calculation.

3.44 What is the specific heat capacity of the liquid which gives the following results with a continuous flow calorimeter?

Current = 2.5 A
Potential difference = 11.0 V
Mass of liquid per second = 3.2 g
Temperature difference between inlet and exit of liquid = 2.0 °C

When the experiment is repeated for the same temperatures,

Current = 2.0 A
Potential difference = 8.7 V
Mass of liquid per second = 2.0 g

3.45 A thermometer has a heat capacity of $40\ J\ °C^{-1}$ and gives a reading of 15.0 °C. It is then completely immersed in 0.40 kg of water and comes to a temperature reading of 45.5 °C. What was the temperature of the water before the thermometer was placed in it?

3.46 A paddle wheel is caused to rotate in water by a mass of 3.00 kg falling through a height of 1.00 m a total of 26 times. The water stirred by the paddle wheel has a mass of 0.300 kg and shows a temperature rise of 0.60 °C. What is the specific heat capacity of of the water if all heat losses are ignored and the vessel containing the water is assumed to have negligible heat capacity? (**Note**: this is a version of an old experiment carried out by Joule.)

Heat capacity of gases

Notes

The specific heat capacity of a gas at constant pressure c_p is the heat needed to raise the temperature of 1 kg of that substance at constant pressure by 1 °C.

The specific heat capacity of a gas at constant volume c_v is the heat needed to raise the temperature of 1 kg of that substance at constant volume by 1 °C.

When heat is supplied at constant volume all the energy goes into increasing the internal energy of the substance. When the heat is supplied at constant pressure, not only is there a change in internal energy, but also work is done as a result of the expansion of the substance. For liquids and solids the differences between the specific heat capacities at constant volume and constant pressure are generally very small. For gases this is not usually the case.

At constant volume, the change in internal energy ΔU is

$$\Delta U = c_v m\theta$$

where m is the mass of gas concerned and θ the change in temperature. If heat is supplied to the same mass of gas at constant pressure and the same temperature rise is produced

$$\text{Heat supplied} = c_p m\theta = \Delta U + p\Delta V$$

The change in internal energy is the same because the change in temperature is the same in both cases.

ΔV is the volume change occurring at the particular gas pressure. For an ideal gas

$$pV = mRT$$

where R is the characteristic gas constant. Hence

$$p(V + \Delta V) = mR(T + \theta)$$

and so

$$c_p m\theta = c_v m\theta + mR\theta$$
$$c_p = c_v + R$$

If the molar heat capacities of the gas are used, i.e. C_v and C_p,

$$C_p = C_v + R_0$$

where R_0 is the universal gas constant.

The ratio γ of the two specific heat capacities, or molar heat capacities, gives information about the atomicity of the gas,

$$\gamma = c_p/c_v = C_p/C_v$$

For a monatomic gas γ is about 1.67, for a diatomic gas about 1.40 and for a polyatomic gas about 1.30.

Units: specific heat capacities – J kg^{-1} °C^{-1} or J kg^{-1} K^{-1}, molar heat capacities – J mol^{-1} °C^{-1} or J mol^{-1} K^{-1}.

Examples

3.47 Which specific heat capacity of a gas will be larger – that at constant pressure or that at constant volume?

Ans The constant pressure value is largest. This is because the internal energy increases as well as the work done. With constant volume there is only a change in internal energy and so less heat is needed per degree change in temperature.

3.48 The molar heat capacity of hydrogen at constant volume is 20.2 J mol^{-1} °C^{-1}. What is the molar heat capacity at constant pressure (universal gas constant is 8.3 J mol^{-1} °C^{-1})?

Ans
$$C_p = C_v + R_0$$
$$= 20.2 + 8.3$$
$$= 28.5 \text{ J mol}^{-1}\ ^\circ\text{C}^{-1}$$

Further problems

Take the universal gas constant to be 8.3 J mol^{-1} °C^{-1}.

3.49 The molar heat capacity of oxygen (molar mass is 32 g) at constant pressure is 29.1 J mol^{-1} °C^{-1}. What is molar specific heat capacity at constant volume?

3.50 If the molar specific heat capacity of hydrogen at constant volume is 20.2 J mol^{-1} °C^{-1}, how much energy is needed to raise the temperature of 1 mole of hydrogen (molar mass is 2 g) under constant volume conditions by 100 °C? How much energy is needed to raise the temperature of 1 kg of hydrogen under constant volume conditions by 100 °C?

3.51 Argon has a molar heat capacity at constant volume of 12.6 J mol^{-1} °C^{-1}. What is the molar heat capacity of argon under constant pressure conditions?

Latent heat

Notes

The specific latent heat of fusion is the heat needed to change 1 kg of a substance from solid to liquid. The specific latent heat of vaporisation is the heat needed to change 1 kg of a substance from liquid to gas. In both cases the change of state involves no change in temperature.

The molar latent heat of fusion is the heat needed to change 1 mole of a substance from solid to liquid. The molar latent heat of vaporisation is the heat needed to change 1 mole of a substance from liquid to gas.

Units: specific latent heat – J kg^{-1}, molar latent heats – J mol^{-1}.

Examples

Take the specific heat capacity of water to be 4200 J kg^{-1} °C^{-1}.

3.52 How much heat is needed to change 200 g of ice at 0 °C to liquid at the same temperature if the specific latent heat of water at this temperature is 0.34 MJ kg^{-1}?

Ans Heat needed $= mL$

where m is the mass of substance and L the specific latent heat capacity. Hence

$$\text{Heat needed} = 200 \times 10^{-3} \times 0.34 \times 10^{6} = 6.8 \times 10^{4}\ \text{J}$$

3.53 A cardboard cup, of negligible heat capacity, contains 200 g of water at a temperature of 40 °C. Small pieces of ice, at 0 °C, are added to the water in the cup and on melting the temperature falls to 12 °C. What is the mass of ice added if all the heat losses, or gains, by the system are neglected (specific latent heat of fusion of water is 0.34 MJ kg^{-1}).

Ans Heat lost by the water in the cup in cooling down to 12°C is $200 \times 10^{-3} \times 4200 \times (40 - 12)$ J. Heat gained by the ice in melting and the resulting water in increasing its temperature from 0 °C to 12 °C is $m \times 0.34 \times 10^{6} + m \times 4200 \times 12$. As the heat gained equals the heat lost, m is 60 g.

3.54 How much energy is needed to take 150 g of water at 20 °C and convert it into steam at 100 °C (specific latent heat of vaporisation of water is 2.27 MJ kg^{-1})?

Ans Heat needed to raise the temperature of 150 g of water from 20 °C to 100 °C without any change of state is $150 \times 10^{-3} \times 4200 \times 80$ J. The heat needed to change 150 g of water at 100 °C to steam at the same temperature is $150 \times 10^{-3} \times 2.27 \times 10^{6}$ J. Hence the total heat needed is 3.9×10^{5} J.

3.55 Why when liquids evaporate do they show a drop in temperature?

Ans For a liquid to evaporate, i.e. change into a vapour, energy is needed. This latent heat is taken from the liquid and so the temperature of the liquid has to decrease. In molecular terms you can consider only the faster moving molecules to be the ones that escape, i.e. evaporate, and so the average kinetic energy, and hence temperature, of the remaining molecules must decrease as only the slower moving ones are left behind. In practice the cooling due to the evaporation may be offset by the gain in heat from the surroundings.

Further problems

3.56 Lead has a specific latent heat of fusion of 25 kJ kg^{-1}. How much heat is needed to change 200 g of lead at its melting point from solid into liquid?

3.57 Water has a specific latent heat of fusion of 0.34 MJ kg^{-1} and a specific heat capacity of 4200 J kg^{-1} °C^{-1}. How much ice, at 0 °C, will be melted when 200 g of water at 60 °C is poured onto some ice?

3.58 Describe methods which could be used in a school laboratory for the determination of (a) the specific latent heat of fusion of water, (b) the specific latent heat of vaporisation of water.

3.59 The molar latent heat of fusion of lead is 4.77 kJ mol^{-1} at the normal melting point of lead (1 mole of which has a mass of 207 g). (a) How much energy is needed to melt 2 moles of lead? (b) How much energy is needed to melt 10 g of lead?

3.60 In an industrial refrigerator ammonia is vaporised in the cooling unit to produce a low temperature. (a) Why should the evaporation of the ammonia reduce the temperature in the refrigerator? (b) How much ammonia must evaporate to freeze 1 kg of water, at 0 °C, if the specific latent heat of fusion of water is 0.34 MJ kg^{-1} and the specific latent heat of vaporisation of ammonia is 1.34 MJ kg^{-1}?

3.61 Trouton's rule states that the molar latent heat of vaporisation divided by the boiling point at atmospheric pressure is a constant. Use data tables to check this and obtain a value for the constant.

3.62 Can you cool a hot kitchen by opening the doors of the refrigerator?

3.63 An electric coffee pot has a heater of power 400 W. How long will it take to boil the water for one cup of coffee if 300 cm^3 of water at 15 °C are used? How long would it take for the coffee pot to boil dry (specific heat capacity of water is 4200 J kg^{-1} °C^{-1}, specific latent heat of vaporisation of water is 0.86 MJ kg^{-1} and density of water is 1000 kg m^{-3})?

3.64 Why is heat needed to change liquid water into a vapour? What is the energy needed for?

Heat transfer

Notes

Heat transfer through a medium with no visible motion of that medium, but which requires the medium, is called conduction. Heat transfer through a medium by virtue of material motion within the medium is called convection. Heat transfer by electromagnetic radiation is called radiation.

For conduction, when steady state conditions occur and when the heat is transferred through a block of material in just one direction, e.g. a lagged bar,

Quantity of heat transmitted per second $= kA \times (\theta_2 - \theta_1)/x$

where k is a constant called the thermal conductivity, A is the cross-sectional area, and $(\theta_2 - \theta_1)/x$ is the temperature gradient.

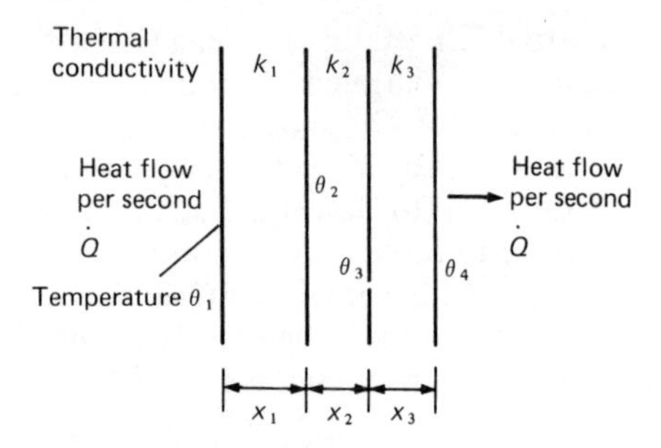

Figure 3.5

If the flow of heat by conduction through a multi-layer structure is considered (*Figure 3.5*) at the steady stage, then for the first layer

Heat flow per second $\dot{Q} = k_1 A \times (\theta_1 - \theta_2)/x_1$

For the second layer,

Heat flow per second $\dot{Q} = k_2 A \times (\theta_2 - \theta_3)/x_2$

For the third layer,

Heat flow per second $\dot{Q} = k_3 A \times (\theta_3 - \theta_4)/x_2$

Rearranging these equations gives

$$\theta_1 - \theta_2 = \frac{\dot{Q}}{A}\frac{x_1}{k_1}$$

$$\theta_2 - \theta_3 = \frac{\dot{Q}}{A}\frac{x_2}{k_2}$$

$$\theta_3 - \theta_4 = \frac{\dot{Q}}{A}\frac{x_3}{k_3}$$

Adding the three equations gives

$$\theta_1 - \theta_4 = \frac{\dot{Q}}{A}\left(\frac{x_1}{k_1} + \frac{x_2}{k_2} + \frac{x_3}{k_3}\right)$$

This equation enables heat transfer by conduction through multi-layer structures to be calculated from a knowledge of the temperatures of just the outer layers.

Convection occurs because an increase in temperature in one part of a fluid results in the fluid having a lower density in that region. This lower density fluid then rises through the colder, higher density fluid and so gives rise to convection currents in which the hotter fluid moves to other regions of the fluid. This is known as natural convection. Forced movement of the fluid by a fan or pump results in what is known as forced convection. For forced convection the rate of loss of heat $\dot{Q}$ per unit area of surface is reasonably proportional to the temperature difference $\Delta\theta$ between the surface and its surroundings. This is known as **Newton's law of cooling.**

$$\frac{\dot{Q}}{A} \propto \Delta\theta$$

For natural convection

$$\frac{\dot{Q}}{A} \propto \Delta\theta^{5/4}$$

The rate of emission of heat radiation from an object is given by **Stefan's law**

$$\frac{\dot{Q}}{A} = e\sigma T^4$$

where T is the temperature of the object on the Kelvin scale, σ is a constant called Stefan's constant and has a value of 56.7×10^{-9} J s^{-1} m^{-2} K^{-4} and e is a quantity called the emissivity of the surface. For a perfect emitter, termed a black body, e has the value of 1. A perfect emitter of radiation is also a perfect absorber of radiation. If an object is at a temperature T in surroundings at a temperature T_s then the net emission of heat from the object is given by

$$\frac{\dot{Q}_{net}}{A} = e\sigma(T^4 - T_s^4)$$

Units: quantity of heat – J, area – m^2, temperature – °C or K, distance – m, thermal conductivity – J s^{-1} m^{-1} °C^{-1} or J s^{-1} m^{-1} K^{-1}, emissivity – no units.

Examples

3.65 What is the rate at which heat will be conducted along a lagged copper rod of length 0.50 m and cross-sectional area 10 cm^2 if the temperature difference between the two ends of the rod is 100 °C (thermal conductivity of copper is 380 J s^{-1} m^{-1} °C^{-1})?

Ans

$$\dot{Q} = kA\frac{\theta_2 - \theta_1}{x}$$

$$= \frac{380 \times 10 \times 10^{-4} \times 100}{0.50}$$

$$= 76 \text{ J/s} = 76 \text{ W}$$

3.66 The cavity wall of a house consists of brick of thickness 120 mm, a 50 mm cavity containing air and then another brick section of thickness 120 mm. The brick has a thermal conductivity of 1.0 J s^{-1} m^{-1} °C^{-1} and the air a thermal conductivity of 0.03 J s^{-1} m^{-1} °C^{-1}. What is the rate at which heat is conducted through a cavity wall of area 15 m^2 when the difference in temperature between the inner and outer surfaces of the cavity wall is 20 °C?

Ans

$$\theta_1 - \theta_4 = \frac{\dot{Q}}{A}\left(\frac{x_1}{k_1} + \frac{x_2}{k_2} + \frac{x_3}{k_3}\right)$$

Hence

$$\dot{Q} = \frac{15 \times 20}{\left(\frac{120}{1.0} + \frac{50}{0.03} + \frac{120}{1.0}\right) \times 10^{-3}}$$

$$= 160 \text{ J/s}$$

3.67 A container of hot liquid in a draughty room is found to be losing heat at the rate of 100 J/s when the liquid is at 60 °C. What is the rate of loss of heat at 30 °C if the surroundings are at 20 °C?

Ans With convection the rate of loss of heat is proportional to the temperature difference between the object concerned and the surroundings, if there is forced convection. When the rate of loss of heat is 100 J/s, the temperature difference is 40 °C. When the container is at 30 °C, the temperature difference is 10 °C. Thus, the rate of loss of heat at this temperature is one quarter of 100 J/s, i.e. 25 J/s.

3.68 An electrical kettle has a thermal capacity of 400 J °C^{-1} and contains 1.0 kg of water. The electrical power to the kettle is switched off when the water reaches boiling point, i.e. 100 °C. After the power is switched off the water is found to cool to 95 °C in 40 s. What was the average rate of loss of heat over the temperature interval 100 °C to 95 °C? How long would the kettle have taken to cool from 100 °C to 90 °C? The surroundings are at 20 °C and the specific heat capacity of water is 4200 J kg^{-1} °C^{-1}.

Ans The rate of cooling at 100 °C to 95 °C is 5/40 °C/s, hence the rate of loss of heat is

$$(400 + 1.0 \times 4200) \times 5/40 = 575 \text{ J/s}$$

The rate of loss of heat will be proportional to the temperature difference between the kettle and its surroundings, if we assume forced convection conditions to occur. Hence

$$575 = h \times (97.5 - 20)$$

where 97.5 °C is the mean temperature of the interval and h is a constant. Hence, for the interval 95 °C to 90 °C,

$$\dot{Q} = h \times (92.5 - 20)$$

Hence

$$\dot{Q} = \frac{72.5 \times 575}{77.5}$$

$$= 538 \text{ J/s}$$

Hence, the time t taken for the temperature to drop from 95 °C to 90 °C is given by

$$(400 + 1.0 \times 4200) \times 5/t = 538$$
$$t = 43 \text{ s}$$

Hence, the time taken for the temperature to drop from 100 °C to 90 °C is 83 s.

3.69 The Sun has a radius of 6.98×10^8 m and behaves as a black body at a temperature of 5800 K. The mean distance of the Sun from the Earth is 1.50×10^{11} m. What is the total power radiated by the Sun and the power received by a surface (normal to the radiation) of area 1.00 m^2 on the Earth if all effects due to the Earth's atmosphere are neglected (Stefan's constant is 56.7×10^{-9} J s^{-1} m^{-2} K^{-4})?

Ans $\dfrac{\dot{Q}}{A} = \sigma T^4$

Hence

Total power $\dot{Q}$
$= 4\pi(6.98 \times 10^8)^2 \times 56.7 \times 10^{-9} \times 5800^4$

The surface area of a sphere is $4\pi r^2$. The total power radiated is thus 3.93×10^{26} W.

This power is dispersed at the Earth's surface to cover a sphere of surface area $4\pi(1.50 \times 10^{11})^2$ m^2. Hence the power received per m^2 is

$$\frac{3.93 \times 10^{26}}{4\pi(1.50 \times 10^{11})^2} = 1.39 \times 10^3 \text{ W m}^{-2}$$

The power received per square metre is called the solar constant.

Further problems

3.70 What is the heat transmitted per second through a rod of lagged steel of length 0.50 m and cross-sectional area 100 cm^2 if there is a temperature difference of 12 °C between the two ends of the rod (thermal conductivity of steel is 54 J s^{-1} m^{-1} °C^{-1})?

3.71 What is the rate of loss of heat through a glass window 2.0 m by 3.0 m and 3.0 mm thick if the temperature difference between the two sides of the window is 30 °C (thermal conductivity of glass is 1.1 J s^{-1} m^{-1} °C^{-1})?

3.72 How would the answer to **3.71** be changed if the window was double glazed? The double glazing involves two sheets of the glass, 3.0 mm thick, separated by an air gap of thickness 5.0 mm and air has a thermal conductivity of 0.026 J s^{-1} m^{-1} °C^{-1}.

3.73 An internal wall in a house consists of a single layer of brick with a layer of plaster on each side of it. If the brick has a thickness of 100 mm and the plaster 15 mm, what will be the heat loss per second through a square metre of such a wall when the temperature difference between the rooms on each side of the wall is 12 °C (thermal conductivity of plaster is 0.5 J s^{-1} m^{-1} °C^{-1} and that of brick is 1.0 J s^{-1} m^{-1} °C^{-1})?

3.74 Describe experiments that could be carried out in a school laboratory for the measurement of the thermal conductivity of (a) a good conductor, (b) a poor conductor of heat. Explain how the results are obtained.

3.75 In one version of the Searle's bar method for the measurement of the thermal conductivity of a good conductor an electric heating coil is wrapped round one end of the sample, a well-lagged bar of cross-sectional area 1.1×10^{-3} m^2, and a coil of copper tubing is wrapped round the other end. Water flows through the copper tubing and temperatures are measured at two points along the bar. If these temperatures are 72.5 °C and 58.0 °C, the distance between the two points is 100 mm and the increase in temperature of the water flowing through the copper tube is 10.8 °C when the flow rate is 14×10^{-4} kg/s, what is the thermal conductivity of the bar (the specific heat capacity of water is 4200 J kg^{-1} °C^{-1})?

3.76 In one version of the Lee's disc experiment for the measurement of the thermal conductivity of a poor conductor a thin disc of the material is sandwiched between two metal discs of the same diameter. An electric heater is used to maintain one of the metal discs at a temperature of 98.9 °C. The other metal disc is at a temperature of 86.8 °C when steady state conditions have been attained. When the current to the heater is switched off the metal disc, initially at 86.8 °C, is found to show an initial rate of cooling of 0.10 °C/s. The heat capacity of the metal disc is 120 J/°C. If the poor conductor has a thickness of 5.0 mm and a diameter of 100 mm, what is its thermal conductivity?

3.77 Explain how the following measures are able to reduce the cost of heating a house. The term thermal conductivity should occur in the answer. (a) Double glazing, (b) covering the loft floor with fibre-glass insulation material and (c) filling the cavity of the cavity walls of a house with a plastic material which includes a large amount of trapped air bubbles.

3.78 A hot object at 80 °C is found to cool by forced convection in surroundings at 20 °C at the rate of 0.6 °C/s. At what temperature will the cooling rate be 0.3 °C/s?

3.79 The planet Venus of radius 6×10^6 m behaves as a black body at a temperature of 240 K. What is the total energy emitted per second by the planet (Stefan's constant is 56.7×10^{-9} J s^{-1} m^{-2} K^{-4})?

3.80 Calculate the apparent temperature of the Sun, if it is assumed to be a black body, from the following data:

Solar constant = 1.40×10^3 W m^{-2}
Ratio of the radius of the Earth's orbit to the radius of the Sun = 216
Stefan's constant = 56.7×10^{-9} J s^{-1} m^{-2} K^{-4}

3.81 A small, blackened sphere is suspended in a vacuum. At what rate must energy be supplied to the sphere to maintain it at 500 K if it has a diameter of 30 mm and the surroundings are at a temperature of 300 K and no conduction occurs along the suspension (Stefan's constant is 56.7×10^{-9} J s^{-1} m^{-2} K^{-4})?

Temperature

Notes

When two systems are each in thermal equilibrium with a third system, they are in thermal equilibrium with each other. This postulate is known as the zeroth law of thermodynamics. When two or more systems are in thermal equilibrium they are said to be at the same temperature.

A temperature scale is specified by selecting some body and a thermometric property of that body, then selecting two fixed points and ascribing values to those points. Assuming a linear relationship between temperatures between these points and the thermometric property gives

$$\theta = \frac{(X_\theta - X_1) \times N}{(X_2 - X_1)}$$

where θ is the temperature on this scale, X_θ is the value of the property at this temperature, X_1 is the value of the property at the first fixed point, X_2 is the value at the second fixed point and N is the number of scale divisions between those two fixed points.

Examples of thermometric properties are:

(i) The length of a liquid column in a glass capillary tube.
(ii) The pressure of a gas which volume is kept constant.
(iii) The vapour pressure of a liquid.
(iv) The resistance of a metal wire or a piece of semiconductor.
(v) The e.m.f. of a thermocouple.
(vi) The colour of the radiation emitted by an object.
(vii) The amount of radiation emitted by an object.

The International Practical Scale of temperature is an internationally agreed scale which specifies a number of fixed points and the type of thermometers to be used to establish the temperatures between those points. The Thermodynamic Kelvin Scale is the fundamental scale. A degree on the Kelvin scale is about the same as a degree on the Celsius scale with the triple point of water (just above the freezing point) being specified as 273.15 K.

Examples

3.82 The length of the mercury thread in a capillary tube is 10 mm at the solidifying point and 250 mm at the boiling point of water. What would be the temperature given by this scale, if the solidifying point is given the scale value of 0 and the boiling point the value 70, when the length of the mercury thread is 100 mm?

Ans $\theta = \dfrac{(100 - 10) \times 70}{(250 - 10)}$

$= 26.25$ (on the scale given)

3.83 A thermocouple with one junction in melting ice gives zero e.m.f. when the other junction is at the melting point of ice and an e.m.f. of 3.50 mV when it is at the boiling point of water. If the melting point is given the value of 0 and the boiling point 100, what will be the temperature on this scale when the e.m.f. is 2.50 mV?

Ans $\theta = \dfrac{(2.50 - 0) \times 100}{(3.50 - 0)}$

$= 71.4$ (on the scale given)

3.84 A constant volume gas thermometer is used to give a temperature scale and uses the value of 0 for the melting point of ice and 100 for the boiling point of water. Why, when it is measuring the same temperature as the thermocouple mentioned in the previous question, is the value of the temperature likely to be different?

Ans Even though the same fixed points are chosen and the same scale values given to them it is unlikely that the scales so specified will agree at all points on those scales. This is because the thermometric properties are not likely to vary in exactly the same way with temperature.

Further problems

3.85 A vapour pressure thermometer gives a pressure reading of 610 N/m^2 at the solidifying point and 101 000 N/m^2 at the boiling point of water. If such a thermometer is used to give a vapour pressure scale of temperature with the solidifying point at 0 and boiling point at 100, what is the room temperature if the vapour pressure at that temperature is 3200 N/m^2?

3.86 Using the Kelvin temperature scale the room temperature for the previous question was 298 K. Why is this result different to that given by the vapour pressure scale of temperature?

3.87 Describe the basic principles of the constant volume gas thermometer, the platinum resistance thermometer and the disappearing filament pyrometer.

3.88 Describe how you would calibrate a mercury-in-glass thermometer for use between 0 °C and 100 °C.

3.89 A resistance thermometer gives a resistance of 52.145 Ω at the solidifying point and 59.361 Ω at the boiling point of water. If such a thermometer is used to give a resistance scale of temperature with the solidifying point at 273 and the boiling point at 373, what is the temperature when the resistance is 55.646 Ω?

The conversion of heat into work

Notes

The work done on or by a sample of material when its volume changes from V_0 to V_1 is given by

$$\text{Work} = \int_{V_0}^{V_1} p\,dV$$

This is the area under the pressure/volume graph for the sample between the V_0 and V_1 ordinates. Heat engines use heat to produce a change in the volume of a gas and hence produce work.

Steam turbines use heat to produce a high pressure or high velocity steam which is then used to cause a turbine to rotate; heat is converted into work.

The internal combustion engine uses a fuel to produce a rise in temperature of a gas and hence expansion; the result is work.

All heat engines can be represented by *Figure 3.6*. There is a heat input Q_h from some high temperature source and waste heat Q_c that flows to a low temperature sink. The difference between these is the work that can be done

$$\text{Work, } W = Q_h - Q_c$$

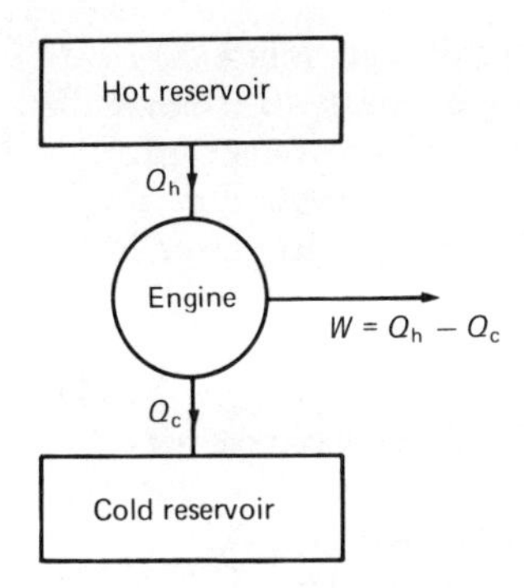

Figure 3.6 The energy flow for a heat engine

The thermal efficiency of an engine is defined as the ratio of the useful work to the heat input to the engine

$$\begin{aligned}\text{Thermal efficiency} &= W/Q_h \\ &= (Q_h - Q_c)/Q_h \\ &= 1 - Q_c/Q_h\end{aligned}$$

The Kelvin temperature scale is defined, in terms of a heat engine, as

$$Q_c/Q_h = T_c/T_h$$

where T_c and T_h are the low temperature sink and high temperature source temperatures expressed on the Kelvin scale. Thus, the efficiency is given by

$$\text{Thermal efficiency} = 1 - T_c/T_h$$

This gives the maximum efficiency an engine can have.

The second law of thermodynamics can be stated as: it is impossible to convert heat energy entirely into work, some of the heat must always be wasted as heat.

Examples

3.90 *Figure 3.7* shows the pressure/volume relationship for an ideal gas. What is the work done when the volume of the gas rises from 1.0×10^{-2} m^3 to 2.0×10^{-2} m^3?

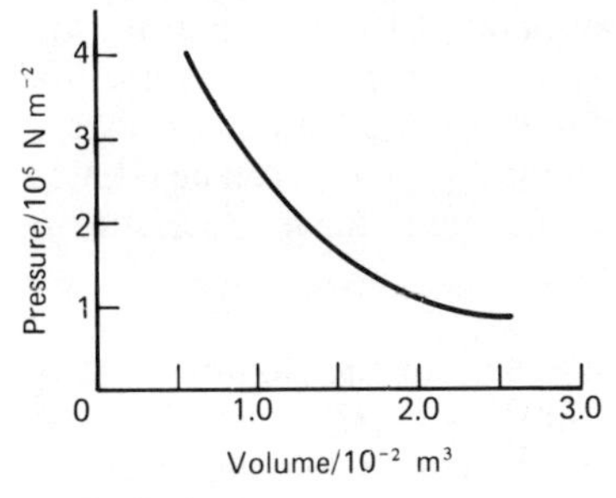

Figure 3.7

Ans The work done is the area under the graph between the specified ordinates which can be estimated from the graph as 2×10^3 J. The answer could be obtained by integration.

3.91 What is the thermal efficiency of an engine that produces, for every 100 J of heat energy input, 20 J of work?

Ans Thermal efficiency $= W/Q_h = 20/100$

The efficiency is 20%.

3.92 What is the maximum efficiency possible for an engine operating between temperatures of 250 °C and 40 °C?

Ans

$$\begin{aligned}\text{Thermal efficiency} &= 1 - T_c/T_h \\ &= 1 - 313/523 \\ &= 0.40\end{aligned}$$

The maximum efficiency is 40%.

Further problems

3.93 What work is done on a gas when its volume is reduced from 5.0×10^{-3} m^3 to 4.8×10^{-3} m^3 if the pressure of the gas remains constant during the change at 6.0×10^5 N/m^2?

3.94 When 1 m^3 of water at 100 °C is converted into steam at that temperature its volume increases to 600 m^3, when the pressure is the normal atmospheric pressure of 1×10^5 N/m^2. What is the work done by the water in this expansion?

3.95 An engine operates between 327 °C and 127 °C. If it absorbs 20 kJ of heat energy at the higher temperature what is (a) the maximum possible work that can be done by the engine, (b) the minimum amount of energy that is wasted as heat?

3.96 What is the maximum thermal efficiency possible for a heat engine operating between 227 °C and 27 °C?

3.97 The efficiency of an engine is 0.20. For every 1000 J heat input to the engine (a) how much work is done by it and (b) how much heat is wasted?

3.98 The thermal efficiency of a power station is 25%. If the power station generates 1 GW of useful energy, what is the waste heat per second?

Energy resources

Notes

Figure 3.8 shows how the world's use of energy is increasing. There is now a very large increase in the energy used each year. Part of the increase is due to the increase in world population, but

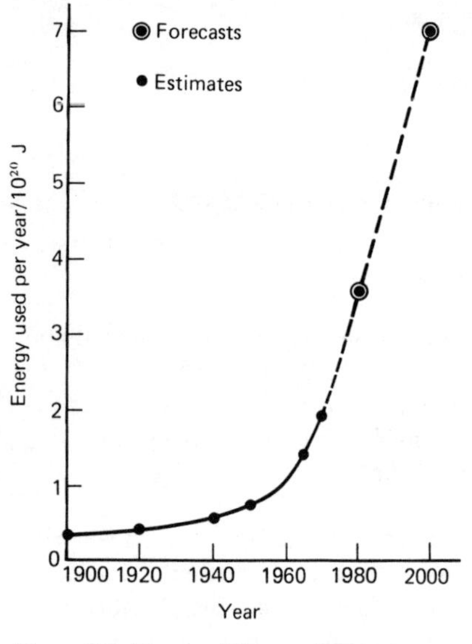

Figure 3.8 The world's use of energy

part is due to the increasing amount of energy needed for each person as a result of the change to a technological society. The world's stock of fossil fuels is limited and estimates place the amount accessible to man as being about 2260×10^{20} J.

Examples

3.99 In the year 1900 the world population was about 1.6×10^9 and in 1950 it was about 2.3×10^9. Use the data given in *Figure 3.8* for the world's use of energy to estimate the amount of energy used, on average, per person (a) in 1900, (b) in 1950. Forecast the amount of energy used per person in the year 2000.

Ans In 1900 the world's use of energy was about 0.3×10^{20} J. This gives an average per person of $0.3 \times 10^{20}/1.6 \times 10^9$ or about 19×10^9 J. In 1950 the world's use of energy had risen to about 0.6×10^{20} J which gives an average per person of about $0.6 \times 10^{20}/2.3 \times 10^9$ or 26×10^9 J. In the first half of the century the population increased by a factor of 1.4. If this same factor is assumed to occur for the second half of the century then the population in the year 2000 would be about 3.2×10^9. The forecast world's use of energy for the year 2000 is about 7×10^{20} J, hence the amount per person might be expected to be about $7 \times 10^{20}/3.2 \times 10^9$ or about 220×10^9 J. As the world population is probably going to increase at a greater rate in the second half of the century this estimate of energy per person is probably an over-estimate. However, it is likely that the figure will certainly be more than 100×10^9 J per person.

Further problems

3.100 Hydroelectricity relies on a cyclic process in which water at a low level is evaporated by the Sun, rises to become a cloud and then falls as rain to a higher level than that from which it started. (a) Describe this process in terms of the sequence of energy transformations that occur. (b) Estimate the energy needed to evaporate 1 kg of water at the low level and so start the cycle. (c) How does the energy used to evaporate the water compare with the possible energy that can be released by that 1 kg of water at the hydroelectric power station?

3.101 In England the Sun supplies, as direct radiation, about 600 W for each square metre when there are no clouds. Suppose solar collectors could be made which had a 50%

efficiency in transforming the radiation into electricity, how big an area of the country would need to be covered with solar collectors to give the same output as a power station generating 1 GW?

3.102 The density of air is 1.2 kg/m^3. Consider a windmill where the air impinging on the sails has a speed of 10 m/s. (a) What mass of air per second hits the sails if the windmill sails have a total area of 30 m^2? (b) What is the total kinetic energy of the air hitting the sails per second? (c) What is the power available to the windmill? (d) If the windmill is 50% efficient, how many such windmills would be needed to replace a power station generating 1 GW?

3.103 The world's reserves of oil have been estimated as about 2×10^{12} barrells (one barrell of oil yields about 6.5×10^9 J). In 1980 oil was used to supply about 1.5×10^{20} J of energy. If oil continued to be used at this rate, how long would it last?

Examination questions

3.104 (a) Outline an experiment to show that linear momentum is conserved in collisions.
(b) In which of the following processes are both momentum and kinetic energy conserved during the impacts?

(i) A collision between two gas molecules.
(ii) A steel ball being dropped onto a stone floor and rebounding to about three quarters of its original height.
(iii) A lump of plasticine being dropped onto a stone floor and not rebounding at all.

For case (ii) discuss the momentum and energy transfers which occur throughout the process.
Sketch two graphs showing how the momentum and kinetic energy of the ball will vary with time from the initial release.

(University of London)

3.105 (a) For a rigid body rotating about a fixed axis explain, with the aid of a suitable diagram, what is meant by angular velocity, kinetic energy and moment of inertia.
(b) In the design of a passenger bus, it is proposed to derive the motive power from the energy stored in a flywheel. The flywheel, which has a moment of inertia of 4.0×10^2 kg m^2, is accelerated to its maximum rate of rotation of 3.0×10^3 revolutions per minute by electric motors at stations along the bus route.

(i) Calculate the maximum kinetic energy which can be stored in the flywheel.
(ii) If, at an average speed of 36 kilometres per hour, the power required by the bus is 20 kW, what will be the maximum possible distance between stations on the level?

(JMB)

3.106 (a) Describe how you would determine the specific heat capacity of a liquid by the continuous flow method.
(b) What are the chief advantages and disadvantages of this method?
(c) What special difficulties would you expect to meet in attempting to use this method for (i) saturated brine and (ii) glycerol?
(d) With a certain liquid, the inflow and outflow temperatures were maintained at 25.20 °C and 26.51 °C, respectively. For a potential difference of 12.0 V and a current of 1.50 A, the rate of flow was 90 g per minute; with 16.0 V and 2.00 A, the rate of flow was 310 g per minute. Find the specific heat capacity of the liquid, and also the power lost to the surroundings.

(Oxford Local Examinations)

3.107 (a) State the principle of conservation of energy. How is this principle applied in a simple experiment to determine the specific heat capacity of a piece of copper? (Experimental details are not required.) Discuss the assumptions which are made.
In your experiment, is the specific heat capacity measured under constant volume (c_v) or constant pressure (c_p)? Would you expect there to be much difference between c_v and c_p? Explain your answer.

(b) Specific latent heats of vaporisation are always much greater than specific latent heats of fusion. Calculate the work done in pushing back the surrounding atmosphere when 1.0 kg of water evaporates and consider whether or not this accounts for the difference in specific latent heat values. What other factors could account for the difference?

Density of steam at 100 °C = 0.60 kg m^{-3}
Specific latent heat of fusion of ice at 0 °C = 3.4×10^5 J kg^{-1}
Specific latent heat of vaporisation of water at 100 °C = 2.3×10^6 J kg^{-1}
Standard atmospheric pressure = 1.0×10^5 Pa

(University of London)

3.108 (a) Define the terms thermal conductivity and specific latent heat of fusion.
(b) Describe a method of determining the latent heat of fusion of ice at 0 °C. Explain how the result is derived from the readings taken and give the precautions which must be taken to minimize experimental errors.
(c) Consider a pond covered with a layer of ice of thickness x, there being a steady temperature difference of θ between the lower and upper surfaces of the ice. If the ice has a density ρ, a thermal conductivity k, a specific latent heat of fusion L and is assumed to have a uniform temperature gradient throughout, derive an expression for the length of time it will take for the thickness of the ice to increase by a small amount δx.

(JMB)

3.109 Outline an experiment to measure the thermal conductivity of a solid which is a poor conductor, showing how the result is calculated from the measurements.

Calculate the theoretical percentage change in heat loss by conduction achieved by replacing a single glass window by a double window consisting of two sheets of glass separated by 10 mm of air. In each case the glass is 2 mm thick. (The ratio of the thermal conductivities of glass and air is 3 : 1.)

Suggest why, in practice, the change would be much less than that calculated.

(University of London)

3.110 (a) Discuss the analogy between the flow of heat under a temperature gradient and the electric current under a potential gradient, and write down the corresponding expressions for thermal conductivity and electrical conductivity.

In the case of a metal, why is thermal conductivity much the more difficult of the two to measure accurately?
(b) Describe how you would measure the thermal conductivity of copper. Mention the chief precautions which must be taken to obtain a reasonably accurate result.

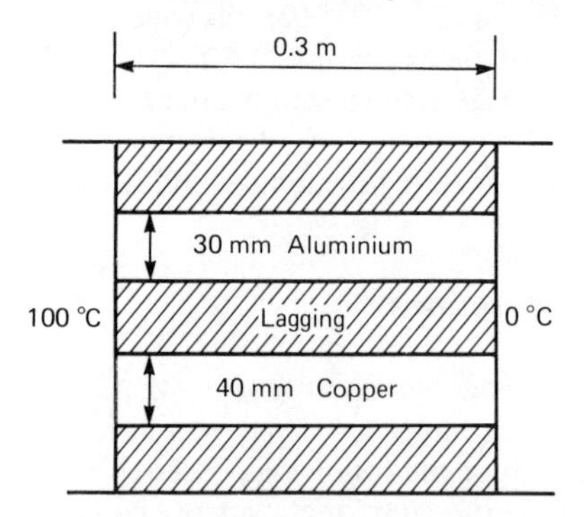

Figure 3.9

(c) *Figure 3.9* shows cylindrical bars of aluminium and copper, each 0.3 m long, which have just been set up in good thermal contact with the flat faces of two large reservoirs maintained at 100 °C and 0 °C, respectively. The diameter of the aluminium bar is 30 mm and that of the copper bar is 40 mm. Both bars are well lagged. The thermal conductivity, specific heat capacity and density of each metal are

Property	*Aluminium*	*Copper*
Thermal conductivity (W m^{-1} K^{-1})	235	380
Specific heat capacity (J kg^{-1} K^{-1})	840	360
Density (kg m^{-3})	2700	8800

(i) In the steady state, calculate the rate of flow of heat down each bar.

(ii) Assuming that both were initially at 15 °C, estimate the energy that has been gained by each bar during the attainment of the steady state.
(iii) Which bar should reach the steady state first? Why?

(Oxford Local Examinations)

3.111 The silica cylinder of a radiant wall heater is 0.6 m long and has a radius of 5 mm. If it is rated at 1.5 kW estimate its temperature when operating. State two assumptions you have made in making your estimate. The Stefan constant is $\sigma = 6 \times 10^{-8}$ W m^{-2} K^{-4}.

(University of London)

Answers

3.9 Ans mgh = force × distance, hence average resistive force = 16 N

3.10 Ans $mgh = 1.3 \times 10^5$ J (or force × distance, both along the incline)

3.11 Ans mgh/t = 18 kW (or force × distance/t, both force and distance along the incline)

3.12 Ans Force along plane = $\mu mg \cos 5^\circ + mg \sin 5^\circ$, force × distance/$t$, both along the incline, gives 99 kW

3.13 Ans $\frac{1}{2}mv^2$ = force × distance, hence force = 0.75 N

3.14 Ans $\frac{1}{2}mv^2$ = force × distance, force = $F + mg \sin 5^\circ$, hence F = 0.32 N

3.15 Ans See *Figure 3.10*

3.16 Ans (a) $Ft = mv$, hence F = 67 N
(b) $\frac{1}{2}mv^2$ = 25 J
(c) 25 J

3.17 Ans Conservation of momentum and kinetic energy gives the velocity of the neutron after the collision as zero, hence the new energy = 0.72 × the initial energy

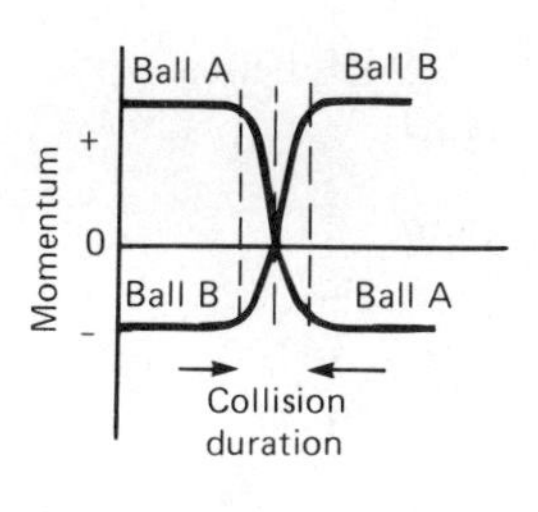

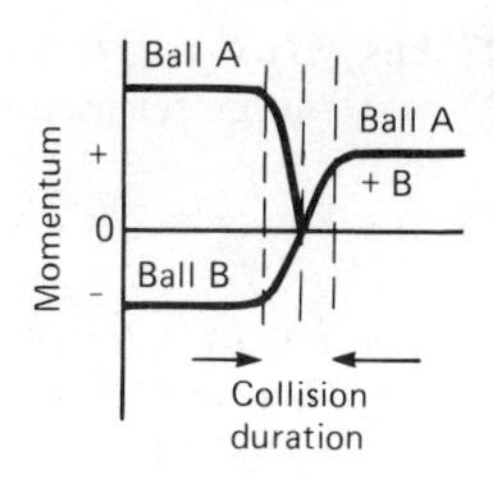

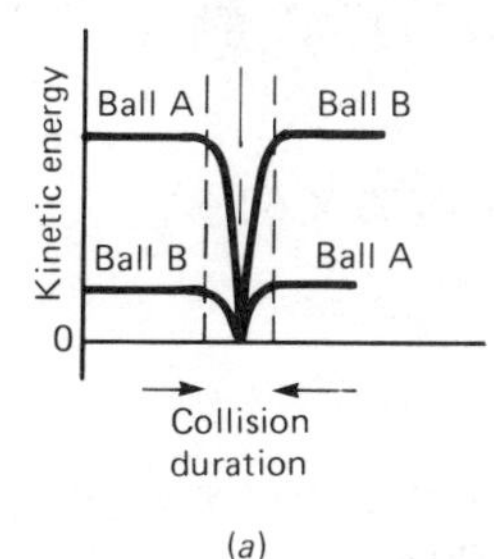

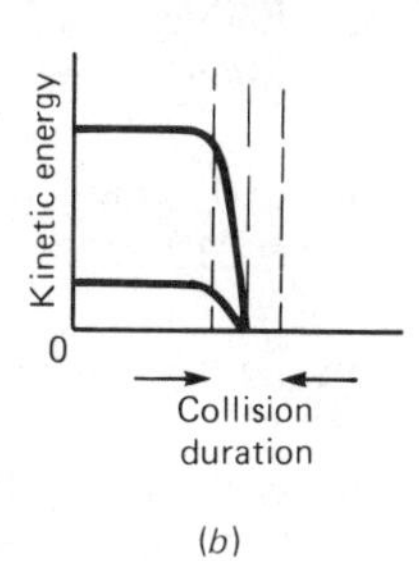

Figure 3.10 (a) Elastic collision; (b) inelastic collision with all kinetic energy lost

3.18 Ans $(mgh + \frac{1}{2}mv^2)/t$ = 2.1 kW if 100% efficient, hence power = 3.5 kW

3.19 Ans Conservation of momentum gives 3.2 m/s, hence loss in K.E. = 12 kJ

3.20 Ans (a) $\frac{1}{2}mv^2 = mgh$, hence v = 7.7 m/s
(b) Conservation of momentum gives 5.5 m/s
(c) 8470 J

3.21 Ans For the rising block $\frac{1}{2}M_w V_w{}^2 = M_w g h_w$, hence V_w = 0.94 m/s (subscript w refers to wooden block), hence conservation of momentum then gives the bullet velocity as 310 m/s

3.22 Ans P.E. = $\frac{1}{2}kx^2 = mgh$, hence h = 5.1 m

3.23 Ans Work = $\frac{1}{2}kx^2$ = 0.020 J, work = $\frac{1}{2}k(40^2 - 20^2) \times 10^{-6}$ = 0.060 J

3.24 Ans Energy = ½ × stress × strain × volume = ½ × stress^2 × volume/E = 34 J

3.25 Ans Energy = $\frac{1}{2}E$ × strain^2 × volume = 0.026 J

3.26 Ans $\frac{1}{2}I\omega^2 = \frac{1}{2}I(2\pi f)^2$ = 95 kJ

3.27 Ans $\frac{1}{2}I\omega^2 = \frac{1}{2}I(2\pi f)^2 = 9.9$ kJ, hence energy released = 7.4 kJ

3.28 Ans Power = $2\pi fT = 1.5$ kW

3.29 Ans $T = I\alpha = I\omega/t = I \times 2\pi f/t = 190$ N m. Gain in K.E. = $\frac{1}{2}I\omega^2 = \frac{1}{2}I(2\pi f)^2$ = 300 kJ

3.30 Ans Energy = $\frac{1}{2}I\omega^2 = \frac{1}{2}I(2\pi f)^2$, hence $I = 51$ kg m^2

3.36 Ans $180 \times 60 + 0.80 \times 4200 \times 60$ = 212 kJ

3.37 Ans $\frac{1}{2}mv^2 = 8000$ J

3.38 Ans $m \times 2100 \times 10 = 150 \times 10^{-3} \times 4200 \times 20$, hence $m = 600$ g

3.39 Ans $200 \times 10^{-3} \times 2400 \times \theta = 150 \times 3 \times 60$, hence $\theta = 56$ °C

3.40 Ans $200 \times 10^{-3} \times 380 \times (\theta - 18) + 120 \times 10^{-3} \times 4200 \times (\theta - 18) = 50 \times 10^{-3} \times 480 \times (100 - \theta)$, hence $\theta = 21.3$ °C

3.41 Ans $\frac{1}{2}mgh = ms\theta$, hence $\theta = 0.041$ °C

3.42 Ans Continuous flow calorimeter could be described with elimination of heat losses by repetition with the same temperatures, but a different flow rate

3.43 Ans As in **3.42**. Heat capacity plays no part because there is no change in temperature of the container

3.44 Ans See **3.33**; $c = 4208$ J kg^{-1} °C^{-1}

3.45 Ans $0.40 \times 4200 \times (\theta - 45.5) = 40 \times (45.5 - 15.0)$, hence $\theta = 46.2$ °C

3.46 Ans $0.300 \times s \times 0.60 = 3.00 \times 9.81 \times 1.00 \times 26$, hence $s = 4250$ J kg^{-1} °C^{-1}

3.49 Ans $C_p = C_v + R_0$, hence $C_v = 20.8$ J mol^{-1} °C^{-1}, and specific heat = 650 J kg^{-1} °C^{-1}

3.50 Ans Energy for 1 mole = 2020 J, energy for 1 kg = 1.01×10^6 J

3.51 Ans $C_p = C_v + R_0$, hence $C_p = 20.9$ J mol^{-1} °C^{-1}

3.56 Ans $mL = 5.0$ kJ

3.57 Ans $mL = Ms\theta$, hence $m = 150$ g

3.58 Ans I leave this to you

3.59 Ans (a) 9.54 kJ
(b) 230 J

3.60 Ans (a) See **3.55**
(b) $m \times 1.34 \times 10^6 = 1 \times 0.34 \times 10^6$, hence $m = 0.25$ kg

3.61 Ans Constant of about 84 J K^{-1}

3.62 Ans No

3.63 Ans $400t = 300 \times 10^{-6} \times 1000 \times 4200 \times 85$, hence $t = 268$ s. To boil dry an extra time t is required, where $400t = 300 \times 10^{-6} \times 1000 \times 0.86 \times 10^6$, i.e. $t = 645$ s, hence total time to boil dry is 913 s

3.64 Ans The energy is needed to break the bonds between the water molecules that occur in the liquid and so allow the water molecules to escape

3.70 Ans $\dot{Q} = kA(\theta_2 - \theta_1)/x = 13$ J/s

3.71 Ans $\dot{Q} = kA(\theta_2 - \theta_1)/x = 6.6 \times 10^4$ J/s

3.72 Ans The rate of heat loss is reduced, $\dot{Q} = A(\theta_2 - \theta_1)/(x_1/k_1 + x_2/k_2 + x_3/k_3)$ = 910 J/s

3.73 Ans $\dot{Q}/A = (\theta_2 - \theta_1)/(x_1/k_1 + x_2/k_2 + x_3/k_3)$ = 75 J s^{-1} m^{-2}

3.74 Ans (a) Describe Searle's bar
(b) Lee's disc is a suitable experiment

3.75 Ans $\dot{Q} = kA(\theta_2 - \theta_1)/x = ms\Delta\theta/t$, hence $k = 398$ J s^{-1} m^{-1} °C^{-1}

3.76 Ans $\dot{Q} = kA(\theta_2 - \theta_1)/x = 120 \times 0.10$, hence $k = 0.63$ J s^{-1} m^{-1} °C^{-1}

3.77 Ans (a) Discuss the effect of an air gap in terms of low thermal conductivity of air trapped between the panes
(b) Fibre-glass has a low thermal conductivity and when in contact with a ceiling reduces the rate of flow of heat through the assembly
(c) The material has a low thermal conductivity. It does, also, prevent convection currents in the air in the cavity

3.78 Ans $\dot{Q} \propto \Delta\theta$, hence temperature is 50 °C

3.79 Ans $\dot{Q} = A\sigma T^4 = 8.5 \times 10^{16}$ J/s

3.80 Ans $\dot{Q} = A\sigma T^4 = 4\pi R^2 \times 1.40 \times 10^3$, hence $T = 5800$ K

3.81 Ans $\dot{Q} = A\sigma(T^4 - T_s^4) = 8.72$ J/s

3.85 Ans $\theta = (X_\theta - X_1) \times N/(X_2 - X_1) = 2.58$

3.86 Ans According to the Kelvin scale the vapour pressure varies non-linearly with temperature. With the temperature scale defined by the vapour pressure the relationship is defined as linear

3.87 Ans Mention the property which varies with temperature and how it is measured

3.88 Ans Mention how it would be calibrated at the freezing and boiling points of water, at standard pressure

3.89 Ans $\theta = (X_\theta - X_1) \times N/(X_2 - X_1) = 48.52$, hence temperature = 321.52

3.93 Ans $6.0 \times 10^5 \times 0.2 \times 10^{-3} = 120$ J

3.94 Ans $1 \times 10^5 \times 599 = 599 \times 10^5$ J

3.95 Ans (a) $6\frac{2}{3}$ kJ
(b) $13\frac{1}{3}$ kJ

3.96 Ans 40%

3.97 Ans (a) 200 J
(b) 800 J

3.98 Ans 3 GW

3.100 Ans (a) Heat gives evaporation which results in potential energy. This then results in heat energy when condensation occurs, then kinetic energy as rain falls. The water does not lose all its potential energy when it falls to a higher level than that from which it evaporated
(b) Energy = specific latent heat = 2.27 MJ/kg. This is the figure at 100 °C, the value at 0 °C is higher
(c) Suppose the 1 kg of water falls through 100 m to hit the turbine at the power station. This would give 980 J, assuming the turbine was 100% efficient. It is much less than the latent heat

3.101 Ans 3.3×10^6 m^2

3.102 Ans (a) 360 kg/s
(b) 18 kW
(c) 18 kW
(d) 111 000
Note: 50% is high

3.103 Ans 87 years

3.104 Ans (a) Possibly a collision between gliders on an air track or trolleys with velocities being determined before and after the collision
(b) (i) K.E. and momentum are conserved if no chemical reaction or excitation of spectra occurs. (ii) Momentum only is conserved. (iii) Momentum only is conserved
In case (ii) velocity after rebound = ¾ of velocity before collision, hence for momentum $mv_0 = MV + \frac{3}{4}mv_0$ and for K.E., $\frac{1}{2}mv_0^2 = \frac{1}{2}MV^2 + \frac{1}{2}m(\frac{3}{4}v_0)^2$, where M is the mass of the Earth and V the velocity it acquires. During the fall the velocity increases to $v = gt$, where g is the acceleration due to gravity. Hence, the momentum increases up to the point of collision. Afterwards the momentum reverses sign, but the kinetic energy does not change sign at the collision (see *Figure 3.11*)

3.105 Ans (a) See chapter 1 and chapter 3 notes
(b) (i) $\frac{1}{2}I\omega^2 = \frac{1}{2}I(2\pi f)^2 = 20$ MJ
(ii) At 20 kW energy lasts for 1000 s, hence distance = 10^4 m

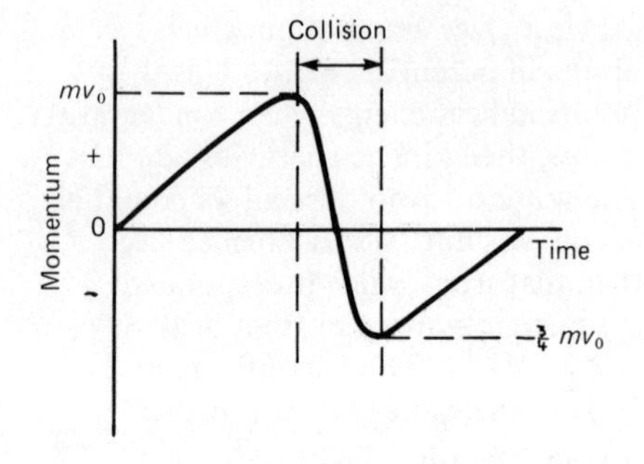

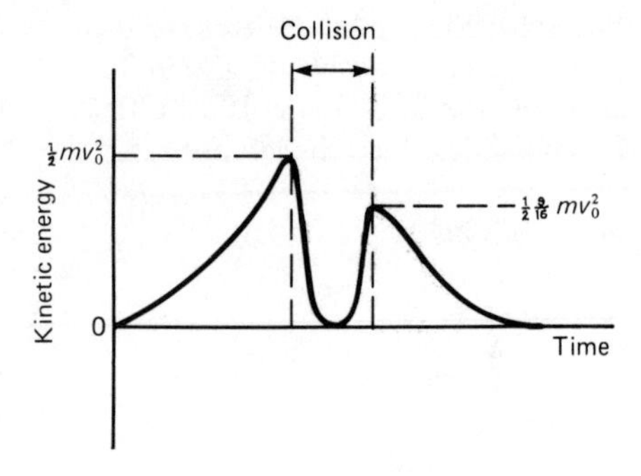

Figure 3.11

3.106 Ans (a) See the notes in this chapter
(b) Advantages: heat losses can be eliminated, heat capacity of apparatus is not needed and it is capable of giving high accuracy. Disadvantages: high viscosity fluids and saturated solutions cannot be used and it requires a large amount of fluid and a constant flow pump
(c) (i) Salt precipitating and also corrosion of the heating element (ii) Difficulty in obtaining a flow through the apparatus
(d) Specific heat capacity = 2915 J kg^{-1} °C^{-1}; power lost = 12.27 W

3.107 Ans (a) See the notes in this chapter. Specific heat capacity is usually at constant pressure (atmospheric pressure). For a solid there is virtually no difference as the expansion is so small and hence the $p\Delta V$ term is small
(b) See **3.94**. Energy is also needed to break the bonds between molecules in the liquid state. Work = $1 \times 10^5 \times (1/0.60) = 1.7 \times 15^5$ J

3.108 Ans (a) See the notes in this chapter
(b) You could describe a method involving an electrical heater placed in melting ice in a funnel, so the amount of ice melted by the heater can be determined. The effect of heat gains from the surroundings can be estimated by having a control experiment in which the heater is not switched on
(c) $\delta Q/\delta t = kA\theta/x$, but $\delta Q = L\delta m = \rho LA\delta x$, hence $\delta t = \rho Lx\delta x/(k\theta)$

3.109 Ans Lee's disc, see **3.76**; double glazing, see **3.72**, 94%. Conduction through the frame of the window and convection currents in the air between the glass lower the change. There are, with both single and double glazing, stagnant air layers at the glass surfaces and these noticeably improve the single glazing figure.

3.110 Ans (a) $I = \sigma AV/L$, $\dot{Q} = kA\Delta\theta/x$, potential gradient V/L compares with thermal gradient $\Delta\theta/x$; current I compares with rate of heat flow $\dot{Q}$. With heat there is difficulty in ensuring uniaxial flow. Also, we cannot measure directly rate of flow of heat, but we can directly measure current
(b) Describe Searle's bar (see **3.75**)
(c) (i) Aluminium, $\dot{Q}$ = 55.4 J/s; copper, $\dot{Q}$ = 159.2 J/s (ii) Aluminium gains 1.68×10^4 J, copper gains 4.18×10^4 J. The average bar temperature has been taken as 50 °C in $Q = mc\Delta\theta$. (iii) Copper, though it requires more energy the flow is faster

3.111 Ans Assume that the heater is a black body, that there is 100% electrical-to-radiation transformation and that the temperature of the surroundings can be ignored. $\dot{Q}/A = \sigma T^4$ and $\dot{Q} = IV$, hence $T = 4800$ K

4 Basic electricity and magnetism

Circuits involving resistors

Notes

The resistance of a circuit component is the potential difference existing across the component divided by the current through it.

$$R = V/I$$

A component is said to obey Ohm's law if the current is directly proportional to the potential difference.

The resistance of metals generally increases with temperature, the following equation reasonably representing this variation.

$$R_\theta = R_0(1 + \alpha\theta)$$

where R_0 is the initial resistance, R_θ the resistance after a temperature change of θ and α the temperature coefficient of resistance. Semiconductors show a non-linear decrease in resistance with an increase in temperature.

The resistance R of a piece of material is related to the length L of the sample and the cross-sectional area A by

$$R = \rho L/A$$

where ρ is the resistivity (in Ω m).

For resistors in series the potential difference across the total group of resistors is equal to the sum of the potential differences across each resistor, the current through each resistor being the same. The series combination has a resistance equal to the sum of the resistances of each individual resistor.

For resistors in parallel the current entering the parallel arrangement is equal to the sum of the currents through each resistor, the potential differences across each resistor being the same. The parallel combination has a resistance R given by

$$\frac{1}{R} = \frac{1}{R_1} + \frac{1}{R_2} + \text{etc.}$$

If a voltmeter is connected across a voltage supply, e.g. a battery, the potential difference indicated will depend upon the current taken from the supply. The value that the potential difference would have if zero current were drawn is known as the electromotive force (e.m.f., E).

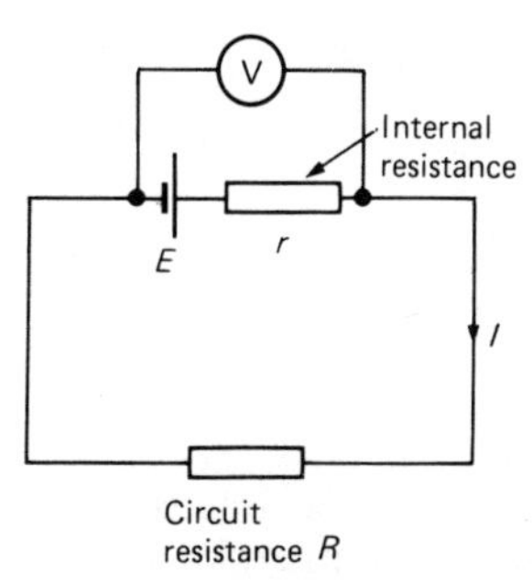

Figure 4.1

The reason for the drop in potential difference when a current is drawn is the existence of an internal resistance to the voltage supply. For the circuit shown in *Figure 4.1*

$$E = Ir + V$$

Units: current – A, potential difference – V, e.m.f. – V, resistance – Ω (ohm).

Examples

4.1 What is the resistance of a system consisting of two resistors in parallel if the resistors have resistances of 10 Ω and 5 Ω?

Ans

$$\frac{1}{R} = \frac{1}{R_1} + \frac{1}{R_2}$$

$$= \frac{1}{10} + \frac{1}{5}$$

$$= 3/10$$

Hence R is $3\frac{1}{3}$ Ω

4.2 A torch battery has an e.m.f. of 1.5 V and an internal resistance of 2 Ω. What is the potential difference across the terminals of the battery when a current of 0.2 A is being taken from it?

Ans $E = Ir + V$

Hence

$$V = 1.5 - 0.2 \times 2$$
$$= 1.1 \text{ V}$$

4.3 Copper has a resistivity of 1.7×10^{-8} Ω m. What is the resistance of a length of copper wire of cross-sectional area 1 mm^2 and length 2 m?

Ans $R = \rho L/A$

$$= \frac{1.7 \times 10^{-8} \times 2}{1 \times 10^{-6}}$$
$$= 3.4 \times 10^{-2} \ \Omega$$

4.4 An electric cable consists of six strands of bare copper wire twisted together. If each wire has a cross-sectional area of 1 mm^2, what will be the resistance of a 20 m length of the cable (resistivity of copper is 1.7×10^{-8} Ω m)?

Ans The resistance of each wire is given by

$$R = \rho L/A$$
$$= \frac{1.7 \times 10^{-8} \times 20}{1 \times 10^{-6}}$$
$$= 0.34 \ \Omega$$

The resistance of the cable is thus that of six resistors each of resistance 0.34 Ω in parallel. Hence

$$1/R = 6/0.34$$
$$R = 0.057 \ \Omega$$

Further problems

Take the resistivity of copper to be 1.7×10^{-8} Ω m and that of aluminium to be 2.8×10^{-8} Ω m.

4.5 Explain how you would determine in the laboratory whether a resistor obeys Ohm's law.

4.6 Resistors having resistances of 50 Ω and 150 Ω are connected in series across a 6.0 V d.c. supply of zero internal resistance. What is the potential difference across each resistor? What is the current through each resistor?

4.7 Resistors having resistances of 20 Ω and 100 Ω are connected in parallel with each other and with a d.c. supply of 2.0 V. What is the current through each resistor?

4.8 What is the maximum resistance that can be obtained by combining resistors of resistances 2 Ω, 4 Ω and 6 Ω?

4.9 What is the resistance of 10 km of copper wire with a cross-sectional area of 1 mm^2?

4.10 A standard 5.00 Ω resistor coil is made of manganin (a copper, manganese and nickel alloy) wire having a resistivity of 4.30×10^{-7} Ω m. How long is the wire in the coil if the wire has a radius of 0.50 mm?

4.11 A copper cable consists of ten strands of bare copper wire, each strand having a cross-sectional area of 2.0 mm^2. What is the resistance of 10 m of such a cable if the copper has a resistivity of 1.7×10^{-8} Ω m?

4.12 How must the radii of a copper wire and an aluminium wire compare if they have the same length, the same potential difference applied across them and the same current is to be passed by each wire?

4.13 Copper (density 8900 kg/m^3) and aluminium (density 2700 kg/m^3) are possible materials for the overhead cables of an electrical railway system. What will be the mass of the cables per kilometre if the resistance per kilometre is to be 0.20 Ω?

4.14 The temperature coefficient of resistance of copper is 0.0043 °C^{-1}. What will be the change in resistance of the copper windings of a motor if the initial resistance is 100 Ω and the temperature rises during the running of the motor by 50 °C?

4.15 A battery of e.m.f. 1.5 V and internal resistance 1.0 Ω is connected to a torch bulb. If the bulb takes 0.30 A current, what will be the potential difference across the bulb?

4.16 A voltmeter, connected across the terminals of a battery, shows a potential difference of 1.3 V when the current taken from the battery is 0.2 A, and 1.0 V when it is 0.4 A. What is (a) the e.m.f., (b) the internal resistance of the battery?

Measurements

Notes

The range of an ammeter can be changed by connecting resistors in parallel with the instrument, such resistors being known as shunts. The range of a voltmeter is changed by putting resistors in series with the instrument, such resistors being known as multipliers.

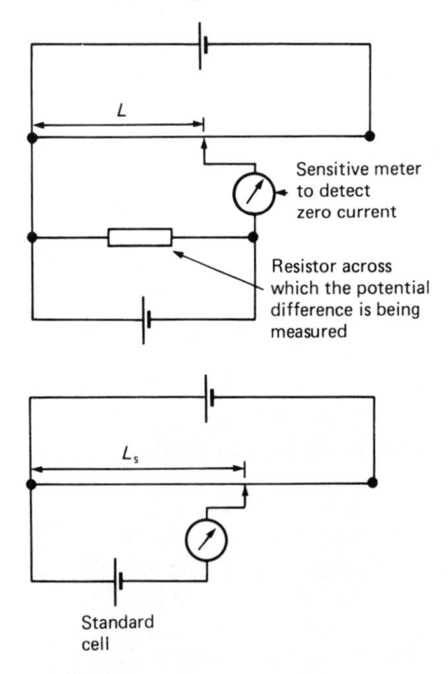

Figure 4.2

Potential differences can be measured by means of a potentiometer circuit (*Figure 4.2*). The unknown potential difference is compared with a variable potential difference, the variable potential difference being adjusted until it is equal and opposite to the one being measured. When this occurs there is no current flow. The variable potential difference is produced by sliding a contact along a length of wire or other linear resistor. The distance (L) moved by the contact along the wire is thus a measure of the potential difference (V). The wire can be calibrated in terms of potential difference by replacing the unknown potential difference with a standard cell of known e.m.f. (V_s).

$$V/V_s = L/L_s$$

The potentiometer is a null method in that the meter is only used to detect when the current is zero. This enables a very sensitive, uncalibrated meter to be used. Because at the zero current condition there is no current being drawn from a cell the e.m.f. of the cell is being compared with the variable potential difference.

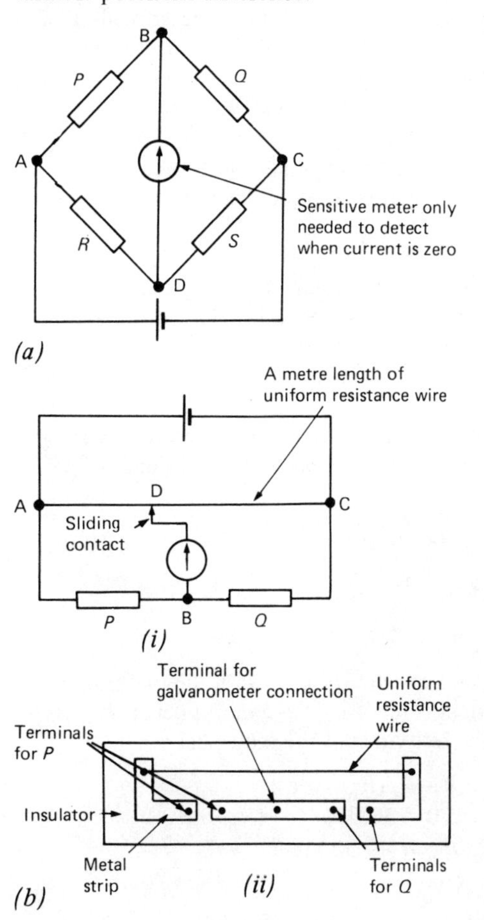

Figure 4.3 (a) The basic Wheatstone bridge circuit; (b) the metre bridge: (i) the circuit, (ii) a simple form of apparatus

By measuring the potential difference across a resistor of known resistance the current flowing through that resistor can be determined. The potentiometer method can thus be used to give an accurate measurement of current.

Figure 4.3(a) shows the basic form of the Wheatstone bridge used for the measurement of resistance. *Figure 4.3(b)* shows a common form of this bridge. When there is no current through the galvanometer

$$P/Q = R/S$$

Examples

4.17 A meter movement has a resistance of 100 Ω and gives a full scale reading with a current of 50.0 μA. (a) How can this meter movement be converted to give an ammeter reading of 500 mA full scale? (b) How can the meter movement be converted to give a voltmeter reading of 5.0 V full scale?

Ans (a) A resistor in parallel with the meter movement is required, i.e. a shunt.

$$I = I_g + I_s$$

where I is the total current, I_g the current through the meter and I_s the current through the shunt.

$$V = I_g R_g = I_s R_s$$

where V is the potential difference across both the meter and the shunt. Hence

$$I = I_g\,(1 + R_g/R_s)$$
$$500 \times 10^{-3} = 50.0 \times 10^{-6}\,(1 + 100/R_s)$$

Hence

$$R_s = 0.010\ \Omega$$

(b) A resistor in series with the meter movement is required, i.e. a multiplier. If R_m is the resistance of the multiplier

$$V = I_g\,(R_m + R_g)$$
$$5.0 = 50.0 \times 10^{-6}(R_m + 100)$$
$$R_m = 99\,900\ \Omega$$

4.18 The e.m.f. of a battery is being determined by means of a slide-wire potentiometer. A standard cell of e.m.f. 1.018 V gives a zero reading on the galvanometer when connected across 160 mm of the slide wire. The battery gives a zero reading when connected across 320 mm. What is the e.m.f. of the battery?

Ans $V/V_s = L/L_s$

Hence

$$V = 1.018 \times 320/160$$
$$= 2.036\ \mathrm{V}$$

4.19 The left-hand gap of a metre bridge (*Figure 4.3(b)*) is closed by a coil having a resistance of 20 Ω and the right-hand gap by an unknown resistance. The zero current position is found to occur when the sliding contact is 250 mm from the left-hand end of the 1000 mm long bridge wire. What is the value of the unknown resistance?

Ans
$$\frac{20}{250} = \frac{R}{(1000 - 250)}$$

Hence

$$R = 60\ \Omega$$

Further problems

4.20 Explain how a basic moving coil meter movement can be used to give both a voltmeter and an ammeter with different ranges.

4.21 An ammeter is to be made from a meter giving a full-scale deflection with 100 μA and having a resistance of 200 Ω. The ammeter is to give a full-scale deflection with 10 A. What resistor should be used for this adaptation?

4.22 A meter movement has a resistance of 50 Ω and gives a full-scale reading with 100 μA. What are the shunt resistances required to convert the movement into a meter with full-scale deflections of (a) 500 μA, (b) 250 mA and (c) 5 A?

4.23 A meter movement has a resistance of 500 Ω and gives a full-scale reading with 40 μA. How can this movement be converted into a voltmeter giving a full-scale reading of 10 V?

4.24 The resistance of the coil of a basic meter movement is 20 Ω and it gives a full-scale deflection with 10 mA. It is converted into a multirange ammeter by the circuit shown in *Figure 4.4*. What are the values of resistances R_1, R_2 and R_3?

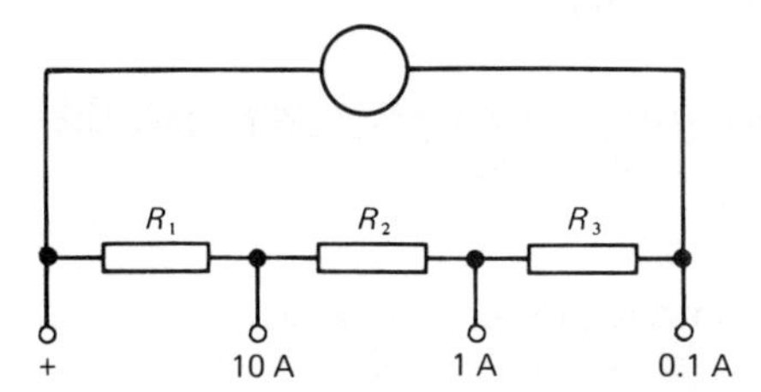

Figure 4.4

4.25 Describe how you would use a potentiometer to determine the e.m.f. of a cell when it has a value of the order of 2 V.

4.26 What are the advantages of a null method?

4.27 A potentiometer used to measure an unknown e.m.f. is found to always give a galvanometer deflection in the same direction, regardless of where the contact is made with the potentiometer wire. What faults could be responsible for this?

4.28 A standard cell of e.m.f. 1.018 V gives a zero reading on the galvanometer in a potentiometer experiment when the contact with the potentiometer wire is at the 340 mm position. A dry battery is found to give a zero reading when the contact is at the 520 mm position. What is the e.m.f. of the battery?

4.29 With a potentiometer a standard cell of e.m.f. 1.018 V is found to give zero current in the galvanometer when the sliding contact on the potentiometer wire is 240 mm from one end of the wire. When the potentiometer is connected across a standard resistance of 2.00 Ω the zero current position is at 150 mm. What is (a) the potential difference across the resistor, (b) the current through it?

4.30 Two resistors are connected in series and the same current thus passes through them. When a potentiometer is used to determine the potential differences across each resistor zero current occurs when the potentiometer wire lengths are 135 mm and 670 mm. What is the ratio of the two resistances?

4.31 How can the simple potentiometer circuit shown in *Figure 4.2* be adapted to enable the e.m.f. of a thermocouple to be determined? A thermocouple has typical e.m.f. values of the order of millivolts.

4.32 A potentiometer wire of length 1.000 m and diameter 0.50 mm is made of wire having a resistivity of 100×10^{-8} Ω m. What is the potential gradient down the wire when it has a battery, of negligible internal resistance and supplying 2.0 V, connected across it? What would be the potential gradient if a 1000 Ω resistor was connected in series with the potentiometer wire?

4.33 Equal lengths of manganin and nichrome (a nickel and chromium alloy) wire are connected in a metre bridge circuit, one length in each of the gaps for resistors. Zero current is found to occur when the balance point on the wire is 400 mm from the end in which the manganin is connected. If the length of the wire is 1000 mm, how do the diameters of the two wires compare (manganin has a resistivity of 43×10^{-8} Ω m and nichrome one of 100×10^{-8} Ω m)?

4.34 A Wheatstone bridge is balanced when a standard resistor of value 1.00 Ω is in one of the bridge arms. When this is replaced by a nominal 1 Ω resistor the balance condition is only obtained when the resistor is shunted with 1.50 m of eureka (a copper–nickel alloy) wire of diameter 0.1 mm and resistivity 49×10^{-8} Ω m. What is the resistance of the nominal 1 Ω resistor?

Energy in a circuit having resistance

Notes

When a current I flows through a resistor of resistance R and gives a potential difference V across it,

Power = energy dissipated per second
$= IV$
$= I^2R$
$= V^2/R$

When the current is in amps and the potential difference in volts, the power is in J/s or watts. The unit used for electricity bills is, however, not the joule but the kilowatt-hour (kWh)

$1 \text{ kWh} = 3.6 \times 10^6 \text{ J}$

With alternating current the mean power developed is $\frac{1}{2}I_m^2R$, where I_m is the maximum value of the current during a cycle. The same power could have been achieved with a steady current of value I, where

$$I^2 = \tfrac{1}{2}I_m^2$$

This equivalent current is referred to as the root mean square current.

$$I_{RMS} = I_m/2^{1/2}$$

Similarly

$$V_{RMS} = V_m/2^{1/2}$$
$$V_{RMS} = I_{RMS}R$$

Units: current – A, potential difference – V, power – W.

Examples

4.35 What is the energy dissipated in 1 minute when an immersion heater of power 5.0 kW is being used?

Ans Energy = power × time
$= 5.0 \times 10^3 \times 60$
$= 3.0 \times 10^5 \text{ J}$

4.36 What power of electric fire can be connected to the 240 V mains supply if the maximum fuse value possible is 13 A?

Ans Power $= IV$
$= 13 \times 240$
$= 3.1 \text{ kW}$

A lower power than this would be used in practice if the fuse is not constantly to be replaced due to slight fluctuations in the voltage supply and the variability of fuses.

4.37 How many units of electricity (kWh) are used when an electric fire element of power 1 kW is used for 4 hours?

Ans One kilowatt for 4 hours is 4 kWh, hence the number of units used is 4.

4.38 The a.c. mains voltage supply has a root mean square voltage of 240 V. What is the maximum voltage?

Ans $V_{RMS} = V_m/2^{1/2}$

Hence

$$V_m = 240 \times 2^{1/2} = 339 \text{ V}$$

Further problems

4.39 The power rating for a 50 Ω resistor is ½ W. What is the maximum current that the resistor can safely pass?

4.40 What power is supplied to an electric heater if the potential difference across it is 12 V and the current through it is 2.0 A?

4.41 Which filament has the higher resistance – that in a 50 W or that in a 100 W lamp bulb when both are operated on the 240 V mains supply?

4.42 What is the resistance of a light bulb that dissipates 100 W when operated at a root mean square voltage of 240 V?

4.43 How many units of electricity (kWh) are used when a tape recorder of power 50 W is run for 5 hours?

4.44 What is the root mean square current passing through a resistor of resistance 50 Ω when the root mean square potential difference across it is 10 V?

Electrolysis

Notes

Compounds which conduct electricity when molten or in solution are called electrolytes and the conductors which conduct the current into the liquid are called electrodes. The electrode connected to the positive side of the d.c. supply is the anode and the one connected to the negative side is the cathode. The process of electrical conduction through a liquid is called electrolysis and takes place as the result of the movement of ions, which are charged compounds, parts of compounds or elements. The current is the rate of movement of charge.

The charge carried by ions is of certain values only, namely integral multiples of 1.6×10^{-19} C (Coulomb). Copper sulphate solution contains positive copper ions carrying two units of charge, i.e. Cu^{2+}, and negative sulphate ions carrying two units of charge, i.e. SO_4^{2-}. Sodium chloride solution contains positive sodium ions carrying one unit of charge, i.e. Na^+, and negative chlorine ions carrying one unit of charge, i.e. Cl^-. The charge required to liberate one mole of a substance is called one Faraday and has the value 96 500 C. Thus, the mass liberated at an electrode during electrolysis is directly proportional to the charge passing between the electrodes, i.e. for a constant current the product of the current and the time. One Faraday can be considered to be the charge carried by one mole of electrons.

Examples

4.45 How much charge Q flows past a point in an electrical circuit in 3 s if there is a current of 0.4 A?

Ans Current = rate of movement of charge
$0.4 = Q/3$
$Q = 1.2$ C

4.46 When a current passes between copper electrodes dipped in copper sulphate solution, copper is deposited on one of the electrodes. (a) On which electrode is the copper deposited? (b) How much charge has to flow to deposit one mole of copper? (c) How long will a current of 5 A have to flow to deposit one mole of copper? (d) How long will a current of 4 A have to flow to deposit 2 g of copper (one mole of copper has a mass of 64 g, one Faraday is 96 500 C and the copper ion is Cu^{2+})?

Ans (a) The cathode
(b) Two Faradays, i.e. 193 000 C
(c) $t = 193\,000/5$
$= 38\,600$ s
(d) 2 g is 2/64 of a mole, hence the charge required is 2/64 th of two Faradays, i.e. 6000 C and hence

$$t = 6000/4 = 1500 \text{ s}$$

Further problems

4.47 Explain the terms electrode, electrolysis and electrolyte.

4.48 One mole of silver is liberated in electrolysis by the passage of 96 500 C. What is the charge carried by the silver ion (Avogadro's number is 6×10^{23})?

4.49 Use the data from the previous question. What is the mass of silver liberated by a current of 2.0 A passing through a silver solution for 5 minutes (one mole of silver has a mass of 108 g)?

4.50 For what time must a current of 10 A be passed through a solution containing copper ions if 10 g of copper is to be deposited on the cathode (one Faraday is 96 500 C, the copper ion is Cu^{2+} and one mole of copper has a mass of 64 g)?

Cells

Notes

Knowledge of the e.m.f. of each metal with respect to some standard electrode allows the e.m.f. of any pair of metals acting as a cell to be computed. The table which gives the e.m.f. of each metal in relation to a hydrogen electrode is known as the electrochemical series.

Electrode	*E.m.f. relative to hydrogen electrode/V*
Magnesium	−2.37
Aluminium	−1.66
Zinc	−0.76
Iron	−0.44
Tin	−0.14
Lead	−0.13
Hydrogen	0.00
Copper	+0.34

With a primary cell the chemicals in the cell are gradually 'used up' during the life of the cell. A secondary cell can, however, be recharged, i.e. the chemical changes reversed, and the cell restored to full working order.

Examples

4.51 What is the e.m.f. for an aluminium–lead cell? Which metal will be the positive terminal of the cell?

Ans The cell has an e.m.f. of

$$-1.66 - (-0.13) = -1.53 \text{ V}$$

The above data was taken from the table given in the notes. The metal which is highest in that table, in the form it is written in this book, is the negative terminal. Thus, the lead is the positive terminal.

Further problems

4.52 Use the data given in the electrochemical series in this chapter. What is the e.m.f. for the following cells? (a) Magnesium–copper, (b) zinc–copper, (c) lead–copper and (d) zinc–lead.

4.53 What feature distinguishes secondary cells from primary cells?

4.54 With a zinc–copper cell, e.g. a Daniell cell, zinc goes into solution as ions from the zinc electrode and copper ions are deposited on the copper electrode. If the zinc electrode has a mass of 200 g for how long will the cell deliver a steady current of 2 A (the zinc ion is Zn^{2+}, one Faraday is 96 500 C and one mole of zinc has a mass of 65 g)?

Capacitors

Notes

If a charge Q is given to an insulated conductor and its potential changes by V then it is said to have a capacitance C, where

$$C = Q/V$$

A parallel plate capacitor consists of two parallel plates separated by an insulator, often referred to as a dielectric. In the uncharged state both of the plates have zero net charge. When a potential difference is connected across the plates they become charged, one plate positively and the other negatively. The net charge on the entire capacitor is zero in that the amount of negative charge exactly balances the amount of positive charge. The charge on one of the plates divided by the potential difference between the plates is the capacitance.

When an uncharged capacitor is connected into a d.c. circuit, a current occurs for a finite period of

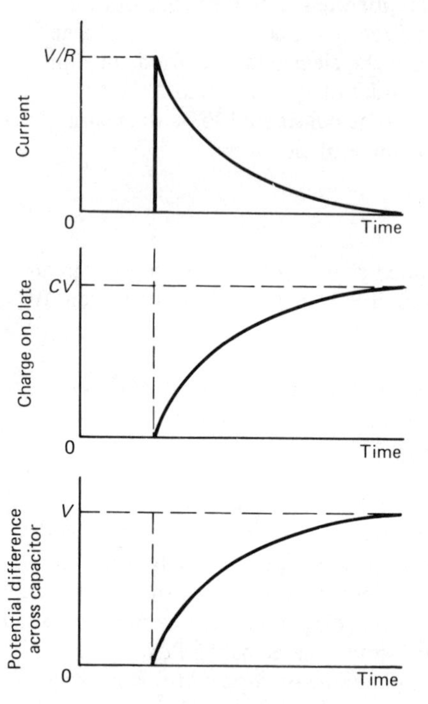

Figure 4.5 Charging a capacitor

time while the capacitor plates are being charged (*Figure 4.5*). When the plates are fully charged the potential difference across the plates has reached the same value as the source potential difference and no current occurs. The time taken to reach this fully charged condition depends on the product CR, where R is the resistance in the circuit. This quantity CR is called the time constant. The bigger the time constant the longer the time taken for the capacitor to become fully charged. When a charged capacitor has its plates connected together via some conducting path, of resistance R, it will discharge. The time taken for the discharge depends on the time constant CR.

If a capacitor is used in a circuit with an alternating potential difference an alternating current occurs at all times.

Units: charge – C (coulomb), potential difference – V, capacitance – F (farad).

Examples

4.55 What is the size of the maximum charge on the plates of a capacitor of capacitance 8 μF when connected to a 2.0 V battery?

Ans $Q = CV$
$= 8 \times 10^{-6} \times 2.0$
$= 16 \times 10^{-6}$ C

4.56 A capacitor of capacitance 16 μF is charged by a 2.0 V d.c. supply connected across it. (a) What is the size of the charge on a plate of the capacitor when fully charged? (b) What is the potential difference between the capacitor plates when the capacitor is fully charged? (c) The capacitor is then discharged through a resistor of resistance 10 kΩ. What is the initial current through the resistor?

Ans (a) $Q = CV$
$= 16 \times 10^{-6} \times 2.0$
$= 32 \times 10^{-6}$ C

(b) Potential difference = 2.0 V
(c) $V = IR$

Hence

$I = 2.0/(10 \times 10^3)$
$= 2.0 \times 10^{-4}$ A

Further problems

4.57 A capacitor of capacitance 16 μF is charged by 12 V connected across it. What is the size of the charge on the capacitor plates?

4.58 A battery of e.m.f. 6 V is used to charge a 8 μF capacitor. What is the charge on the plates when the capacitor is fully charged? What is the potential difference across the plates in this condition?

4.59 A fully charged capacitor of capacitance 8 μF has a potential difference of 8 V across its plates. It is discharged through a resistance of 4 kΩ. (a) What is the initial charge on the capacitor? (b) What is the initial current through the resistor?

4.60 What potential difference has to be applied to a capacitor of capacitance 4 μF if the size of the charge on the plates is to be 200 μC when the capacitor is fully charged?

The magnetic effect of a current

Notes

When a current passes through a wire, or any other conductor, a magnetic field is produced. The direction indicated by the North-seeking end of a compass needle is taken to be the direction of the magnetic field. For a long, straight wire the field pattern is one of circles concentric with the wire (*Figure 4.6*). For a solenoid the field pattern within the solenoid is one of parallel lines, parallel to the axis of the solenoid.

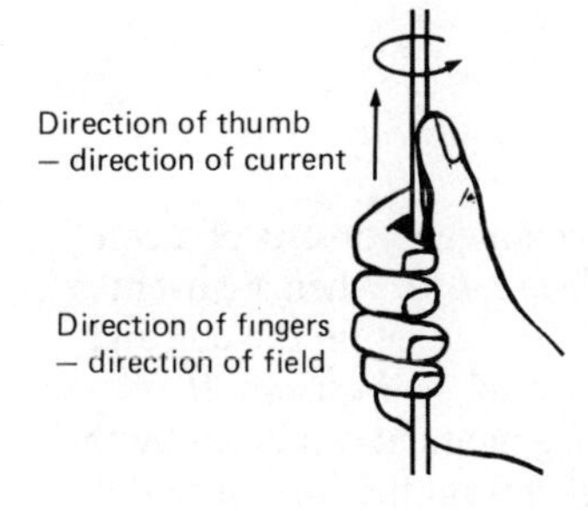

Figure 4.6 Simple aid to remembering the relation between current and field direction

A current-carrying conductor experiences a force when in a magnetic field.

Force, $F = BIL$

where I is the current, L the length of the current-carrying conductor in the field and B the magnetic flux density component at right angles to the current carrying conductor (*Figure 4.7*).

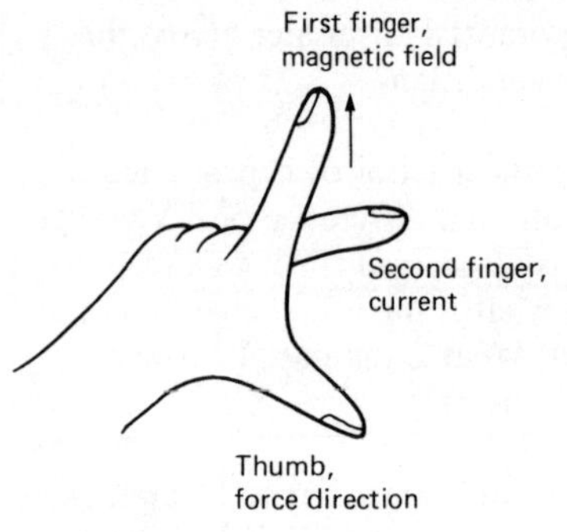

Figure 4.7

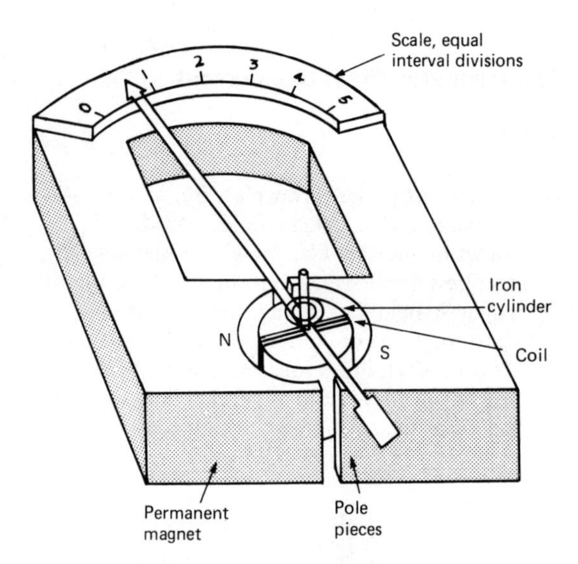

Figure 4.8 The basic meter movement

The basic moving coil meter consists of a coil in a magnetic field (*Figure 4.8*). When a current flows through the coil, forces act on the coil and cause it to rotate against the hair springs. It rotates until the turning moment produced by the forces on the coil arising from the current in the coil is balanced by the opposing torque produced by the springs. For the radial magnetic field shown in *Figure 4.8*, the field is always at right angles to the coil and thus at equilibrium.

$k\theta = NBIA$

where k is a constant for the springs used, θ the angle through which the coil has rotated, I the current, B the radial flux density, A the coil area and N the number of turns on the coil. The current sensitivity of the movement is θ/I.

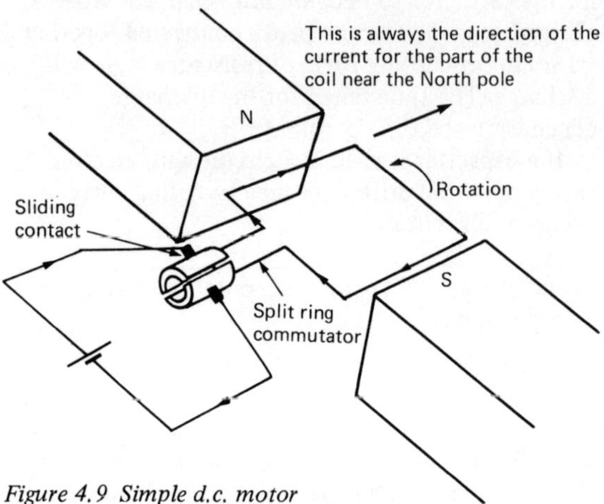

Figure 4.9 Simple d.c. motor

Figure 4.9 shows the basic principles of a simple d.c. motor. When a current passes through the coil, forces are produced which cause the coil to rotate. To obtain continuous rotation of the coil the current through the coil is reversed every time the plane of the coil is at right angles to the magnetic field.

Units: force – N, current – A, length – m, flux density – $N\,A^{-1}\,m^{-1}$ or Wb/m^2 or T (Wb is the unit of flux, the weber, $1\,N\,A^{-1}\,m^{-1} = 1\,Wb/m^2 = 1\,T$ (Tesla)).

Examples

4.61 A wire carrying a current of 2 A has a length of 100 mm in a uniform magnetic field of $0.8\,N\,A^{-1}\,m^{-1}$ (or Wb/m^2). What is the force acting on the wire if the field is (a) at right angles to the wire, (b) parallel to the wire and (c) at 60° to the wire?

Ans $F = BIL$

where B is the field component at right angles to the wire.

(a) Thus

$$F = 0.8 \times 2 \times 100 \times 10^{-3}$$
$$= 0.16 \text{ N}$$

(b) There is no field component at right angles to the wire and so there is no force
(c) The field component at right angles to the wire is $B \sin \theta$. Thus

$$F = 0.8 \times \sin 60° \times 2 \times 100 \times 10^{-3}$$
$$= 0.14 \text{ N}$$

4.62 A plane, rectangular coil measures 20 mm by 20 mm, has 20 turns of wire and carries a current of 100 mA. What is the turning moment (torque) acting on the coil when it is in a magnetic field of 80×10^{-3} N A^{-1} m^{-1} (or Wb/m^2) and the field is parallel to the plane of the coil?

Ans Turning moment = Fw

where w is the width of the coil. But

$$F = NBIL$$

where L is the length of the wire in the coil of N turns. Hence

$$\text{Turning moment} = NBILw$$
$$= NBIA$$

where B is the flux density at right angles to the wires down the sides of the coil, i.e. parallel to the plane of the coil.

$$\text{Turning moment} = 20 \times 80 \times 10^{-3} \times 100 \times 10^{-3} \times 20 \times 10^{-3} \times 20 \times 10^{-3}$$
$$= 6.4 \times 10^{-5} \text{ N m}$$

Further problems

4.63 What is the force acting on a wire carrying a current of 1.5 A when it has a length of 200 mm in a uniform magnetic field of 1.2 N A^{-1} m^{-1} and the field is (a) at right angles to the wire, (b) makes an angle of 30° with the wire?

4.64 A plane, rectangular coil with 25 turns of wire has a height of 25 mm, a width of 12 mm and carries a current of 100 mA. What is the torque acting on the coil when it is in a uniform magnetic field of 2.0 Wb/m^2 and the field is parallel to the plane of the coil?

4.65 A wire is suspended so that it hangs between the poles of a bar magnet. What is the force acting on the wire if it carries a current of 500 mA and has a length of 12 cm at right angles to the magnetic field of strength 8 N A^{-1} m^{-1}?

4.66 Describe how a current balance can be used to measure the strength of a magnetic field.

4.67 Explain the basic principle of operation of a moving coil galvanometer.

4.68 The plane of a rectangular coil of wire 40 mm by 60 mm is parallel to a magnetic field of flux density 0.20 Wb/m^2. If the coil has 10 turns of wire what is the torque acting on it when a current of 2.0 A flows through the coil? How does the torque change as the angle between the plane of the coil and the magnetic field changes?

Electromagnetic induction

Notes

When a magnetic flux linked to a circuit is changing, an e.m.f. is induced in the circuit. The magnitude of the induced e.m.f. is proportional to the rate of change of the linked flux. These two statements are known as **Faraday's laws.**

$$\text{Induced e.m.f.} = -\frac{d\phi}{dt}$$

where ϕ is the flux and is equal to the product of the flux density B and the area A of the circuit loop concerned,

$$\phi = BA$$

The minus sign in the induced e.m.f. equation arises because the direction of the induced e.m.f. is such that it tends to opposte the change producing it and does oppose it if an induced current flows. The direction of an induced current is such that its effect opposes the change in magnetic flux which gave rise to the current. This is known as **Lenz's law.**

A changing current in one coil of wire produces an induced e.m.f. in another coil nearby.

This effect is called mutual inductance. The frequency of the induced e.m.f. is the same as the frequency of the primary alternating current, though out of phase with it. When the rate of change of the primary current is zero, i.e. when

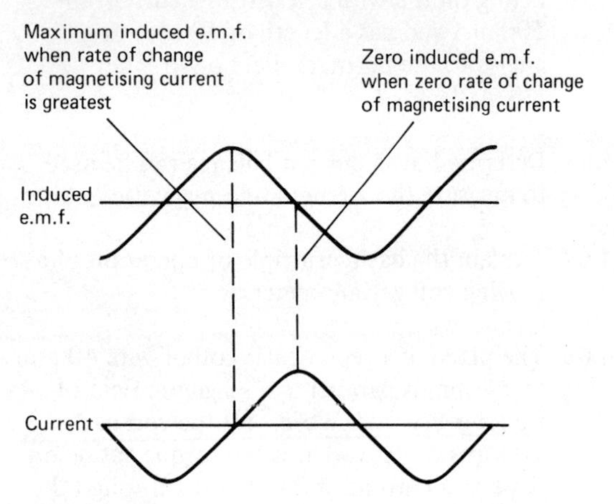

Figure 4.10

the current is passing through its maximum or minimum values, then the induced e.m.f. is zero (*Figure 4.10*).

$$\text{Induced e.m.f.} = -M\frac{dI}{dt}$$

where M is a constant for a pair of coils and is called the mutual inductance.

The transformer consists of two coils wound on an iron core. When an alternating current is applied to one coil an alternating voltage is produced in the other coil.

For an ideal transformer in which all the magnetic flux produced by the primary coil links the secondary coil and no current is taken from the secondary coil, i.e. it is an open circuit, the flux linked by the entire secondary coil of N_s turns is $N_s\phi$, where ϕ is the flux produced by the primary coil. The induced e.m.f. in the secondary coil E_s is thus $-N_s\, d\phi/dt$. But the changing magnetic flux links both the coils. Hence, the induced e.m.f. E_p in the primary coil is $-N_p\, d\phi/dt$. If no current is taken from the secondary coil and we assume conservation of energy for the two coils, then as no energy is taken from the secondary coil the induced e.m.f., E_{in}, in the primary coil must cancel out the input e.m.f. in that coil, i.e.

$$E_{in} - E_p = 0$$

Thus

$$E_{in}/E_s = N_p/N_s$$

If there is a current in the secondary coil and there is assumed to be no energy loss, e.g. heating, then the input power equals the output power and so

$$E_pI_p = E_sI_s$$

$$I_p/I_s = N_s/N_p$$

A changing current in a conductor produces a changing magnetic field which can then induce an e.m.f. in the conductor itself. The effect is called self inductance. The e.m.f. opposes the current producing it and, thus, if the primary current is increasing the e.m.f. slows down the rate of increase. The greater the rate of change of the current the greater

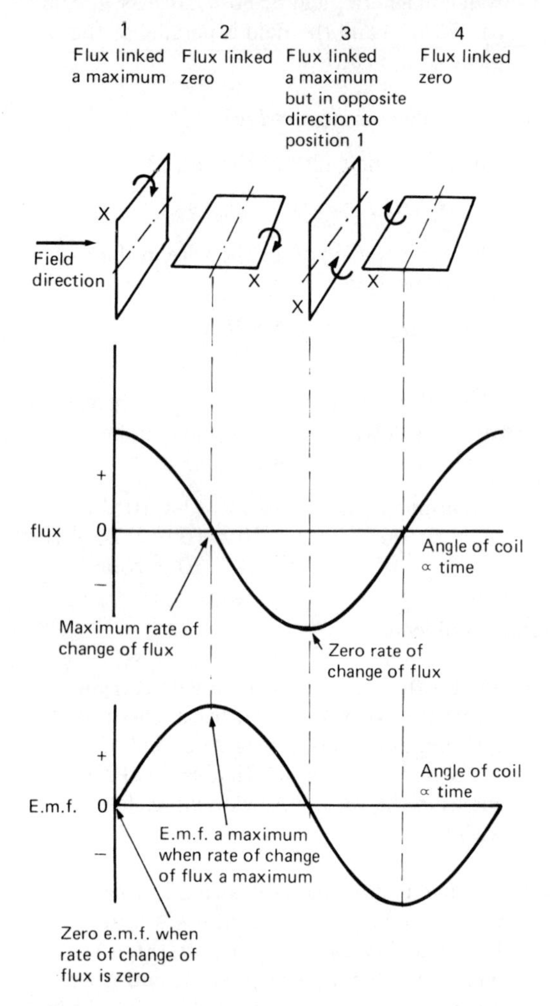

Figure 4.11

this opposing e.m.f., or back e.m.f. as it is often termed.

Back e.m.f. $= -L\, dI/dt$

where L is called the inductance.

An alternating current generator consists essentially of just a coil rotating in a magnetic field. As the coil rotates the flux linked by the coil continually changes and so an alternating e.m.f. is induced in the coil (*Figure 4.11*).

Units: e.m.f. – V, flux – Wb (weber), time – s, mutual inductance – H (Henry), self inductance – H.

Examples

4.69 A plane, rectangular coil 20 mm by 40 mm has 50 turns of wire. What is the flux linked by the coil if the flux density at right angles to the plane of the coil is 1×10^{-3} Wb/m^2? What would be the flux linked if the field was parallel to the plane of the coil?

Ans Flux linked per turn of wire $= \phi = BA$

where B is the flux density at right angles to the plane of the coil. Hence total flux linked is

$$50 \times 20 \times 10^{-3} \times 40 \times 10^{-3} \times 1 \times 10^{-3}$$
$$= 5 \times 10^{-5} \text{ Wb}$$

When the field is parallel the flux linked is zero.

4.70 What is the e.m.f. induced in a coil of wire when the flux linked by the coil changes at a constant rate from 2×10^{-3} Wb/m^2 to 5×10^{-3} Wb/m^2 in 2 s?

Ans Induced e.m.f. $= -\dfrac{d\phi}{dt}$

$$= (-3 \times 10^{-3})/2$$
$$= -1.5 \times 10^{-3} \text{ V}$$

4.71 A small coil has 40 turns of wire and an area of 100 mm^2. What is the average e.m.f. induced in the coil when it is moved in a magnetic field in such a way that the flux density component normal to the plane of the coil changes from 2 Wb/m^2 to 0 in 0.1 s?

Ans Induced e.m.f. $= -\dfrac{d\phi}{dt}$

$$= (-40 \times 100 \times 10^{-6} \times 2)/0.1$$
$$= -8 \times 10^{-2} \text{ V}$$

4.72 What turns ratio is required for a transformer which has an input of an alternating voltage with a root mean square value of 120 V and is required to give an output of 6 V across a high resistance load?

Ans $E_{in}/E_s = N_p/N_s$

Hence

Turns ratio $N_p/N_s = 120/6$
$= 20/1$

4.73 A simple a.c. generator consists of a coil rotating in a magnetic field of flux density B. (a) If the plane of the coil makes an angle θ with the field direction what is the flux linked by the coil of area A and having N turns? (b) How does the flux linked by the coil vary with time if the coil rotates with a uniform angular velocity? How does the induced e.m.f. vary with time? (c) What is the maximum value of the induced e.m.f.?

Ans (a) $NAB \sin \theta$
(b) $\phi = NAB \sin \omega t$
where ω is the angular velocity

Hence

$$E = -\frac{d\phi}{dt} = -\frac{NAB d(\sin \omega t)}{dt}$$

$$E = -NAB\, \omega \cos \omega t$$

(c) $NAB\omega$

4.74 A metal disc of radius 100 mm rotates around an axis through its centre at 50 revolutions per second. If the disc is in a magnetic field of flux density 0.8 Wb/m^2 at right angles to the plane of the disc, what is the potential difference produced between the axis of the disc and its rim?

Ans $E = -d\phi/dt$

The problem can be considered in terms of a

radius arm sweeping round and cutting flux at the rate of BA in each revolution when A is the area of the disc. As there are f revolutions every second the rate of change of flux linked is

$$E = -fBA$$

As

$$A = \pi r^2$$

then

$$E = -\pi f B r^2$$

Thus

$$E = -\pi \times 50 \times 0.8 \times (100 \times 10^{-3})^2 = 1.26 \text{ V}$$

Further problems

4.75 Describe how you would demonstrate the laws of electromagnetic induction in the school laboratory.

4.76 A coil has an area of 200 mm^2 and 30 turns of wire. What are the changes in flux linkage occurring in the following cases? (a) The flux perpendicular to the plane of the coil changes from 5×10^{-5} Wb to 3×10^{-5} Wb. (b) The flux density at right angles to the plane of the coil is changed from 2×10^{-3} Wb/m^2 to 8×10^{-3} Wb/m^2. (c) The flux density perpendicular to the plane of the coil is 6×10^{-3} Wb/m^2. The coil is removed from this situation to one where the flux density is zero. (d) The flux density perpendicular to the plane of the coil is 8×10^{-3} Wb/m^2. The coil is rotated from this position to one at right angles to its initial position. (e) The flux density perpendicular to the plane of the coil is 5×10^{-2} Wb/m^2. The coil is rotated from this position to one where the field is at 60° to the plane of the coil.

4.77 Calculate the average induced e.m.f. in each of the situations described in **4.76**, if the changes each take place in 0.01 s.

4.78 An aircraft has a wing span of 20 m and is moving horizontally with a velocity of 400 m/s. What is the potential difference produced between the tips of the wings if the Earth's magnetic field has a vertical component of 50×10^{-6} Wb/m^2?

4.79 A rectangular coil of 50 turns is 100 mm by 120 mm. What is the maximum e.m.f. when the coil is rotated about an axis through the midpoint of its short sides at a rate of 10 revolutions per second (the magnetic field is 6.0×10^{-3} Wb/m^2 and is perpendicular to the axis of rotation)?

4.80 A search coil has 40 turns of wire and an area of 120 mm^2. An e.m.f. of 20 V was found to be induced in the coil when it was moved in 0.12 s from a magnetic field to a region where there was no field. What was the strength of the magnetic field?

4.81 A coil has a self inductance of 0.01 H. What is the e.m.f. induced in the coil when the current in it is changing as the rate of 4 A/s?

4.82 The mutual inductance between coils X and Y is 0.2 H. What is the e.m.f. induced in coil Y when the current in coil X is changing at the rate of 20 A/s?

4.83 A transformer has a primary coil of 800 turns and a secondary coil of 100 turns. What is the voltage in the secondary coil if the root mean square voltage applied to the primary is 240 V? Assume that the secondary coil is connected to a high resistance load.

4.84 A transformer used with the a.c. mains of 240 V root mean square voltage applied to the primary coil gives a secondary output of 6.0 V. What is the turns ratio?

4.85 A copper disc of radius 60 mm rotates at 4.0 revolutions per second about an axis which is parallel to a magnetic field of 2.0×10^{-3} Wb/m^2. What is the potential difference produced between the axis of the disc and its rim?

4.86 Describe the basic principles of a simple a.c. generator.

4.87 A coil of area 500 mm^2 and having 60 turns rotates at a constant rate of 50 revolutions per second in a magnetic field of 1.2 Wb/m^2. (a) What is the frequency of the e.m.f. induced in the rotating coil? (b) What is the maximum e.m.f. induced in the coil?

Examination questions

4.88 (a) Explain what is meant by the electromotive force and the terminal potential difference of a battery.
(b) A bulb is used in a torch which is powered by two identical cells in series each of e.m.f. 1.5 V. The bulb then dissipates power at the rate of 625 mW and the p.d. across the bulb is 2.5 V. Calculate (i) the internal resistance of each cell and (ii) the energy dissipated in each cell in one minute.

(JMB)

4.89 (a) Define resistivity, and temperature coefficient of resistance.
(b) Outline briefly how you would measure the temperature coefficient of resistance of copper in the laboratory.
(c) The temperature at which the tungsten filament of a 12 V, 36 W lamp operates is 1750 °C. Taking the temperature coefficient of resistance of tungsten to be 6×10^{-3} K^{-1}, find the resistance of the filament at room temperature, 20 °C.
(d) The table gives readings for two filament lamps A and B of different ratings.

Current, I/A	0	0.05	0.10	0.15	0.20
P.d. across lamp A, V_A/V	0	0.40	1.1	2.8	6.5
P.d. across lamp B, V_B/V	0	1.25	2.6	5.0	9.1

(i) On the same graph, with I as y-axis (ordinate), draw the graph of I against V for each lamp.
(ii) The two lamps A and B are connected in parallel. Find and tabulate the corresponding values of the current I and the p.d. across the lamps up to 6 V, and draw the I/V graph (on the same graph as for (i)).

(Oxford Local Examinations)

4.90 (a) Explain, with the aid of circuit diagrams, how a potentiometer can be used to measure

(i) a current known to be of the order of 10 A,
(ii) an e.m.f. known to be of the order of 5 mV.

You may assume that any apparatus you require is available.
(b) In an experiment to measure the e.m.f. of a thermocouple, a potentiometer is used in which the slide wire is 2.000 m long and the resistance of the wire is 6.000 Ω. The current through the wire is 2.000 mA and the balance point is 1.055 m from one end. Calculate the e.m.f. of the thermocouple.

(AEB)

4.91 (a) Draw a fully labelled diagram showing the structure of a transformer capable of giving an output of 12 V from a 240 V a.c. supply. State two sources of power loss in the transformer and describe how they may be minimized.
(b) Explain why, when the secondary is not delivering a current, the transformer consumes very little power from the supply.
(c) Why will the transformer not work with a direct current?
(d) A factory requires power of 144 kW at 400 V. It is supplied by a power station through cables having a total resistance of 3 Ω. If the power station were connected directly to the factory,

(i) show that the current through the cables would be 360 A,
(ii) calculate the power loss in the cables,
(iii) calculate the generating voltage which would be required at the power station,
(iv) calculate the overall efficiency.

If the output from the power station provided a 10 000 V input to a transformer at the factory, the transformer having an efficiency of 96%,

(v) show that the current through the cables would be 15 A,

(vi) calculate the power loss in the cables,
(vii) calculate the generating voltage which would be required at the power station,
(viii) calculate the overall efficiency.

(AEB)

4.92 Describe, with the aid of any necessary diagram, the construction of a moving coil galvanometer.

A moving coil galvanometer has a coil of 80 turns, each of area 50 mm^2, suspended in a radial magnetic field of 0.3 T by a wire of torsional constant 6×10^{-9} N m rad^{-1}. The resistance of the coil is 20 Ω. Calculate the angular deflection of the coil produced by (a) a current of 1 μA, (b) an applied potential difference of 1 μV.

The galvanometer described above is modified by rewinding the coil using 160 turns of thinner wire having half the cross-section area of that previously employed. The turns on the new coil have each the same area as those on the old one. What effect have these modifications on the angular deflections produced by (a) a current of 1 μA, (b) a potential difference of 1 μV?

(Southern Universities)

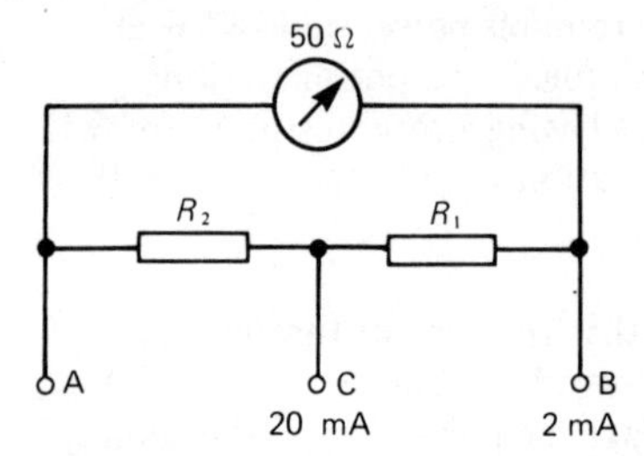

Figure 4.12

4.93 The diagram (*Figure 4.12*) shows a dual-range d.c. milliammeter which has a common negative terminal A and positive terminals B and C. The indicator is a moving-coil meter of resistance 50 Ω which gives a full-scale deflection with a current of 1.0 mA.
(a) When the instrument is connected into a circuit using terminals A and B, full-scale deflection is obtained with a current of 2.0 mA. Find the total shunt resistance $R_1 + R_2$.
(b) When connection is made with terminals A and C, a current of 20 mA gives full-scale deflection. Find the resistance R_2.

(University of Cambridge)

Answers

4.5 Ans Take readings of the potential difference across the resistor for different currents through it. If Ohm's law is obeyed a graph of the potential difference against the current will be a straight line passing through the origin

4.6 Ans $V = IR$, hence $6.0 = 200\,I$, and $I = 0.03$ A,
$V = IR = 0.03 \times 50 = 1.5$ V,
$V = IR = 0.03 \times 150 = 4.5$ V

4.7 Ans For each resistor $V = 2.0$ V, hence
$I = V/R = 2.0/20 = 0.1$ A and $I = 2.0/150 = 0.013$ A

4.8 Ans In series to give 12 Ω

4.9 Ans $R = \rho L/A = 170\ \Omega$

4.10 Ans $R = \rho L/A$, hence $L = 9.13$ m

4.11 Ans For one strand $R = \rho L/A = 0.085\ \Omega$.
For the ten strands in parallel $R = 0.0085\ \Omega$

4.12 Ans ρ/r^2 = a constant, hence radius of copper wire is 0.78 times that of the aluminium wire

4.13 Ans $A = \rho L/R$, hence for copper $A = 8.5 \times 10^{-5}$ m^2 and for aluminium $A = 1.4 \times 10^{-4}$ m^2. Mass per kilometre = density × volume, hence for copper the mass per kilometre is 756.5 kg and for aluminium it is 378 kg

4.14 Ans Change in resistance $= R_0\alpha\theta = 21.5\ \Omega$

4.15 Ans $V = E - Ir = 1.2$ V

4.16 Ans (a) $V = E - Ir$, thus $1.3 = E - 0.2r$ and $1.0 = E - 0.4r$; hence $E = 1.6$ V
(b) $r = 1.5\ \Omega$

4.20 Ans For the voltmeter by introducing different resistors in series with the meter and for the ammeter by introducing different resistors in parallel with the meter

4.21 Ans See **4.17**, $I = I_g\ (1 + R_g/R_s)$, hence resistance of shunt = 0.002 Ω

4.22 Ans (a) See **4.17**, $I = I_g(1 + R_g/R_s)$, hence $R_s = 12.5\ \Omega$
(b) $R_s = 0.020\ \Omega$
(c) $R_s = 0.0010\ \Omega$

4.23 Ans See **4.17**, $V = I_g(R_m + R_g)$, hence $R_m = 2.5 \times 10^5\ \Omega$

4.24 Ans For the 0.1 A range, $I = I_g(1 + R_g/R_s)$ and thus $R_1 + R_2 + R_3 = 2.22\ \Omega$. For the 1 A range R_3 is in series with the meter, thus $I = I_g\,[1 + (20 + R_3)/(R_1 + R_2)]$ and hence $99(R_1 + R_2) = 20 + R_3$. For the 10 A range $I = I_g\,[1 + (20 \times R_2 + R_3)/R_1\,]$ and hence $1000R_1 = 20 + R_2 + R_3$, hence $R_3 = 2.00\ \Omega$, $R_2 = 0.21\ \Omega$ and $R_1 = 0.0022\ \Omega$

4.25 Ans The driver cell must produce a potential difference across the wire greater than 2 V

4.26 Ans A very sensitive, uncalibrated galvanometer can be used

4.27 Ans The driver cell and the cell being tested may not be connected positive side to positive side, or negative to negative. There could be a break in the circuit connecting the driver cell to the potentiometer wire. The driver cell could be supplying a potential difference smaller than that of the test cell

4.28 Ans $1.018 \times 520/340 = 1.557$ V

4.29 Ans (a) $1.018 \times 150/240 = 0.636$ V
(b) $V = IR$, hence $I = 0.318$ A

4.30 Ans 135/670

4.31 Ans Effectively increasing the length of the wire by putting a resistor in series with the wire. The standard cell can be connected across the resistor plus part of the wire and the thermocouple only connected across part of the wire. Instead of length of wire in the calculation the resistance is used. See **4.32**

4.32 Ans $R = \rho L/A = 5.093\ \Omega$. Without the 1000 Ω resistor the potential is 2.0 V/m. With the 1000 Ω resistor the 2.0 V is across a total resistance of 1005.093 Ω. This is equivalent to a potentiometer wire of length 1005.093/5.093 = 197.35 m, hence the potential gradient is 0.01013 V/m

4.33 Ans $R = \rho L/A$, hence $R \propto \rho/d^2$ and the diameter of nichrome is 1.25 times that of manganin

4.34 Ans $R = \rho L/A = 93.6\ \Omega$;
$1/R = 1/R_1 + 1/R_2$, hence resistance = 1.01 Ω

4.39 Ans $P = I^2R$, hence $I = 0.1$ A

4.40 Ans $P = IV = 24$ W

4.41 Ans $P = V^2/R$, hence the 50 W bulb has the highest resistance

4.42 Ans $P = V^2/R$, hence $R = 576\ \Omega$

4.43 Ans 0.25 kWh

4.44 Ans $V = IR$, hence $I_{RMS} = 0.2$ A

4.47 Ans See the notes

4.48 Ans $96\ 500/(6 \times 10^{23}) = 1.6 \times 10^{-19}$ C (this is 1 electron)

4.49 Ans Charge moved = $2.0 \times 5 \times 60 = 600$ C, hence mass liberated = $600 \times 108/96\ 500 = 0.67$ g

4.50 Ans Charge required = $2 \times 96\ 500 \times 10/64 = 30\ 200$ C, hence $t = 3020$ s

4.52 Ans (a) 2.71 V
(b) 1.10 V
(c) 0.47 V
(d) 0.63 V

4.53 Ans Secondary cells can be recharged

4.54 Ans Charge required $= 2 \times 96\,500 \times 200/65 = 5.9 \times 10^5$ C, hence $t = 3.0 \times 10^5$ s

4.57 Ans $Q = CV = 1.92 \times 10^{-4}$ C

4.58 Ans $Q = CV = 4.8 \times 10^{-5}$ C; 6 V

4.59 Ans (a) $Q = CV = 6.4 \times 10^{-5}$ C
(b) $V = IR$, hence $I = 2$ mA

4.60 Ans $V = Q/C = 50$ V

4.63 Ans (a) $F = BIL = 0.36$ N
(b) $F = BIL \sin\theta = 0.18$ N

4.64 Ans Torque $= NBIA = 15 \times 10^{-3}$ N m

4.65 Ans $F = BIL = 0.48$ N

4.66 Ans The end wire of the balance must be at right angles to the field being measured and the field must also be at right angles to the direction in which the wire is to move

4.67 Ans See the notes

4.68 Ans Torque $= NBIA = 9.6 \times 10^{-3}$ N m; torque $= NBIA \sin\theta$, where θ is the angle between the plane of the coil and the field direction.

4.75 Ans You could describe what happens when a loop of wire connected to a galvanometer is moved near a magnet; also, when a.c. is passed through a coil wound on an iron core and the e.m.f. induced in the secondary coil is monitored with a cathode ray oscilloscope.

4.76 Ans (a) 60×10^{-5} Wb
(b) $30 \times 6 \times 10^{-3} \times 200 \times 10^{-6} = 3.6 \times 10^{-5}$ Wb
(c) $6 \times 10^{-3} \times 30 \times 200 \times 10^{-6} = 3.6 \times 10^{-5}$ Wb
(d) $30 \times 8 \times 10^{-3} \times 200 \times 10^{-6} = 4.8 \times 10^{-5}$ Wb
(e) $30 \times (5 \times 10^{-2} - 5 \times 10^{-2} \times \sin 60^\circ) \times 200 \times 10^{-6} = 4.0 \times 10^{-5}$ Wb

4.77 Ans (a) 60×10^{-3} V
(b) 3.6×10^{-3} V
(c) 3.6×10^{-3} V
(d) 4.8×10^{-3} V
(e) 4.0×10^{-3} V

4.78 Ans Area swept out in 1 s $= 20 \times 400$ m^2. Rate of change in flux linked $= 20 \times 400 \times 50 \times 10^{-6}$ Wb/s, hence e.m.f. $= 0.4$ V

4.79 Ans See 4.73. Maximum e.m.f. $= NAB\omega = 50 \times 100 \times 120 \times 10^{-6} \times 6.0 \times 10^{-3} \times 2\pi \times 10 = 0.23$ V

4.80 Ans $B = 20 \times 0.12/(40 \times 120 \times 10^{-6}) = 500$ Wb/m^2

4.81 Ans 0.04 V

4.82 Ans 4 V

4.83 Ans $E_{in}/E_s = N_p/N_s$, hence $E_s = 30$ V

4.84 Ans $N_p/N_s = 240/6.0 = 40$

4.85 Ans See 4.74. $E = -\pi fBr^2 = 9.0 \times 10^{-5}$ V

4.86 Ans Essentially a coil rotating in a magnetic field

4.87 Ans (a) 50 Hz
(b) See 4.73. $E_{max} = NAB\omega = 60 \times 500 \times 10^{-6} \times 1.2 \times 2\pi \times 50 = 11.3$ V

4.88 Ans (a) See the notes in this chapter
(b) $P = IV$, hence $625 \times 10^{-3} = 2.5\,I$. Thus $I = 0.25$ A, $2E = V + I \times 2r$ and thus $r = 1\ \Omega$. Energy dissipated in each cell $= I^2 r = 0.0625$ W

4.89 Ans (a) See the notes in this chapter
(b) You could use the metre bridge, see the notes in this chapter. Quite an amount of fine copper wire would be needed in order to give a significant enough resistance
(c) $P = V^2/R$, hence $R = 12^2/36 = 4.0\ \Omega$, this is the resistance at 1750 °C. $R_{1750} = R_0(1 + 1750\alpha)$ and

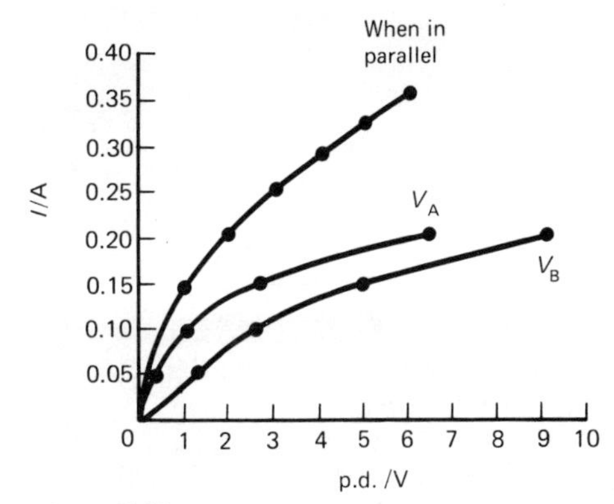

Figure 4.13

$R_{20} = R_0(1 + 20\alpha)$, hence the resistance at 20 °C is 0.389 Ω
(d) (i) See *Figure 4.13*. (ii) For the lamps in parallel the p.d. will be the same for both. The current through the arrangement is the sum of the currents through each lamp. Thus, for 1 V across each lamp the currents are 0.08 A and 0.06 A, i.e. a total of 0.14 A. For 2 V the total current is 0.20 A. For 3 V it is 0.25 A, for 4 V it is 0.29 A, for 5 V it is 0.33 A, for 6 V it is 0.35 A. The current/p.d. graph is obtained by adding, at a particular p.d. value, the current values for the two lamps

4.90 Ans (a) (i) Pass the current through a resistor and measure the potential difference across the resistor. The resistor would need to be of the order of a few ohms and capable of withstanding the power. (ii) A resistor could be used in series with the potentiometer wire and the e.m.f. connected across just a segment of the wire while the standard cell is used across the resistor plus a segment of the wire
(b) The potential drop across the wire $= IR = 2.000 \times 10^{-3} \times 6.000$, hence the potential drop per metre of wire is 6.000×10^{-3} V/m. The e.m.f. of the thermocouple is thus 6.330×10^{-3} V

4.91 Ans (a) The diagram should show the primary and secondary coils and the iron former on which the coils are both wound. The primary coil will have 20 times the number of turns on the secondary coil. Losses occur due to heating in the coil windings, eddy currents in the iron core, not all the magnetic flux produced by the primary linking with the secondary coil, hysteresis losses in the magnetisation of the core. Eddy currents are minimized by laminating the iron core. The use of an iron core and its shape minimizes the loss of magnetic flux
(b) The back e.m.f. in the primary coil is virtually opposite and equal to the input e.m.f. There is thus little net current in the primary coil
(c) A changing current is required in the primary coil in order to produce a changing magnetic flux linking the secondary coil and hence an induced e.m.f.
(d) (i) $P = IV$, hence $I = 144 \times 10^3/400$ $= 360$ A. (ii) Power loss $= I^2R = 360^2 \times 3$ $= 390$ kW. (iii) $P = IV = (390 + 144) \times 10^3$ $= 360$ V, hence $V = 1.5$ kV. (iv) Efficiency = power supplied/total power produced = 27%. (v) $P = IV = 100 \times 144 \times 10^3/96 = 10\,000\,I$, hence $I = 15$ A. (vi) Power loss $= I^2R = 15^2 \times 3$ $= 675$ W. (vii) $P = IV = 144 \times 10^3 + 675$ $= 15$ V, hence $V = 9.6$ kV. (viii) Efficiency = 99.5%

4.92 Ans See the notes. $k\theta = NBIA$
(a) 0.2 radians
(b) $V = IR$, hence $I = 10^{-6}/20$ A, thus angle = 0.01 radians
Doubling N and halving A has no effect on (a) and the result is still 0.2 radians. $R = \rho L/A$ and the resistance is increased by a factor of 4; thus for (b) the current is decreased by a factor of 4 and so the angle is 0.0025 radians

4.93 Ans (a) $50\,I_g = (R_1 + R_2)I_s$ and $I = I_g + I_s = 2 \times 10^{-3}$ A. Thus as $I_g = 1.0 \times 10^{-3}$ A, $(R_1 + R_2) = 50$ Ω
This could have been argued from half the total current going through the meter and the other half through the shunt. They must therefore have equal resistances.
(b) $(50 + R_1)\,I_g = R_2 I_s$ and $I = I_g + I_s = 20 \times 10^{-3}$ A; $R_2 = 5$ Ω

5 Fields

Gravitational fields

Notes

The gravitational field strength g at a point is defined as the force per unit mass at that point.

$$g = F/m$$

The gravitational field strength at a point is numerically equal to the acceleration of a freely falling object at that point.

Newton's law of universal gravitation states that any two objects with mass attract each other with a force given by

$$F = (Gm_1 m_2)/d^2$$

where d is the distance between the centres of objects of masses m_1 and m_2 and G is a constant having the value 6.67×10^{-11} N m^2 kg^{-2}. Note that F is sometimes shown as negative, and hence g is negative, as the force is attractive.

Kepler's laws for planetary orbits are:

(i) The orbit of each planet is an ellipse with the Sun at one focus.
(ii) The line joining the Sun to a given planet sweeps out equal areas in equal times.
(iii) The square of a planet's year, i.e. orbit time, divided by the cube of its mean distance from the Sun is the same for all planets.

Units: force – N; mass – kg, field strength – N/kg.

Examples

Take G to be 6.67×10^{-11} N m^2 kg^{-2}.

5.1 The Moon orbits the Earth once every 27.3 days, the radius of the orbit being 3.85×10^5 km. What is the mass of the Earth?

Ans

$$F = (Gm_1 m_2/r^2) = (m_2 v^2)/r$$

where m_1 is the mass of the Earth and m_2 the mass of the Moon.

$$v = (2\pi r)/T$$

where T is the time for one orbit. Thus

$$m_1 = \frac{4\pi^2 r^3}{GT^2}$$

The above equation demonstrates **Kepler's third law**, i.e.

$$T^2/r^3 = \text{a constant}$$

$$m_1 = \frac{4\pi^2 \times (3.85 \times 10^8)^3}{(27.3 \times 24 \times 3600)^2 \times 6.67 \times 10^{-11}}$$

$$= 6.07 \times 10^{24} \text{ kg}$$

5.2 The field strength at the surface of the Earth is 9.81 N/kg. If the Earth has a radius of 6400 km, what is the mass of the Earth?

Ans

$$F = m_2 g = (Gm_1 m_2)/r^2$$

Hence

Field strength $g = (Gm_1)/r^2$

$$m_1 = \frac{9.81 \times (6400 \times 10^3)^2}{6.67 \times 10^{-11}}$$

$$= 6.02 \times 10^{24} \text{ kg}$$

Further problems

Take G to be 6.67×10^{-11} N m^2 kg^{-2}.

5.3 What is the Earth's orbital speed about the Sun if the Sun has a mass of 2×10^{30} kg and it is 1.5×10^{11} m from the Earth?

5.4 The Moon has a mass of 7.40×10^{22} kg and a radius of 1.74×10^6 m. What is the gravitational field strength at the surface of the Moon and the acceleration due to gravity at the surface?

5.5 In an experiment to measure the universal gravitational constant G a small sphere of mass 2.0 g is placed near a 500 g sphere, the distance between their centres being 50 mm. What is the force between the spheres that has to be measured?

5.6 The mass of the Moon is 1/81 and its radius is 1/4 that of the Earth. If the acceleration due to gravity at the surface of the Earth is 9.81 m/s^2, what is the value at the surface of the Moon?

5.7 If G were to be measured on the Moon, would the value be the same as that obtained on Earth?

5.8 There is a force of attraction between the Sun and the Earth. Why do the two bodies not move together and collide?

5.9 Explain how the mean density of the Earth can be estimated from the values of the acceleration due to gravity at the Earth's surface and the radius of the Earth. Assume a knowledge of G.

5.10 The planet Neptune takes 165 years to orbit the Sun, i.e. 165 times longer than the Earth takes. Show that the radius of Neptune's orbit is approximately 30 times that of the Earth's orbital radius.

5.11 The satellite Phobos takes 2.75×10^4 s to complete a nearly circular orbit of radius 9.7×10^6 m round Mars. What is the mass of Mars?

Gravitational potential

Notes

The potential energy of a system of particles is equal to the work that must be done to assemble the system, starting from some initial reference configuration. In the case of an object being lifted off the ground the work done in moving it to a height h above the ground is mgh and thus this is the gravitational potential energy of the Earth–object system if we take the potential energy to have been zero when the object was on the ground.

Potential is defined as the potential energy per unit mass. The zero potential value is taken to be a point at infinity for theoretical derivations. The gravitational potential at a point, with reference to this zero definition, is the work required to bring unit mass from infinity to the point concerned.

Because gravitational forces are always attractive, potential values are always negative.

An equipotential surface is one on which the potential is the same at all points. Thus, no work is done in moving a mass from one point on the surface to another point on the same surface. The gravitational field strength is at right angles to the equipotentials.

Gravitational field strength,

$$g = -\text{(potential gradient)}$$

$$g = -\mathrm{d}V/\mathrm{d}r$$

For a spherical object of mass M the potential at a point a distance r away from its centre can be obtained by calculating the work done W in bringing a test mass m from infinity to that point.

$$W = \int_{\infty}^{r} F\,\mathrm{d}x$$

$$= \int_{\infty}^{r} \frac{GMm}{x^2}\,\mathrm{d}x$$

$$= -\frac{GMm}{r}$$

Thus, the potential V is

$$V = W/m$$

$$= -GM/r$$

The term escape velocity v_e is used for the minimum velocity which an object must have in order to escape from some sphere, e.g. the Earth. The potential energy of an object of mass m at the surface of a sphere of radius r and mass M is

$$Vm = -GMm/r$$

To escape to infinity the object must have a kinetic energy plus potential energy at the Earth's surface sufficient to get it to infinity, i.e.

$$\tfrac{1}{2}mv_e^2 + (-GM/r) = 0$$

and the potential energy must be zero at infinity.

Hence

$$v_e = (2GM/r)^{1/2}$$

But the field strength g at the surface of the sphere is GM/r^2. Hence

$$v_e = (2gr)^{1/2}$$

Units: potential energy – J, mass – kg, potential – J/kg.

Examples

5.12 The Earth has a radius of 6400 km. How does the potential at the following points compare with that at the surface of the Earth?
(a) 6400 km above the Earth's surface,
(b) 12 800 km above the Earth's surface.
Both distances are measured along a radius.

Ans $V = -GM/r$

where r is measured from the centre of the sphere. Thus, doubling r to 6400 km halves the potential. Trebling r reduces the potential to a third and so (a) halved; (b) a third.

5.13 Given a graph showing how the gravitational potential varies with distance away from the Earth, how could the graph of gravitational field strength with distance be derived?

Ans The field strength is the potential gradient. It can thus be obtained from the potential/distance graph by drawing tangents to the graph.

5.14 The Apollo 11 spacecraft on its journey from the Earth to the Moon had a velocity of 5374 m/s when 26 306 km from the centre of the Earth. What would have been its velocity at 50 000 km from the Earth's centre if the spacecraft did not fire its motors during the journey (take the value of GM to be 4×10^{14} N m^2 kg^{-1})?

Ans $V = -GM/r$

At 26 306 km

$$V = -4 \times 10^{14}/(26\,306 \times 10^3)\ \text{J/kg}$$

At 50 000 km

$$V = -4 \times 10^{14}/(50\,000 \times 10^3)\ \text{J/kg}$$

The change in potential is

$$-4 \times 10^{14}\left(\frac{1}{50\,000 \times 10^3} - \frac{1}{26\,306 \times 10^3}\right)$$
$$= 7.206 \times 10^6\ \text{J/kg}$$

There is a gain in potential, and hence potential energy mV. The gain in potential energy results in a decrease in kinetic energy. Therefore

Change in kinetic energy
$$= -7.206 \times 10^6 \times m$$
$$\tfrac{1}{2}mv^2 - \tfrac{1}{2}mu^2 = -7.206 \times 10^6 \times m$$

Thus

$$v^2 = u^2 - 14.412 \times 10^6$$
$$= 5374^2 - 14.412 \times 10^6$$

Hence

New velocity $v = 3804$ m/s

5.15 Calculate the escape velocity from the surface of the Earth if the acceleration due to gravity at the surface is 9.8 m/s^2 and the radius of the Earth is 6.4×10^6 m.

Ans Escape velocity $= (2gr)^{1/2}$
$$= (2 \times 9.81 \times 6.4 \times 10^6)^{1/2}$$
$$= 11.2 \times 10^3\ \text{m/s}$$

Further problems

5.16 At what distance from the Earth's centre are the following half their values at the surface: (a) the Earth's field strength and (b) the Earth's potential?

5.17 Calculate the gravitational potential at (a) $2R$, (b) $3R$, (c) $4R$, (d) $5R$ from the centre of the Earth if R is the radius of the Earth and has the value 6400 km. Take GM to have the value 4.0×10^{14} N m^2 kg^{-1}.

5.18 What is the value of the potential gradient at the surface of the Earth ($g = 9.8$ N/kg)?

5.19 The Apollo 11 spacecraft (mass 4.4×10^4 kg) on its journey from the Earth to the Moon had a velocity of 3630 m/s when 54.4×10^6 m from the Earth and 2620 m/s when 95.7×10^6 m away. If the spacecraft did not fire its motors between these two points, what is the gravitational potential difference between them?

5.20 The controllers of the Apollo 11 spacecraft needed the spacecraft, on its journey from the Earth to the Moon, to have a velocity of 2620 m/s at 95.7×10^6 m from the Earth. If they did not want to fire the rocket motors after 11×10^6 m from the Earth, what must have been the velocity of the spacecraft at this distance (take GM to be 4.0×10^{14} N m^2 kg^{-1}?

5.21 How does the escape velocity from the Moon compare with that from the surface of the Earth (mass of the Moon is 0.012 that of the Earth and its radius is 0.27 that of the Earth)?

Electric fields

Notes

The electric field strength E is defined as the force per unit charge at a point.

$$E = F/Q$$

The direction of the field is the direction of the force that would be experienced by a positively charged object. An electric field can be represented pictorially by lines of force. The tangent to a line of force at a point on the line represents the direction of the electric field at that point.

Coulomb's law:

The force between two charged objects is proportional to the charges, Q_1 and Q_2, carried by the bodies and inversely proportional to the square of the distance d between their centres.

$$F = \frac{Q_1 Q_2}{4\pi\epsilon_0 d^2}$$

where ϵ_0 is a constant known as the permittivity of free space, having the value 8.85×10^{-12} C^2 N^{-1} m^{-2}. The above equation refers to the force in a vacuum. In any other medium the constant $\epsilon_r \epsilon_0$ has to be used where ϵ_r is the relative permittivity of the medium (sometimes referred to as the dielectric constant).

There is no electric field inside any charged shell. This means that objects can be shielded from electric fields by enclosing them in a metal case.

Units: force – N, charge – C, electric field strength – N/C, distance – m, relative permittivity – no units, a ratio.

Examples

5.22 What is the electric field at a point in a vacuum 100 mm from a small object carrying a charge of 2×10^{-6} C ($\epsilon_0 = 8.85 \times 10^{-12}$ C^2 N^{-1} m^{-2})?

Ans If a charge Q_2 is placed at a point then the electric field strength E at that point is

$$E = F/Q_2$$

But

$$F = \frac{Q_1 Q_2}{4\pi\epsilon_0 r^2}$$

Hence

$$E = \frac{Q_1}{4\pi\epsilon_0 r^2}$$

$$= 1.8 \times 10^6 \text{ N/C}$$

5.23 The electric field between a pair of parallel charged plates has a constant value of 1000 N/C. What would be the force experienced by a small, charged oil drop carrying a charge of 1.6×10^{-19} C?

Ans $E = F/Q$

Hence

$$F = 1000 \times 1.6 \times 10^{-19}$$
$$= 1.6 \times 10^{-16} \text{ N}$$

Further problems

Take ϵ_0 to be 8.85×10^{-12} C^2 N^{-1} m^{-2}.

5.24 Can electric lines of force cross?

5.25 An object having a mass of 2.0 g and a charge of 3.0×10^{-5} C is placed in a uniform electric field of strength 400 N/C. What is (a) the force acting on the object due to the electric field, (b) the net force acting on the object if the force due to the electric field is vertically upwards, (c) the acceleration of the object?

5.26 What is the electric field strength a distance of 50 mm from a small charged object carrying a charge of 2×10^{-6} C in a vacuum?

5.27 Two point charges of $+1.0 \times 10^{-6}$ C and -2.0×10^{-6} C are 100 mm apart. What are the forces acting on each charge?

5.28 A spherical metal shell has a radius of 20 mm. What is the electric field strength at distances of (a) 10 mm, (b) 30 mm, from the centre of the sphere if the sphere carries a charge of 5.0×10^{-7} C?

5.29 Two small balls are suspended by insulating threads from a common point. Each ball has a mass of 0.20 g and the suspension threads are 1.0 m long. When the balls are given equal quantities of positive charge each suspension thread is found to make an angle of 7° with the vertical. What are the charges carried by the two balls?

Electric potential

Notes

The change in electric potential energy resulting from a movement of a charged object of charge Q, a distance d in the direction of an electric field E is given by

$$\text{Work} = \text{force} \times \text{distance} = -QEd$$

Hence

$$\text{Resulting change in potential energy} = -QEd$$

The electric potential at a point is the electric potential energy per unit charge at that point. Thus, in a uniform electric field E, when an object with a charge Q is moved through a distance d in the direction of the field,

$$\text{Change in potential, } \Delta V = \text{change in potential energy}/Q = -QEd/Q$$

Hence

$$E = -\Delta V/d$$

The electric field strength is the potential gradient. The minus signs indicate that the electric field direction is in the opposite direction to that in which the potential or potential energy increases.

The zero potential value is taken to be a point at infinity for theoretical derivations. The electric potential at a point is thus the work required to bring unit charge from infinity to the point concerned.

An equipotential surface is one on which the potential is the same at all points. Thus, no work is done in moving a charge from one point on the surface to another on the same surface. The electric field is at right angles to the equipotentials.

Consider moving a +1 charge from infinity to some point a distance r from the centre of a charged sphere carrying a charge $+Q$. The work done is given by

$$\text{Work} = -\int_{\infty}^{r} F\,dr = -\int_{\infty}^{r} \frac{Q \times 1}{4\pi\epsilon_0 r^2}\,dr = \frac{Q}{4\pi\epsilon_0 r}$$

This is the potential at the point concerned.

Units: Work – J, force – N, distance – m, electric potential energy – J; potential – N m C^{-1} or J/C or V (where 1 N m C^{-1} = 1 J/C = 1 V), electric field strength – N/C or V/m (where 1 N/C = 1 V/m).

Examples

Take ϵ_0 to be 8.85×10^{-12} $C^2\ N^{-1}\ m^{-2}$.

5.30 What is the potential at a distance of 200 mm from a point charge of 3.0×10^{-7} C in a vacuum?

Ans $V = \dfrac{Q}{4\pi\epsilon_0 r}$

$= \dfrac{3.0 \times 10^{-7}}{4\pi \times 8.85 \times 10^{-12} \times 200 \times 10^{-3}}$

$= 1.35 \times 10^4$ V

5.31 A uniform electric field is created between a pair of parallel metal plates by producing a potential difference of 1500 V between them. If they are 200 m apart, what is the electric field?

Ans Electric field = –potential gradient

$= -\dfrac{1500}{200 \times 10^{-3}}$

$= -7500$ N/C

5.32 A metal sphere of radius 100 mm carries a charge of 4.0×10^{-8} C. (a) What is the electric field at the surface of the sphere? (b) What is the electric potential at the surface of the sphere? (c) What is the potential gradient at the surface of the sphere?

Ans (a) $E = \dfrac{Q}{4\pi\epsilon_0 r^2}$

$= \dfrac{4.0 \times 10^{-8}}{4\pi \times 8.85 \times 10^{-12} \times (100 \times 10^{-3})^2}$

$= 3.6 \times 10^4$ N/C

(b) $V = \dfrac{Q}{4\pi\epsilon_0 r}$

$= \dfrac{4.0 \times 10^{-8}}{4\pi \times 8.85 \times 10^{-12} \times 100 \times 10^{-3}}$

$= 3.6 \times 10^3$ V

(c) Electric field = –potential gradient

Hence

Potential gradient = -3.6×10^4 V/m

Further problems

Take ϵ_0 to be 8.85×10^{-12} C^2 N^{-1} m^{-2}.

5.33 A spherical metal shell has a charge of $+2 \times 10^{-8}$ C. If the sphere has a radius of 20 mm what is the potential at (a) 10 mm, (b) 20 mm and (c) 30 mm from the centre of the sphere?

5.34 What is the electric field between a pair of parallel metal plates if they are 20 mm apart and a potential difference of 500 V is maintained between them?

5.35 A pair of parallel metal plates are 20 mm apart and a potential difference of 1200 V is maintained between them. What is (a) the electric field in the space between the plates, (b) the force acting on a small particle carrying a charge of 1.9×10^{-19} C and lying between the plates?

5.36 What is the kinetic energy gained by an electron in being accelerated through a potential difference of 200 V (charge on the electron $= -1.6 \times 10^{-19}$ C)?

5.37 How much work has to be done to move a charge of $+2.0 \times 10^{-10}$ C from a point where the potential is 3.0×10^6 V to one where it is 5.0×10^6 V?

5.38 Is it possible for two equipotential surfaces to intersect?

5.39 If the electric field at some point is zero, must the potential at that point also be zero?

5.40 How much work has to be done to bring a small sphere with a charge of 4×10^{-10} C from infinity to a point 30 mm from another small sphere having a charge of 2×10^{-10} C?

5.41 The electric field strength at the surface of the Earth is about 130 N/C. The Earth can be considered to be a conducting sphere of radius 6.4×10^6 m. What is (a) the charge carried by the Earth, (b) the potential at the surface of the Earth?

Capacitors

Notes

A charged, isolated, conducting sphere has a surface potential given by

$$V = \frac{Q}{4\pi\epsilon_0 r}$$

It thus has a capacitance given by

$$C = Q/V$$

$$= 4\pi\epsilon_0$$

The capacitance of a parallel plate capacitor with plates of area A and separation d is given by

$$C = (\epsilon_r \epsilon_0 A)/d$$

where ϵ_r is the relative permittivity of the medium between the plates. For air ϵ_r is generally considered to have the value 1.

The electric field between the parallel plates is given by

$$E = V/d$$

where V is the potential difference between the two plates and

$$E = \frac{\sigma}{\epsilon_r \epsilon_0}$$

where σ is the charge density, i.e. the charge per unit area (Q/A).

The total capacitance of two capacitors in parallel is

$$C = C_1 + C_2$$

while the capacitance for two in series is

$$\frac{1}{C} = \frac{1}{C_1} + \frac{1}{C_2}$$

The energy involved in moving a charge ΔQ through a constant potential difference V is $V\Delta Q$. When a capacitor is charged the charge on the plates increases at the same time as the potential difference across the plates increases, the charge at any instant being proportional to the potential difference at that instant ($Q = CV$). The total energy involved in charging a capacitor from zero charge to a charge Q is thus

$$\text{Energy} = \int_0^Q V \, \mathrm{d}Q$$

$$= \int_0^Q \frac{Q}{C} \, \mathrm{d}Q$$

$$= \frac{1}{2} \frac{Q^2}{C}$$

$$= \tfrac{1}{2} QV$$

This energy is the area under the potential difference/charge graph up to the Q ordinate.

Units: capacitance – F, area – m^2, distance – m, relative permittivity – no units, potential difference – V, electric field strength – N/C or V/m, charge density – $\mathrm{C/m^2}$, energy –J.

Examples

Take ϵ_0 to be $8.85 \times 10^{-12}\ \mathrm{C^2\ N^{-1}\ m^{-2}}$.

5.42 What is the capacitance of a parallel plate capacitor having plates of area $1.0\ \mathrm{m}^2$ and a plate separation of 3.0 mm (the space between the plates contains only air)?

Ans

$$C = (\epsilon_r \epsilon_0 A)/d$$

$$= \frac{1 \times 8.85 \times 10^{-12} \times 1.0}{3.0 \times 10^{-3}}$$

$$= 3.0 \times 10^{-9}\ \mathrm{F}$$

5.43 What is (a) the electric field between the plates and (b) the charge density on the plates of the capacitor described in **5.42** when a potential difference of 500 V is connected across the plates?

Ans

(a) $E = V/d$

$$= \frac{500}{3.0 \times 10^{-3}}$$

$$= 1.7 \times 10^5\ \mathrm{N/C}$$

(b) Charge density σ

$$= \epsilon_r \epsilon_0 E$$

$$= 8.85 \times 10^{-12} \times 1.7 \times 10^5$$

$$= 1.5 \times 10^{-6}\ \mathrm{C/m^2}$$

5.44 The Earth is a sphere of radius 6.4×10^6 m. What is its capacitance?

Ans

$$C = 4\pi\epsilon_0 r$$

$$= 4\pi \times 8.85 \times 10^{-12} \times 6.4 \times 10^6$$

$$= 7.1 \times 10^{-4}\ \mathrm{F}$$

5.45 What is the capacitance of a system consisting of an 8 μF capacitor in series with a 16 μF capacitor?

Ans $\frac{1}{C} = \frac{1}{C_1} + \frac{1}{C_2}$

$= \frac{1}{8 \times 10^{-6}} + \frac{1}{16 \times 10^{-6}}$

Hence

$C = 5.3\ \mu F$

5.46 A capacitor C_1 is charged until there is a potential difference of V_0 across it. The charging battery is then removed and an uncharged capacitor C_2 connected across C_1. What is the resulting potential difference of C_1?

Ans The charge on C_1 when there is the potential difference V_0 across it is $Q = C_1 V_0$. When the other capacitor is connected across it this charge becomes distributed across both capacitors.

$Q = Q_1 + Q_2$

Thus

$C_1 V_0 = C_1 V + C_2 V$

where V is the new potential difference. Thus

$V = \frac{V_0 C_1}{C_1 + C_2}$

5.47 A capacitor of capacitance 8 μF is fully charged by a 6 V battery connected across it. What is the energy stored in the charged capacitor?

Ans Energy $= \frac{1}{2}CV^2$
$= \frac{1}{2} \times 8 \times 10^{-6} \times 6^2$
$= 1.4 \times 10^{-4}$ J

5.48 A parallel plate air capacitor has plates of area 500 mm^2 and separation 10 mm. The capacitor is charged to a potential difference of 6.0 V. The charging battery is then disconnected and the plates pulled apart until their separation is 20 mm. What is (a) the initial capacitance, (b) the new potential difference between the plates and (c) the work needed to separate the plates?

Ans (a) $C = (\epsilon_0 A)/d$

$= \frac{8.85 \times 10^{-12} \times 500 \times 10^{-6}}{10 \times 10^{-3}}$

$= 44 \times 10^{-14}$ F

(b) When the plates are pulled apart the charge on the plates remains constant. But doubling the separation halves the capacitance. Thus, as

$Q = CV$

the potential difference must have doubled to 12 V.
(c) The initial energy of the charged capacitor is

$\frac{1}{2}QV = \frac{1}{2}CV^2$

This is thus $\frac{1}{2} \times 44 \times 10^{-14} \times 6.0^2$ J. After the plates are pulled apart, and the capacitance is halved and the potential difference doubled, the energy stored is $\frac{1}{2} \times 22 \times 10^{-14} \times 12^2$ J. The work done is thus the difference in these two energies, i.e. 7.92×10^{-12} J.

Further problems

Take ϵ_0 to be $8.85 \times 10^{-12}\ C^2\ N^{-1}\ m^{-2}$.

5.49 A raindrop in air has a radius of 2.0 mm. What is its capacitance?

5.50 A parallel plate capacitor has plates, each of area 100 cm^2, and separated by an air gap of 20 mm. What is (a) the capacitance, (b) the charge on the plates when the potential difference between them is 200 V and (c) the electric field between the plates of the charged capacitor?

5.51 Three capacitors of capacitances 4 μF, 8 μF and 16 μF are connected (a) in series, (b) in parallel. What is the total capacitance in each case?

5.52 What is the capacitance of a capacitor consisting of plates, each of area 400 mm^2, separated by an insulator of thickness 1.0 mm if the insulator has a dielectric constant of 2.6?

5.53 What is the energy stored in a capacitor of

capacitance 8 μF when it is charged to a potential difference of 12 V?

5.54 A pacemaker requires pulses of energy 2×10^{-4} J. If these pulses are supplied by a capacitor of capacitance 12 μF, what is the potential difference that must be applied to the capacitor?

5.55 A battery of e.m.f. 12 V is used to charge fully a capacitor of capacitance 20 μF. What is the energy stored in the capacitor?

5.56 A capacitor of capacitance 8 μF is charged until there is a potential difference of 12 V across it. The charging battery is then removed and an uncharged capacitor of capacitance 4 μF is connected across the charged capacitor. (a) What is the new potential difference across the 8 μF capacitor? (b) What is the new charge on the 8 μF capacitor? (c) What is the charge on the 4 μF capacitor?

5.57 A parallel plate capacitor has two plates, each of area 4000 mm^2, separated by an air gap of 30 mm. The capacitor is permanently connected to a battery of e.m.f. 50 V.
(a) What is the charge on the capacitor plates?
(b) What is the energy stored by the capacitor?
(c) The separation of the capacitor plates is then increased to 60 mm, without disconnecting the battery. What is the new charge on the plates and the energy stored by the capacitor?

5.58 What would the answers to **5.57** have been if the battery had been disconnected after the initial charging and before the plates were moved further apart?

5.59 An air capacitor consists of two parallel plates and has a capacitance of 10 pF.
(a) What is the potential difference between the capacitor plates if the charge on the plates is 2×10^{-7} C? (b) What will be the potential difference between the plates if the separation of the plates is doubled, without any change in the charge on the plates? (c) How much work is needed to produce this increased separation?

Magnetic fields

Notes

$$\text{Magnetic flux density, } B = \frac{F}{IL}$$

where F is the force acting on a current element of length L carrying a current I.

$$\text{Magnetic flux } \phi = \text{flux density } B \times \text{area } A$$

where B is the flux density component at right angles to the area concerned.

Magnetic fields can be explored and the flux density determined by a number of methods.

(i) A current balance (*Figure 5.1*) in which the carrying wire in the magnetic field is measured.
(ii) A search coil (*Figure 5.2*) in which a magnetic flux, linked by a small coil, changes and an e.m.f. is induced in the coil. This can be

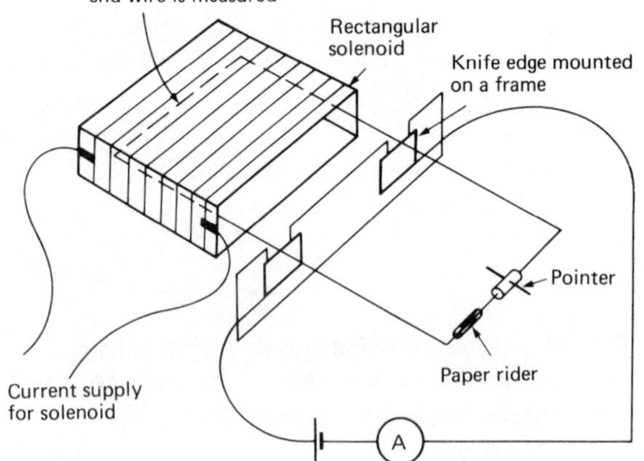

Figure 5.1 A current balance being used to measure the flux density inside a solenoid

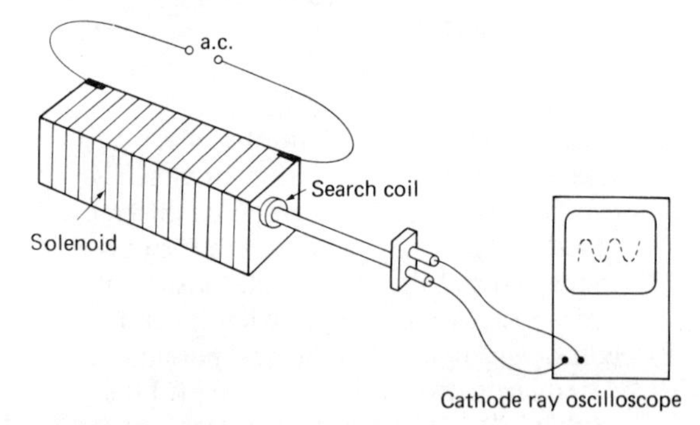

Figure 5.2 A search coil

detected by a cathode ray oscilloscope or a ballistic galvanometer. The flux change can occur as a result of suddenly removing the coil from the magnetic field or, in the case of an electromagnet, using alternating current for the electromagnet.

(iii) A Hall probe (*Figure 5.3*) which consists of a thin slice of a semiconducting material through which a current is passed. In the absence of a magnetic field component at right angles to the surface of the slice there is no current in a direction at right angles to the supplied current. In practice some adjustment is needed to ensure this. When there is a magnetic field a current can be detected, and measured, in a direction at right angles to the applied current, this current being proportional to the magnetic flux density.

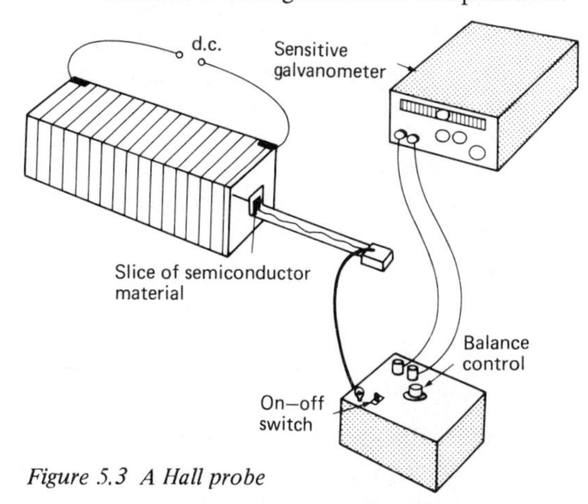

Figure 5.3 A Hall probe

A ballistic galvanometer is one which has a large period of oscillation. When a short-duration current pulse is applied to such an instrument the pulse is over before the galvanometer coil has appreciably deflected. The initial deflection θ is proportional to the total charge Q passed through the coil of the instrument.

$$\theta = bQ$$

where b is a constant called the charge sensitivity. The induced e.m.f. in a search coil of N turns is given by

$$\text{Induced e.m.f.} = -NA\frac{dB}{dt}$$

where A is the area of the coil. If the search coil circuit has a resistance R then the current I flowing is given by

$$I = \frac{V}{R}$$

$$= -\frac{NA}{R}\frac{dB}{dt}$$

But

$$I = dQ/dt$$

thus

$$\frac{dQ}{dt} = -\frac{NA}{R}\frac{dB}{dt}$$

Thus, if the search coil suffers a flux density change of B_1 to B_2

$$Q = \frac{NA}{R}(B_1 - B_2)$$

and

$$\text{Galvanometer deflection, } \theta = \frac{bNA}{R}(B_2 - B_1)$$

The flux density B at a distance r from a long straight wire carrying a current I is given by

$$B = \frac{\mu_0 I}{2\pi r}$$

where μ_0 is a constant called the permeability of free space and has the value $4\pi \times 10^{-7}$ N/A^2.

The flux density inside a long solenoid having n turns per unit length and carrying a current I is given by

$$B = \mu_0 nI$$

For any closed path, the value of the sum of the length of each small segment in the path multiplied by the component of the flux density along that segment is the total circulation of line integral of the flux density round the closed path. This circulation is equal to $\mu_0 NI_0$ where N is the number of conductors carrying the current I which are enclosed by the closed path. This statement is known as **Ampere's circuital law.**

Forces occur between two parallel current-carrying conductors, each wire being a current-carrying conductor in the magnetic field of the other. Thus, the flux density at wire 2 due to wire 1 is

$$B = \mu_0 I_1/(2\pi a)$$

where a is the separation of the two wires. But

$$F = BI_2L$$

Thus

$$F = \frac{\mu_0 I_1 I_2 L}{2\pi a}$$

The ampere is defined as that current which, if maintained in two straight parallel conductors of infinite length and negligible circular cross-section placed 1 metre apart in a vacuum, will produce between these conductors a force equal to 2×10^{-7} N per metre length.

Examples

Take μ_0 to be $4\pi \times 10^{-7}$ N/A^2.

5.60 An air-cored solenoid has 1000 turns of wire in its length of 200 mm and carries a current of 2.0 A. What is the magnetic flux density along the axis of the solenoid in the centre?

Ans $B = \mu_0 nI$

$$= 4\pi \times 10^{-7} \times \frac{1000}{200 \times 10^{-3}} \times 2.0$$

$$= 1.3 \times 10^{-2} \text{ Wb/m}^2$$

5.61 What is the force acting per unit length of two parallel wires carrying a current of 2.0 A if they are situated 40 mm apart?

Ans $$\frac{\text{Force}}{\text{Length}} = \frac{\mu_0 I^2}{2\pi a}$$

$$= 2.0 \times 10^{-5} \text{ N/m}$$

5.62 A search coil is placed with its axis parallel to a magnetic field produced in a solenoid. The search coil has a cross-sectional area of 200 mm^2 and has 100 turns of wire. It is connected to a ballistic galvanometer having a charge sensitivity of 200 mm/μC. The ballistic galvanometer circuit has a total resistance of 1500 Ω. When the coil is quickly removed from the magnetic field to a region where the magnetic field is virtually zero, the ballistic galvanometer gives a deflection of 50 mm. What was the magnetic flux density in the solenoid?

Ans $$\theta = \frac{bNA}{R}(B_1 - B_2)$$

Hence

$$B_1 = \frac{\theta R}{bNA}$$

$$= \frac{50 \times 1500}{200 \times 10^6 \times 100 \times 200 \times 10^{-6}}$$

$$= 1.9 \times 10^{-2} \text{ Wb/m}^2$$

5.63 Use Ampere's circuital law to derive an equation giving the flux density at a radial distance r from a long straight wire carrying a current I.

Ans Consider the closed loop to be a circle with a centre through which the wire passes and which has a radius r. The circumference of the circle is $2\pi r$ and the flux density is constant at all points round the circle. Thus, the product of the flux density and the length of the path is $2\pi rB$. Hence

$$2\pi rB = \mu_0 I$$

$$B = \frac{\mu_0 I}{2\pi r}$$

5.64 An alternating current of peak value 1.2 A and of frequency 50 Hz is passed through a solenoid. A search coil with its axis parallel to the axis of the solenoid is situated inside the solenoid and is connected to a cathode ray oscilloscope. The screen of the oscilloscope shows an alternating voltage having a peak value of 20 mV. What would be the peak voltage value indicated on the oscilloscope screen if (a) the peak current value through the solenoid was doubled to 2.4 V at 50 Hz, (b) the peak current was kept at the same value of 1.2 A but the frequency increased to 100 Hz?

Ans The current variation with time is given by

$$I = I_{max} \sin 2\pi ft$$

where f is the frequency and I_{max} the peak value of the flux density. As the flux density is proportional to the current then

$$B = kI_{max} \sin 2\pi ft$$

The induced e.m.f. in the search coil is

$$V = -NA\,\mathrm{d}B/\mathrm{d}t$$

Thus

$$V = -2\pi fNAI_{max} \cos 2\pi ft$$

The maximum voltage is $2\pi fNAI_{max}$.
(a) Doubling the maximum current doubles the maximum induced voltage.
(b) Doubling the frequency doubles the maximum induced voltage.

Further problems

Take μ_0 to be $4\pi \times 10^{-7}$ N/A^2.

5.65 What is the magnetic flux density at a point 40 mm from a long straight wire carrying a current of 4.0 A?

5.66 Two long parallel wires are 150 mm apart. If there is a current of 2.0 A in one of the wires and 5.0 A in the same direction in the other what is the flux density at a point half way between the wires?

5.67 What would the answer to **5.66** be if the currents were in opposite directions?

5.68 What is the force per unit length of wire for the situation described in **5.66**?

5.69 What is the flux density along the axis of a solenoid with 5000 turns per metre and carrying a current of 2.0 A?

5.70 A solenoid has a length of 200 mm and a cross-sectional area of 300 mm^2. If it has 600 turns and carries a current of 2.0 A what is (a) the flux density parallel to the axis of the solenoid and near its centre, (b) the flux in the core of the solenoid?

5.71 A search coil has 20 turns of wire and an area of 140 mm^2. What is the average e.m.f. induced in the coil if it is moved in 20 s from a region where the flux density is zero to one where it is 2.0 Wb/m^2?

5.72 Consider a solenoid in the form of a spring and assume that the individual turns do not come into electrical contact. (a) A current is passed through the spring. What happens to the magnetic flux density inside the solenoid if the spring is stretched to double its length? (b) What happens to the magnetic flux within the solenoid if the contacts for passing the current through the spring are changed from across the full length of the spring to across only half the length of the spring? Assume that changing to half the length of the spring doubles the current.

5.73 For the current balance arrangement shown in *Figure 5.1* the length of the current balance wire that is in the magnetic field, and at right angles to it, is 50 mm, the current through the balance wire being 1.5 A. A paper rider of mass 0.040 g is used to restore balance to the same position as when no current flowed through the balance. What is the flux density of the magnetic field?

5.74 The ampere can be established with the aid of a spring balance and a metre rule. Explain this statement.

5.75 Describe an experiment by which you could establish the relationship between the current passed through a solenoid and the magnetic flux density in that solenoid.

Magnetic circuits

Notes

The term magnetic circuit is used for a closed loop of magnetic flux; this compares with a closed loop of electrical current for an electrical circuit. For a magnetic circuit

Magnetomotive force
= magnetic flux × total circuit reluctance

This compares with, for an electrical circuit

Electromotive force (e.m.f.)
= current × total circuit resistance

The magnetomotive force (m.m.f.) is the source of the magnetic flux. In the case of a coil of N turns carrying a current I,

$$\text{m.m.f.} = NI$$

The reluctance R_m is given by

$$R_m = \frac{L}{\mu_r \mu_0 A}$$

where L is the length of flux path, A its cross-sectional area and μ_r the relative permeability. This is a numerical factor which expresses how much better a material is at producing flux in, say, a coil rather than a vacuum. The magnetomotive circuit equation can be therefore written as

$$NI = \frac{\theta L}{\mu_r \mu_0 A}$$

This compares with, for an electrical circuit,

$$E = \frac{I\rho L}{A}$$

for a circuit composed of material having a resistivity ρ, length L and cross-sectional area A.

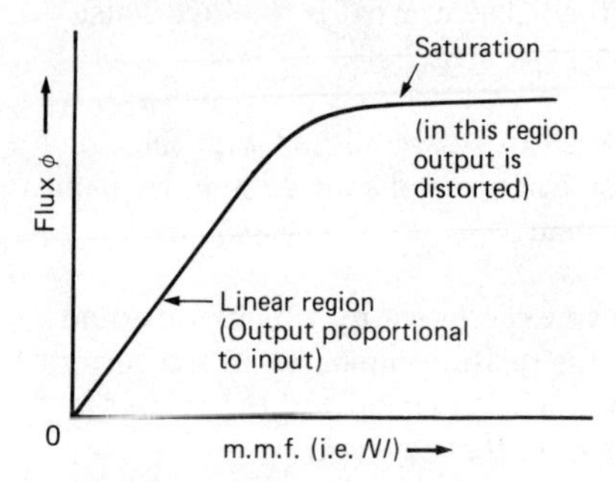

Figure 5.4 A magnetisation curve

The graph in *Figure 5.4* shows how the flux in a material varies with the m.m.f. and is called the magnetisation curve. Instead of m.m.f. the quantity B_0 is sometimes plotted against B instead of ϕ, where

$$B_0 = \mu_0 NI/L$$

i.e. the flux density that would exist if there was a vacuum. The quantity B is the flux density in the material, i.e. $\phi/(\mu_r A)$.

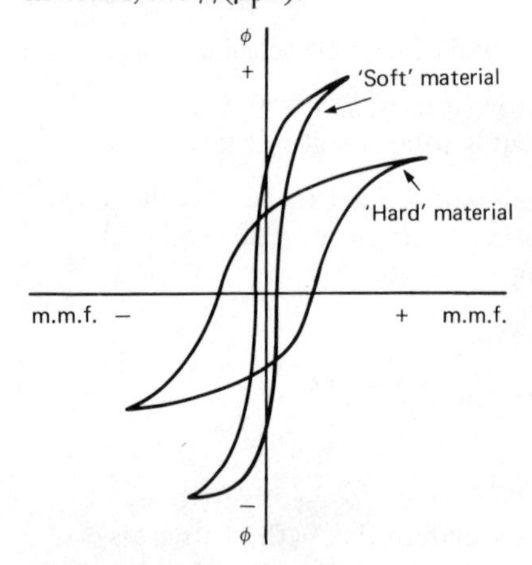

Figure 5.5 The hysteresis loop

If, after magnetising the material, the m.m.f. is decreased and then reversed a hysteresis loop (*Figure 5.5*) is produced. The area enclosed by the loop is proportional to the energy expended each time the magnetising current goes through a complete cycle. Materials termed soft are easily magnetised and demagnetised; such materials are used for transformer cores. Materials termed hard are more difficult to magnetise and retain a higher percentage of their magnetism.

Units: magnetomotive force – A (often called ampere-turns), flux – Wb, reluctance – A/Wb.

Examples

5.76 A toroid has a mean circumferential length of 450 mm and a cross-sectional area of 500 mm^2. The toroid consists of 400 turns of wire wound on an iron core and carries a current of 0.15 A. (a) What is the magnetomotive force (m.m.f.) in the toroid? (b) What is the reluctance of the magnetic circuit? (c) What is the flux in the toroid (relative permeability can be taken to have a value of 700 and $\mu_0 = 4\pi \times 10^{-7}$ N/A^2)?

Ans (a) M.m.f. $= NI$
$= 400 \times 0.15$
$= 60$ A

(b)
$$R_m = \frac{L}{\mu_r \mu_0 A}$$
$$= \frac{450 \times 10^{-3}}{700 \times 4\pi \times 10^{-7} \times 500 \times 10^{-6}}$$
$$= 1.0 \times 10^6 \text{ A/Wb}$$

(c) M.m.f. $= \phi R_m$

Hence

$$\phi = 60/(1.0 \times 10^6)$$
$$= 6.0 \times 10^{-5} \text{ Wb}$$

Further problems

5.77 How does a magnetic circuit compare with an electrical circuit?

5.78 A toroid consists of an iron core wound with 600 turns of wire carrying a current of

200 mA. The toroid has a mean circumferential length of 300 mm and a cross-sectional area of 400 mm^2. What is (a) the magnetomotive force in the toroid, (b) the reluctance of the circuit and (c) the flux in the toroid ($\mu_r = 600$ and $\mu_0 = 4\pi \times 10^{-7}$ N/A^2)?

5.79 A magnetic circuit is designed to be a solid piece of iron in the form of a ring. Unfortunately, the ring contains an air gap and a continuous ring of iron does not occur. What is the effect of such an air gap on the flux developed inside the iron?

5.80 Explain the significance of the hysteresis loop for a magnetic material.

5.81 Describe how you could obtain the hysteresis loop for a material in the form of a ring.

Charge

Notes

Charge comes in packets of size 1.6×10^{-19} C. Only multiples, either positive or negative, of this fundamental charge have yet been found. The classic experiments on the determination of this charge value were carried out by R. Millikan. *Figure 5.6* shows the fundamentals of the apparatus used by him.

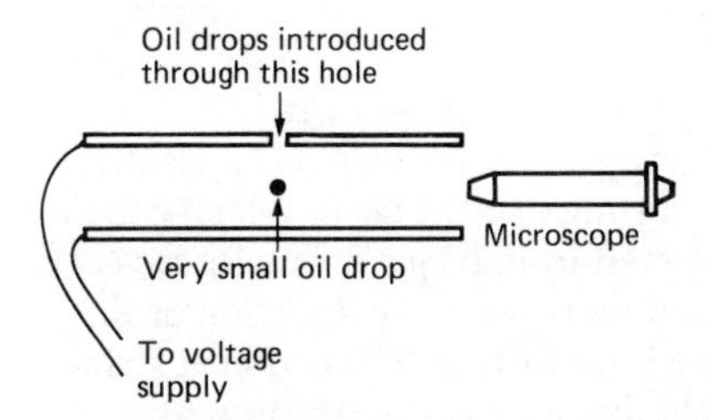

Figure 5.6 The Millikan oil drop experiment

Oil drops were produced by an atomiser and drifted into the space between the two parallel plates. The gravitational force acting on an oil drop is

$$mg = \frac{4}{3}\pi r^3 \rho g$$

where ρ is the density of the oil and r the drop radius. The upthrust force (see chapter 2) is $\frac{4}{3}\pi r^3 \sigma g$, where σ is the density of the air. Thus, the net downward force is $\frac{4}{3}\pi r^3 g(\rho - \sigma)$. The viscous drag on the falling drop is $6\pi\eta r v$ (see chapter 2). When the viscous force equals the net downward force (i.e. weight minus upthrust), the drop falls with a constant, terminal velocity.

By applying a potential difference V between plates of separation d and electric field

$$E = V/d$$

is produced. This can be adjusted so that it is opposite and equal to the net downward force and thus the drop is held stationary in space. When this happens

$$Vq/d = \frac{4}{3}\pi r^3 g(\rho - \sigma)$$

and so

$$q = \frac{4\pi r^3 gd(\rho - \sigma)}{3V}$$

Examples

Take g to be 9.8 m/s^2.

5.82 An oil drop of mass 3.0×10^{-14} kg and carrying a charge of $2 \times 1.6 \times 10^{-19}$ C is held stationary between the horizontal parallel plates in a Millikan experiment. What is the size of the electric field (neglect any effects due to the density of the air)?

Ans Gravitational force = mg

Electric force = Eq

Thus

$$E = \frac{mg}{q} = \frac{3.0 \times 10^{-14} \times 9.8}{3.2 \times 10^{-19}}$$

$$= 9.8 \times 10^5 \text{ V/m}$$

5.83 In a Millikan experiment a charged drop is found to fall through 2.0 mm in a time of 54.8 s with a constant velocity when no electric field is present. When an electric field of 2.37×10^4 V/m is applied between the horizontal parallel plates the oil drop ceases to fall and remains stationary. (a) What is the radius of the oil drop? (b) What is the charge carried by the oil drop? (c) How many excess electrons does the drop carry (density of the oil is 824 kg/m^3, that of air is 1.29 kg/m^3, the viscosity of air is

1.80×10^{-5} N s m^{-2}, and charge on electron = 1.60×10^{-18} C)?

Ans (a) When the drop is freely falling with the terminal velocity

$$\frac{4}{3}\pi r^3 g(\rho - \sigma) = 6\pi\eta r v$$

Hence

$$r^2 = \frac{9\eta v}{2g(\rho - \sigma)}$$

Hence

$$r = 6.052 \times 10^{-7} \text{ m}$$

(b) When the drop is held stationary by the electric field,

$$Eq = \frac{4}{3}\pi r^3 g(\rho - \sigma)$$

Hence

$$q = 3.161 \times 10^{-19} \text{ Q}$$

(c) $1.60 \times 10^{-19}\ N = 3.161 \times 10^{-19}$
$N = 1.98$

The answer is 2 electrons.

Further problems

Take g to be 9.8 m/s^2 and the charge on an electron to be 1.6×10^{-19} C.

5.84 Explain how Millikan's results with his oil drop experiment led to the belief that charge comes in packets.

5.85 An oil drop has a mass of 2.0×10^{-12} kg and carries 4 excess electrons. (a) If the density of the oil is 820 kg/m^3 and the density of air 1.3 kg/m^3, what will be the terminal velocity of the oil drop when freely falling in the absence of any electric field (viscosity of air is 1.8×10^{-5} N s m^{-2})? (b) What size electric field will be needed to make the drop cease falling and become stationary? (c) If the parallel plates are 15.0 mm apart, what is the potential difference that has to be applied between the plates to produce the field specified by (b)?

5.86 An oil drop is held stationary in a Millikan experiment with horizontal parallel plates. If the oil drop has a radius of 1.64×10^{-6} m and the electric field is 1.92×10^5 V/m, what number of excess electrons is carried by the drop (density of the oil is 850 kg/m^3, that of air is 1.30 kg/m^3)?

5.87 The following are some of Millikan's results (converted into the units used in this book). What can be deduced from them with regard to the basic size of the charge packet? Observed charges/10^{-19} C: 6.55, 8.20, 9.87, 11.49, 13.13, 14.81, 16.47, 17.97.

Charged particles in electric and magnetic fields

Notes

When a charged particle carrying a charge q is accelerated by a potential difference V, then the energy gained by the particle is qV and this results in a gain in kinetic energy of $\frac{1}{2}mv^2$. Thus

$$\tfrac{1}{2}mv^2 = qV$$

A charged particle carrying a charge q and moving in a magnetic field which has a flux density component B at right angles to its path experiences a force of Bqv, where v is the velocity of the particle. If the flux density is uniform and at right angles to the path then the charged particle will move in a circular path of radius r, and

$$mv^2/r = Bqv$$

where m is the mass of the charged particle.

If a beam of charged particles enters an electric field which is at an angle to the beam velocity then the beam is deflected from its path. The force acting on the charged particles in the direction of the field is Eq. There is no force in a direction at right angles to the field direction and so the initial velocity component in that direction remains unchanged. The situation is similar to that of a projectile in the Earth's gravitational field (see chapter 1). For a beam starting with a velocity v at right angles to the electric field, the distance x covered in a time t in the initial velocity direction is

$$x = vt$$

The displacement y of the beam in the electric field direction after a time t is

$$y = \tfrac{1}{2}at^2$$

where

$$F = EQ = ma$$

and so

$$y = \frac{Eqt^2}{2m}$$

The equation of the path described by the beam while in the electric field can be obtained by eliminating t from the y and x equations.

$$y = \frac{Eqx^2}{2mv^2}$$

It is the equation of a parabola.

Examples

Take the charge on an electron to be 1.6×10^{-19} C and the mass of an electron to be 9.1×10^{-31} kg.

5.88 What is the velocity of electrons after they have been accelerated through a potential difference of 400 V?

Ans $\frac{1}{2}mv^2 = Vq$

Hence

$$v^2 = \frac{2Vq}{m}$$

$$= \frac{2 \times 400 \times 1.6 \times 10^{-19}}{9.1 \times 10^{-31}}$$

$$v = 1.2 \times 10^7 \text{ m/s}$$

5.89 A beam of electrons moves with a velocity of 4.0×10^6 m/s into a uniform magnetic field of flux density 2.0×10^{-3} Wb/m^2 at right angles to the beam's velocity. What is the radius of the circular path of the electron beam in the magnetic field?

Ans $Bqv = mv^2/r$

Hence

$$r = \frac{mv}{Bq}$$

$$= \frac{9.1 \times 10^{-31} \times 4.0 \times 10^6}{2.0 \times 10^{-3} \times 1.6 \times 10^{-19}}$$

$$= 1.1 \times 10^{-2} \text{ m}$$

5.90 A beam of electrons moves with a velocity of 2.0×10^7 m/s along the axis of a cathode ray oscilloscope before passing between a pair of parallel metal plates 30 mm long and 20 mm apart. If there is a potential difference of 500 V between the plates and the beam is initially parallel to the plates, with what angle to the axis will the beam leave the space between the plates? Assume that the electric field between the plates is perfectly uniform up to the edges of the plates.

Ans See *Figure 5.7*. The horizontal velocity v_x of the electrons remains unchanged at 2.0×10^7 m/s. The vertical velocity v_y is

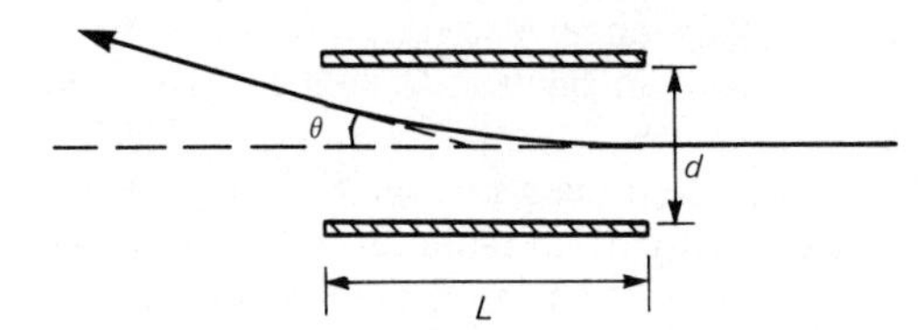

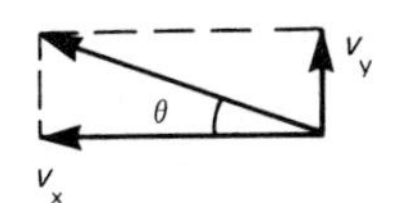

Figure 5.7

zero on entering the plates. During the passage of the beam between the plates the electrons are acted on by a force of Eq, where

$$E = V/d$$

This causes an acceleration in the y direction given by

$$F = ma$$

Thus

$$a = \frac{F}{m} = \frac{Vq}{dm}$$

The time for which the electrons are between the plates is given by

$$t = L/v_x$$

Thus, as this is the time for which the electrons are being accelerated in the y direction,

$$v_y = u + at$$

$$v_y = 0 + \frac{VqL}{v_x dm}$$

The resultant velocity of v_x and v_y on leaving the plates is in a direction given by

$$\tan\theta = \frac{v_y}{v_x}$$

$$= \frac{VqL}{v_x^2 dm}$$

$$= \frac{500 \times 1.6 \times 10^{-19} \times 30 \times 10^{-3}}{(2.0 \times 10^7)^2 \times 20 \times 10^{-3} \times 9.1 \times 10^{-31}}$$

Hence

$$\theta = 18^\circ$$

Further problems

Take the charge on an electron to be 1.6×10^{-19} C and the mass of an electron to be 9.1×10^{-31} kg.

5.91 A proton of mass 1.67×10^{-27} kg moves in a circular path of radius 300 mm in a magnetic field, at right angles to the proton path, of 1.80×10^{-2} Wb/m^2. What is the speed of the proton (charge on the proton $= 1.6 \times 10^{-19}$ C)?

5.92 A beam of electrons is introduced into a vacuum chamber between the poles of a large electromagnet. If the electrons have a speed of 2.0×10^7 m/s in a direction at right angles to the flux density of 2.0×10^{-2} Wb/m^2, what will be the time taken for an electron to complete a circular orbit?

5.93 An electron is liberated from a filament with zero velocity. Between the filament and the anode, a metal plate, there is a potential difference of 100 V. With what velocity will the electron hit the anode?

5.94 Describe an experiment by which the charge to mass ratio for electrons can be determined.

5.95 A beam of electrons has a velocity of 2.0×10^6 m/s in a horizontal direction. It then passes through a vertical electric field of 3.0×10^5 V/m. What magnetic flux density should be applied over the same region of space if the beam of electrons is not to be deviated from its straight line horizontal path?

5.96 Sketch the main features of a cathode ray tube and explain the functions of each element.

Current and charge movement

Notes

If there are in a conductor n charge carriers per unit volume, each carrying a charge q and moving with a velocity v along the conductor, then in a time t a charge carrier will cover a distance vt. Thus, in a time t all those charge carriers in a volume vtA will have passed through some particular cross-section of the conductor. Thus, in time t the number of charge carriers passing through the section is $nvtA$. Current is the rate of movement of charge and thus is $nvtAq/t$, hence

$$I = nvAq$$

When charged particles are moving in a conductor and the conductor is in a magnetic field a force acts on the charged particles and results in a potential difference transverse to the direction of the charged particles motion. This effect is known as the Hall effect.

The force on a charge carrier is Bqv when there is a flux density B at right angles to the current (*Figure 5.8*). This causes negative charge to drift

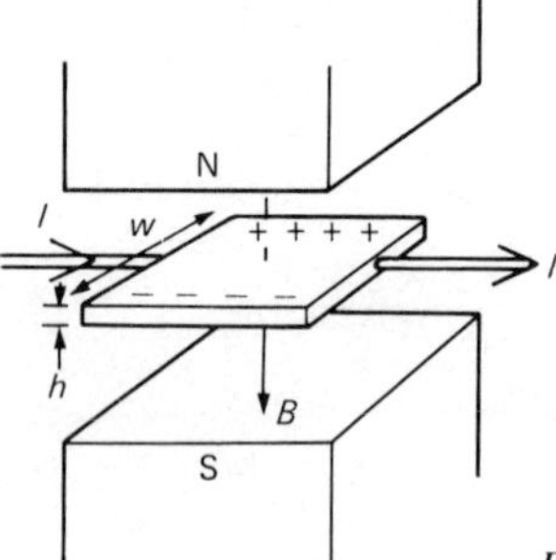

Figure 5.8 The Hall effect

to one edge of the slab of conductor and positive charge to the other. An electric field E is produced. This field exerts a force Eq on the charged particles and equilibrium is reached when

$$Eq = Bqv$$

Thus

$$E = Bv$$

The electric field means a potential difference V, where

$$V = Ew$$

Hence

$$V/w = Bv$$

As

$$I = nvAq$$

and

$$A = wh$$

then

$$V = \frac{BI}{nhq}$$

If a transverse current I' is taken through a resistance R, then $V = I'R$ and so

$$I' = \frac{BI}{Rnhq}$$

Examples

5.97 One mole of copper has a mass of 63.6 g. If copper has a density of 8930 kg/m^3 and each copper atom donates one electron for conduction purposes, what is the number of charge carriers per unit volume of copper? What is the velocity of the charge carriers when a copper wire of cross-sectional area 1.0 mm^2 carries a current of 200 mA (Avogadro's number is 6.0×10^{23} and the charge on the electron $= 1.6 \times 10^{-19}$ C)?

Ans Number of atoms in 1 mole $= 6.0 \times 10^{23}$
Number of electrons in 1 mole $= 6.0 \times 10^{23}$
1 m^3 of copper has a mass of 8930 kg
One mole of copper has a mass of 63.6×10^{-3} kg

Therefore 1 m^3 of copper is $8930/(63.6 \times 10^{-3})$ mole

Therefore

Number of charge carriers in 1 m^3

$$= \frac{8930 \times 6.0 \times 10^{23}}{63.6 \times 10^{-3}}$$

$$= 8.42 \times 10^{28}$$

$$I = nvAq$$

Therefore

$$v = \frac{200 \times 10^{-3}}{8.42 \times 10^{28} \times 1.0 \times 10^{-6} \times 1.6 \times 10^{-19}}$$

$$= 1.48 \times 10^{-5} \text{ m/s}$$

5.98 What is the Hall voltage produced in a silver ribbon 1.0 mm thick when a current of 10 A flows through it and the ribbon is in a magnetic field of flux density 1.4 Wb/m^2 at right angles to the surface of the ribbon (number of charge carriers per cubic metre is 7.4×10^{28}, each carrying a charge of 1.6×10^{-19} C)?

Ans $$V = \frac{BI}{nhq}$$

$$= \frac{1.4 \times 10}{7.4 \times 10^{28} \times 1.0 \times 10^{-3} \times 1.6 \times 10^{-19}}$$

$$= 1.2 \times 10^{-6} \text{ V}$$

Further problems

5.99 If there are about 10^{29} free electrons in each cubic metre of copper, what is the drift velocity of the electrons in a copper wire of cross-sectional area 1 mm^2 when carrying a current of 5 A (charge on the electron $= 1.6 \times 10^{-19}$ C)?

5.100 Explain how the Hall effect can be used to determine the sign of the charge carriers in a material.

5.101 A wafer of n-type semiconductor material is found to give a Hall voltage when a magnetic field of flux density 0.18 Wb/m^2 is applied normally to the face of the slice and a current of 2.0 mA is passed through it. The slice has a thickness of 0.50 mm. If the number of charge carriers per cubic metre of material is 6.0×10^{20} and each carries a charge of 1.6×10^{-19} C, what is the Hall voltage?

5.102 A blood-flow meter operates on the principle of the Hall effect. What is the velocity of the blood if a magnetic field of flux density 2.0 Wb/m^2 generates a potential difference of 0.60 mV between two electrodes transverse to the blood flow direction and separated by 1.0 mm?

5.103 A Hall voltage of 4.5 μV is found to occur when a current of 10 A passes through a strip of copper of thickness 1.0 mm in a magnetic field of flux density 5.0 Wb/m^2 normal to the slice. What is the number of charge carriers per cubic metre of copper (assume each charge carrier to carry a charge of 1.6×10^{-19} C)?

Examination questions

5.104 Explorer 38, a radio-astronomy research satellite of mass 200 kg, circles the Earth in an orbit of average radius $3R/2$ where R is the radius of the Earth. Assuming the gravitational pull on a mass of 1 kg at the Earth's surface to be 10 N, calculate the pull on the satellite.

(University of London)

5.105 (a) Give an account of one method of measuring the value of the gravitational constant, G, explaining how the result is calculated from the observations made.
(b) Kepler's third law of planetary motion, as simplified by taking the orbits to be circles round the Sun, states that if r denotes the radius of the orbit of a particular planet and T denotes the period in which that planet describes its orbit, then r^3/T^2 has the same value for all the planets.

The orbits of the Earth and of Jupiter are very nearly circular with radii of 150×10^9 m and 778×10^9 m, respectively, while Jupiter's period round the Sun is 11.8 years.

(i) Show that these figures are consistent with Kepler's third law.
(ii) Taking the value of the gravitational constant, G, to be 6.67×10^{-11} N m^2 kg^{-2}, estimate the mass of the Sun.

(Oxford Local Examinations)

5.106 (a) Define gravitational field strength and gravitational potential, stating the relationship between them. Explain what is meant by the term uniform field and discuss to what extent the gravitational field of the Earth can be considered to be uniform by considering two points on the surface
(i) separated by a distance of about 10 km,
(ii) at opposite ends of a diameter. Assume that the Earth is a homogeneous sphere.
(b) Write down an expression for the gravitational potential at the surface of the Earth in terms of its mass M, radius R and the gravitational constant G. Sketch a graph showing the variation of potential with position along a line passing through the centre of the Earth and point out the important features of the graph. (Only consider points external to the surface and in one direction only.)
(c) Derive an expression for the escape velocity, v, at the surface of a planet in terms of the radius, r, of the planet and the acceleration of free fall, g_p, at the surface of the planet.

(JMB)

5.107 (a) Define capacitance.
Describe an experiment to show that the capacitance of an air spaced parallel plate capacitor depends on the separation of the plates. State three desirable properties of a dielectric material for use in a high quality capacitor.
(b) Two plane parallel plate capacitors, one with air as dielectric and the other with mica as dielectric, are identical in all other respects. The air capacitor is charged from a 400 V d.c. supply, isolated, and then connected across the mica capacitor which is initially uncharged. The potential difference across this parallel combination becomes 50 V. Assuming the relative permittivity of air to be 1.00, calculate the relative permittivity of mica.
(c) Compare the energy stored by the single charged capacitor in part (b) with the energy stored in the parallel combination.
Comment on these energy values.

(AEB)

5.108 Give an expression for the electric field strength at a distance d from an isolated point charge Q. Define the term potential at any point in the field. Sketch a graph showing how the potential changes with distance from the charge.

One plate of a parallel plate capacitor is earthed and the other plate has a positive charge. For the region between the plates sketch a graph of potential against distance from the earthed plate if the plates are close together.

How could you find the electric field strength at any point from these graphs? Justify your statement by deducing the relationship between field strength and potential gradient.

A parallel plate capacitor with air as a dielectric has plates of area 4.0×10^{-2} m^2 which are 2.0 mm apart. It is charged by connecting it to a 100 V battery. It is then disconnected from the battery and connected in parallel with a similar, uncharged capacitor with plates of half the area which are twice the distance apart. Calculate the final charge on each capacitor. Edge effects may be neglected.
Permittivity of air = 8.8×10^{-12} F m^{-1}.

(University of London)

5.109 The capacitance of a certain variable capacitor may be varied between limits of 1×10^{-10} F and 5×10^{-10} F by turning a knob attached to the movable plates. The capacitor is set to 5×10^{-10} F, and is charged by connecting it to a battery of e.m.f. 200 V.
(a) What is the charge on the plates?
The battery is then disconnected and the capacitance changed to 1×10^{-10} F.
(b) Assuming that no charge is lost from the plates, what is now the potential difference between them?
(c) How much mechanical work is done against electrical forces in changing the capacitance?

(University of Cambridge)

5.110 Write down expressions for the electric field E at the surface and the potential V of an isolated spherical conductor of radius R carrying a charge q.

If air ionises in an electric field greater than 3.0×10^6 V m^{-1}, what is the maximum operating potential of a Van de Graaff generator with a spherical dome of radius 6.0 cm?

(University of Cambridge)

5.111 In a measurement of the electron charge by Millikan's method, a potential difference of 1.5 kV can be applied between horizontal, parallel metal plates 12 mm apart. With the field switched off, a drop of oil of mass 10^{-14} kg is observed to fall with constant velocity 400 μm/s. When the field is switched on, the drop rises with a constant velocity of 80 μm/s. How many electron charges are there on the drop? (You may assume that the air resistance is proportional to the velocity of the drop and that air buoyancy may be neglected and the charge on an electron = 1.6×10^{-19} C.)

(Southern Universities)

5.112 A beam of protons is accelerated from rest through a potential difference of 2000 V and then enters a uniform magnetic field which is perpendicular to the direction of the proton beam. If the flux density is 0.2 T calculate the radius of the path which the beam describes. Proton mass = 1.7×10^{-27} kg and electronic charge = -1.6×10^{-19} C.

(University of London)

5.113 Explain the origin of the Hall effect. Include a diagram showing clearly the directions of the Hall voltage and other relevant vector quantities for a specimen in which electron conduction predominates.

A slice of indium antimonide is 2.5 mm thick and carries a current of 150 mA. A magnetic field of flux density 0.5 T, correctly applied, produces a maximum Hall voltage of 8.75 mV between the edges of the slice. Calculate the number of free charge carriers per unit volume, assuming that they each have a charge of -1.6×10^{-19} C. Explain your calculation clearly.

What can you conclude from the observation that the Hall voltage in different conductors can be positive, negative or zero?

(University of Cambridge)

Answers

5.3 Ans $m_1v^2/r = Gm_1m_2/r^2$, hence $v = 2.98 \times 10^4$ m/s

5.4 Ans $m_1g = Gm_1m_2/r^2$, hence $g = 1.63$ N/kg and acceleration due to gravity = 1.63 m/s^2

5.5 Ans $F = Gm_1m_2/r^2$, hence $F = 2.67 \times 10^{-11}$ N

5.6 Ans $m_1g = Gm_1m_2/r^2$, hence $g_M/g_E = M_M r_E^2/M_E r_M^2$, thus g_M=1.94 m/s^2

5.7 Ans It would be the same

5.8 Ans Because the force is at right angles to the velocity of the Earth in its orbit about the Sun. The force provides the centripetal force needed to keep the Earth in its orbit. The Earth can be considered to be always falling towards the Sun but constantly missing it and so ending up in an orbit.

5.9 Ans $m_1g = Gm_1m_2/r^2$, hence $m_2 = gr^2/G$; but

$$m = \frac{4}{3}\pi r^3\rho$$

thus $\rho = 3g/(4\pi rG)$

5.10 Ans $m_1v^2/r = Gm_1m_2/r^2$, hence $r_N^3/r_E^3 = T_N^2/T_E^2$ and r_N/r_E is approximately 30

5.11 Ans $m_1v^2/r = Gm_1m_2/r^2$, hence $m_2 = 4\pi^2r^3/T^2 = 7.14 \times 10^{23}$ kg

5.16 Ans (a) $2^{1/2} \times$ Earth radius
(b) $2 \times$ Earth radius

5.17 Ans (a) -3.125×10^6 J/kg
(b) -2.08×10^6 J/kg
(c) -1.56×10^6 J/kg
(d) -1.25×10^6 J/kg

5.18 Ans $g = -dV/dr = -9.8$ N/kg

5.19 Ans Change in potential energy $= \frac{1}{2}m \times 3630^2 - \frac{1}{2}m \times 2620^2$. Change in potential = change in potential energy/mass = 3.156×10^6 J/kg

5.20 Ans Potential difference $= -GM(1/r_1 - 1/r_2) = -3.23 \times 10^7$ J/kg. Hence potential energy difference $= -3.23 \times 10^7 m$ J. Kinetic energy change balances this, hence the velocity = 8454 m/s

5.21 Ans $v_E = (2GM/r)^{1/2}$, hence escape velocity from the Moon = 0.21 that from the Earth

5.24 Ans No. A point cannot have two field directions

5.25 Ans (a) $E = F/Q$, hence $F = 1.2 \times 10^{-2}$ N
(b) $mg - EQ = 0.0076$ N
(c) $F = ma$, hence $a = 3.8$ m/s^2

5.26 Ans $E = Q/(4\pi\epsilon_0 r^2) = 7.2 \times 10^6$ N/C

5.27 Ans $F = Q_1Q_2/(4\pi\epsilon_0 r^2) = -1.8$ N; this is the force of attraction on each charge

5.28 Ans (a) Zero, inside the shell
(b) $E = Q/(4\pi\epsilon_0 r^2) = 5.0 \times 10^6$ N/C

5.29 Ans Distance apart of two balls $= 2 \times 1.0 \sin 7° = 0.24$ m; $F = Q^2/(4\pi\epsilon_0 r^2) = T \sin 7°$, where T is the tension in the thread. But $T \cos 7° = 0.02 \times 10^{-3} \times 9.8$ and hence $Q = 1.2 \times 10^{-8}$ C

5.33 Ans (a) The potential at every point inside the shell is the same as that on the shell, i.e. $V = Q/(4\pi\epsilon_0 r) = 9.0 \times 10^3$ V. The constant potential inside the sphere means a zero electric field
(b) Same as (a), i.e. 9.0×10^3 V
(c) 6.0×10^3 V

5.34 Ans $E = -$ potential gradient $= -2.5 \times 10^4$ N/C

5.35 Ans (a) $E = -$ potential gradient $= -60 \times 10^3$ N/C
(b) $F = EQ = 1.1 \times 10^{-14}$ N

5.36 Ans Gain in energy $= QV = 3.2 \times 10^{-17}$ J

5.37 Ans Work $= QV = 4 \times 10^{-4}$ J

5.38 Ans No

5.39 Ans No, the potential needs only to be some constant value

5.40 Ans Work $= QV = 4 \times 10^{-10} \times 2 \times 10^{-10}/(4\pi\epsilon_0 \times 30 \times 10^{-3}) = 2.4 \times 10^{-20}$ J

5.41 Ans (a) $E = Q/(4\pi\epsilon_0 r^2)$, hence $Q = 5.9 \times 10^5$ C
(b) $V = Q/(4\pi\epsilon_0 r) = 8.3 \times 10^8$ V

5.49 Ans $C = 4\pi\epsilon_0 r = 2.2 \times 10^{-13}$ F

5.50 Ans (a) $C = \epsilon_0 A/d = 4.4 \times 10^{-12}$ F
(b) $Q = CV = 8.85 \times 10^{-12}$ C
(c) $E = V/d = 10\,000$ N/C

5.51 Ans (a) $1/C = 1/4 + 1/8 + 1/16$, hence $C = 2.3\ \mu$F
(b) $C = 4 + 8 + 16 = 28\ \mu$F

5.52 Ans $C = \epsilon_r\epsilon_0 A/d = 9.2 \times 10^{-12}$ F

5.53 Ans Energy $= \frac{1}{2}CV^2 = 0.58 \times 10^{-3}$ J

5.54 Ans Energy $= \frac{1}{2}CV^2$, hence $V = 5.8$ V

5.55 Ans Energy $= \frac{1}{2}CV^2 = 1.4 \times 10^{-3}$ J

5.56 Ans See **5.46**
(a) $V = V_0 C_1/(C_1 + C_2) = 8$ V
(b) $Q = C_1 V = 6.4 \times 10^{-5}$ C
(c) $Q = C_2 V = 3.2 \times 10^{-5}$ C

5.57 Ans (a) $C = \epsilon_0 A/d = 1.2 \times 10^{-12}$ F; hence $Q = CV = 6.0 \times 10^{-11}$ C
(b) Energy $= \frac{1}{2}CV^2 = 1.5 \times 10^{-9}$ J
(c) V remains constant and C decreases to 0.6×10^{-12} F. Thus $Q = CV = 3.0 \times 10^{-11}$ C and energy $= \frac{1}{2}CV^2 = 7.5 \times 10^{-10}$ J

5.58 Ans **5.57 Ans** (c) would become: Q remains constant and C decreases to 0.6×10^{-12} F. Thus $Q = 6.0 \times 10^{-11}$ C and energy $= \frac{1}{2}Q^2/C = 3.0 \times 10^{-9}$ J

5.59 Ans (a) $V = Q/C = 20$ kV
(b) C is halved, thus V is doubled to 40 kV
(c) $\frac{1}{2}Q(V_1 - V_0) = 2 \times 10^{-3}$ J

5.65 Ans $B = \mu_0 I/(2\pi r) = 2.0 \times 10^{-5}$ Wb/m²

5.66 Ans $B = \mu_0 \times 5.0/(2\pi \times 75 \times 10^{-3}) - \mu_0 \times 2.0/(2\pi \times 75 \times 10^{-3}) = 8.0 \times 10 = 8.0 \times 10^{-6}$ Wb/m²

5.67 Ans 18.7×10^{-6} Wb/m²

5.68 Ans $F/L = \mu_0 I_1 I_2/(2\pi a) = 1.3 \times 10^{-5}$ N/m

5.69 Ans $B = \mu_0 nI = 1.26 \times 10^{-2}$ Wb/m²

5.70 Ans (a) $B = \mu_0 nI = 7.54 \times 10^{-3}$ Wb/m²
(b) $\phi = BA = 2.26 \times 10^{-6}$ Wb

5.71 Ans $E = -NA\ \mathrm{d}B/\mathrm{d}t = -2.8 \times 10^{-2}$ N

5.72 Ans (a) n is halved and so B is halved
(b) Only the current is doubled and so B is doubled

5.73 Ans $F = 0.040 \times 10^{-3} \times 9.8 = BIL$, hence $B = 5.2 \times 10^{-3}$ Wb/m²

5.74 Ans See the notes on the definition of the ampere

5.75 Ans You could use a search coil inside a solenoid supplied by an alternating current, or a current balance with the end wire inside the solenoid, or a Hall probe inside a solenoid

5.77 Ans See the notes

5.78 Ans (a) m.m.f. $= NI = 120$ A
(b) $R_m = L/(\mu_r\mu_0 A) = 9.9 \times 10^5$ A/Wb
(c) m.m.f. $= \phi R_m$, hence $\phi = 1.2 \times 10^{-4}$ Wb

5.79 Ans The relative permeability of air is 1. Thus such an air gap considerably increases the reluctance and hence the flux in the circuit is reduced

5.80 Ans See the notes

5.81 Ans *Figure 5.9* illustrates a possible method

5.84 Ans Millikan only obtained results for the charges carried by oil drops that were integral values of 1.6×10^{-19} C

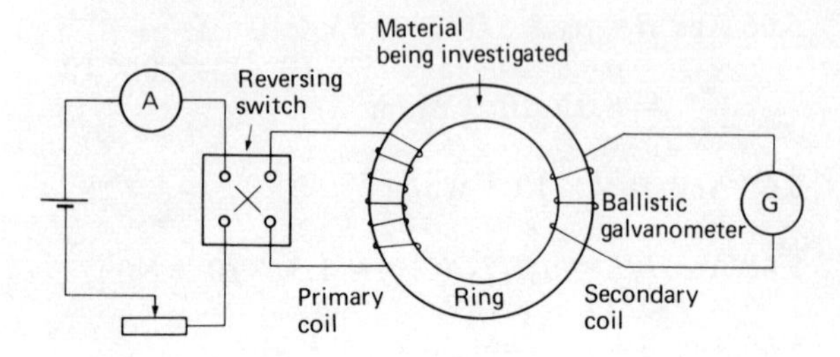

Figure 5.9

5.85 Ans (a) $\frac{4}{3}\pi r^3 g(\rho - \sigma) = 6\pi\eta r v$. Hence as $\frac{4}{3}\pi r^3 \rho = m$, then $v = 6.9 \times 10^{-3}$ m/s
(b) $\frac{4}{3}\pi r^3 g(\rho - \sigma) = Eq$, hence $E = 3.1 \times 10^7$ V/m
(c) $E = V/d$, hence $V = 4.6 \times 10^5$ V

5.86 Ans $\frac{4}{3}\pi r^3 g(\rho - \sigma) = Eq$, hence $q = 8.01 \times 10^{-19}$ C and $1.6 \times 10^{-19}N = q$, hence $N = 5$

5.87 Ans The common factor is 1.64×10^{-19} C which gives N as 4, 5, 6, etc

5.91 Ans $mv^2/r = Bqv$, hence $v = 5.17 \times 10^5$ m/s

5.92 Ans $mv^2/r = Bqv$ and $v = 2\pi r/T$, hence $T = 1.79 \times 10^{-9}$ s. This is independent of the speed

5.93 Ans $\frac{1}{2}mv^2 = Vq$, hence $v = 5.93 \times 10^6$ m/s

5.94 Ans There are a number of possible experiments. Many of them involve the measurement of the radius of the arc, or circle, in which a beam of electrons moves when in a magnetic field, the velocity of the electrons having been obtained from the potential difference through which they were accelerated

5.95 Ans $Eq = Bqv$, hence $B = 0.15$ Wb/m^2

5.96 Ans See *Figure 5.10*

5.99 Ans $I = nvAq$, hence $v = 3.1 \times 10^{-4}$ m/s

5.100 Ans The direction of the Hall potential difference across the width of the material depends on the sign of the charge carriers

5.101 Ans $V = BI/(nhq) = 7.5 \times 10^{-3}$ V

5.102 Ans $E = Bv$, hence $v = E/B = V/Bd = 0.30$ m/s

5.103 Ans $V = BI/(nhq)$, hence $n = 6.9 \times 10^{28}$

5.104 Ans $g = GM/R^2$, hence $g_s/g_E = R/(9R/4)$ and so $g_s = 40/9$ N/kg where g_s = field strength at satellite height and g_E = field strength at surface of the Earth. Hence force $= mg_s = 4.4 \times 10^3$ N

5.105 Ans (a) Cavendish's experiment involving a torsion balance can be described
(b) (i) $(150 \times 10^9)^3/1^2$ and $(778 \times 10^9)^3/11.8^2$ are reasonably equal
(ii) $mv^2/r = GMm/r^2$ and $v = 2\pi r/T$; hence $M = 4\pi^2 R^3/(GT^2)$. Using the data for the Earth $M = 2.01 \times 10^{30}$ kg

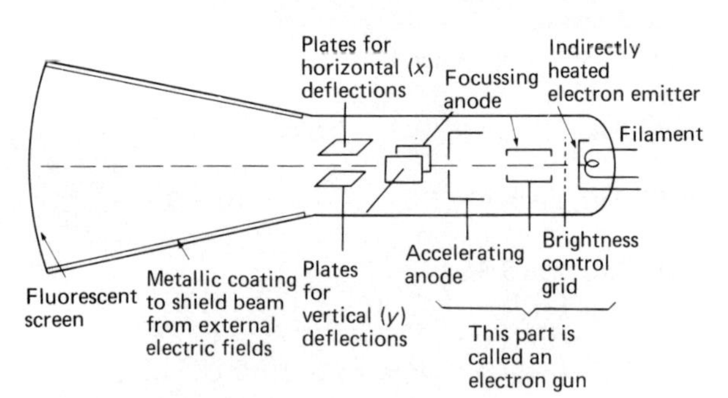

Figure 5.10 The cathode ray tube

5.106 Ans (a) See the notes for the definitions. Field strength is a vector quantity and thus a uniform field strength must mean not only a constant value of the field strength but the same direction. For points only 10 km apart on the Earth the direction of the field strength does not significantly change; for points at the opposite ends of a diameter the directions of the field strength are in completely opposite directions. The size of the field strength at all points on the surface of a constant diameter, homogeneous Earth is the same only if the spin of the Earth is neglected. Taking the spin into account gives different values at the poles and the equators

(b) $V = -GM/r$. See **5.12**
(c) See the notes

5.107 Ans (a) Define capacitance in terms of $C = Q/V$. An electrometer could be used to measure the charge on a parallel plate capacitor after it has been charged by the application of a potential difference. The experiment can be repeated with different plate separations and different areas of plates. Another possibility is to use a reed switch. A dielectric should have a very high resistivity, a generally high permittivity, not absorb water moisture and be capable of withstanding high electric fields without breakdown
(b) $C_m = \epsilon_r C_a$ and $Q = Q_1 + Q_2$, hence $400C_a = 50C_a + 50C_m = 50C_a + 50\epsilon_r C_a$ and hence $\epsilon_r = 7$
(c) Energy $= \frac{1}{2}CV^2$; with a single capacitor, energy $= \frac{1}{2}C_a 400^2$; with the two capacitors, energy $= \frac{1}{2}C_a 50^2 + \frac{1}{2}C_a \times 7 \times 50^2 = \frac{1}{2}C_a \times 8 \times 50^2$ and so the loss in energy $= \frac{1}{2}C_a(160\,000 - 20\,000)$ J. Only 1/8 of the energy of the single capacitor remains stored in the two capacitors. The rest of the energy is dissipated in the connecting wires when the charge flows from one capacitor to the other.

5.108 Ans See the notes. For a point charge the potential varies as the reciprocal of the distance from the charge. For a parallel plate capacitor the potential is proportional to the distance from the earthed plate. The electric field strength is the potential gradient. Electric force on a point charge Q in an electric field E is given by $F = EQ$. The work done in moving this charge a distance dx in the same direction as E is $dW = -EQ\,dx$. Thus

$$E = -\frac{1}{Q}\frac{dW}{dx} = -\frac{dV}{dx}$$

and $C = \epsilon_0 A/d = 1.76 \times 10^{-10}$ F. The second capacitor has $C = 0.49 \times 10^{-10}$ F. Initial charge $= CV = 1.76 \times 10^{-8}$ C; $Q = Q_1 + Q_2 = C_1 V' + C_2 V'$. Thus $V' = 80$ V and the charges are therefore $80 \times 1.76 \times 10^{-10} = 1.41 \times 10^{-8}$ C and $80 \times 0.49 \times 10^{-10} = 3.92 \times 10^{-9}$ C

5.109 Ans (a) $Q = CV = 1.0 \times 10^{-7}$ C
(b) $V = Q/C = 1000$ V
(c) Initial energy $= \frac{1}{2}CV^2 = 2.0 \times 10^{-5}$ J; final energy $= \frac{1}{2}CV^2 = 5.0 \times 10^{-5}$ J. The work done is thus 3.0×10^{-5} J

5.110 Ans $E = Q/(4\pi\epsilon_0 r^2)$; $V = Q/(4\pi\epsilon_0 r)$. Thus $E = V/r$ and so $V = 1.8 \times 10^5$ V

5.111 Ans With no field: $mg = kv$, hence $k = mg/v = 2.45 \times 10^{-10}$. With the field $mg + kv' = NeV/d$, hence $N = 5.9$ and the number is six

5.112 Ans $Vq = \frac{1}{2}mv^2$ for the acceleration. In the magnetic field $mv^2/r = Bqv$; hence $r = 0.033$ m

5.113 Ans See the notes. $V = BI/(nhq)$, hence $n = 2.14 \times 10^{22}$ m^{-3}. The different sign of the Hall voltage indicates that different sign charge carriers predominate for the current. A zero value would indicate that either there are no free charge carriers or that there are equal numbers of positive and negative charge carriers contributing to the current

6 Rays, waves and oscillations

Reflection and refraction

Notes

Laws of reflection:
The incident ray, the normal to the reflecting surface and the reflected ray all lie in the same plane. The angle of incidence equals the angle of reflection.

For a plane mirror the image is as far behind the mirror as the object is in front. The image is virtual.

An image is said to be real if light rays actually pass through it. A virtual image is when no light actually passes through the image, the light rays only appear to come from the image.

Laws of refraction:
The incident ray, the normal to the refracting surface and the refracted ray all lie in the same plane. For light of a given colour (frequency) and for a given pair of media

$$\frac{\sin i}{\sin r} = \text{a constant}$$

This is **Snell's law**. The value of the constant, known as the refractive index n, depends on the media concerned and the colour of the light. The term absolute refractive index is used for a medium when light is considered to be passing from a vacuum into the medium concerned. In all other cases the two media, 1 and 2, concerned need to be specified.

$${}_1n_2 = \frac{1}{{}_2n_1}$$

The angle of incidence for a particular pair of media for which the angle of refraction is 90° is called the critical angle C. For angles of incidence greater than this the rays are not refracted, but internally reflected. Thus, for water and air

$${}_{\text{water}}n_{\text{air}} = \frac{\sin C}{\sin 90^\circ}$$

For light refracted by a prism the angle through which the light is deviated has a minimum value $D_{\min}$ when the light passes symmetrically through the prism. When this occurs

$$n = \frac{\sin\,[(A + D_{\min})/2]}{\sin\,(A/2)}$$

where A is the apex angle of the prism.

The different constituent colours of white light give rise to different refractive indices for a pair of media. White light passing through a prism is spread out into a spectrum, this spreading out being called dispersion.

Examples

6.1 Water has a refractive index of 4/3 for light passing into it from air. (a) for an angle of incidence of 30° what is the angle of refraction? (b) What is the critical angle for light passing from water into air?

Ans (a) $\dfrac{\sin i}{\sin r} = \dfrac{4}{3}$

Hence

$$\sin r = \frac{3}{4}\sin 30^\circ$$

$$r = 22^\circ$$

(b) $\sin C = {}_{\text{water}}n_{\text{air}}$

$$= \frac{1}{{}_{\text{air}}n_{\text{water}}}$$

Thus

$$\sin C = \frac{3}{4}$$

$$C = 49^\circ$$

6.2 What is the angle of minimum deviation for

a prism with an apex angle of 60° and a refractive index of 1.6?

Ans
$$n = \frac{\sin[(A + D_{min})/2]}{\sin(A/2)}$$

$$1.6 = \frac{\sin[(60° + D_{min})/2]}{\sin 30°}$$

Thus

$$\sin[(60° + D_{min})/2] = 1.6 \times \sin 30° = 0.80$$
$$(60° + D_{min})/2 = 53°$$
$$60° + D_{min} = 106°$$
$$D_{min} = 46°$$

6.3 An optical fibre is made of glass with a refractive index of 1.50. What is the largest angle of incidence a ray of light can enter the plane end of the fibre and still be totally internally reflected from the sides of the fibre?

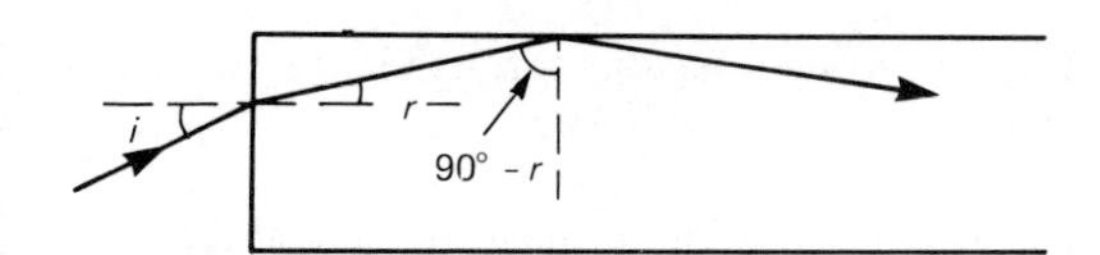

Figure 6.1

Ans See *Figure 6.1*

$$\sin C = {}_{glass}n_{air}$$
$$= \frac{1}{{}_{air}n_{glass}}$$
$$= \frac{1}{1.50}$$

Thus

$$C = 41.8°$$

$$\frac{\sin i}{\sin r} = {}_{air}n_{glass} = 1.50$$

The minimum value of the angle at which the ray must meet the side of the fibre is C, if reflection is to occur. The angle of incidence of the ray on the side of the fibre is $(90° - r)$. When this equals C, then r has a maximum value of

$$90° - 41.8° = 48.2°$$

The maximum value of i for which this is the angle of refraction is thus

$$\sin i = 1.50 \sin 48.2°$$
$$\sin i = 1.13$$

This means that there is no angle of incidence for which internal reflection will not occur.

6.4 A ray of light is incident at an angle of incidence of 30° on a prism with an apex angle of 60° and a refractive index of 1.50. By how much will the ray be deviated in passing through the prism?

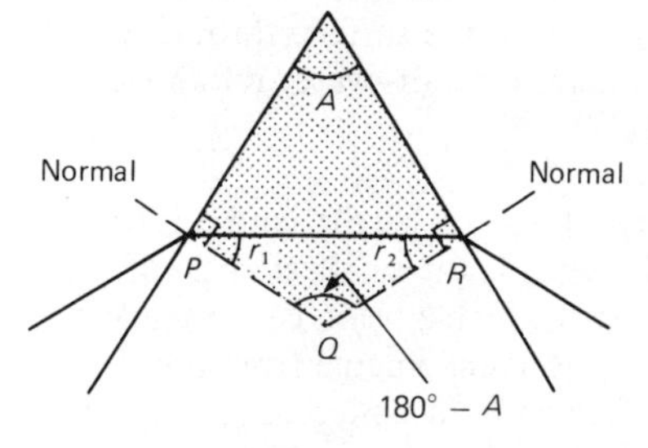

Figure 6.2

Ans See *Figure 6.2.*

Deviation at first prism face $= i_1 - r_1$
Deviation at second prism face $= i_2 - r_2$
Total deviation $= (i_1 - r_1) + (i_2 - r_2)$

In the quadrilateral $APQR$ the angles must add up to 360°. Thus, angle PQR must be $180° - A$. For the triangle PQR the sum of the angles must be 180°. Thus

$$r_1 + r_2 = A$$

i_1 is known and thus, as the refractive index is known, r_1 can be calculated.

$$\frac{\sin i_1}{\sin r_1} = 1.50$$

$$r_1 = 19.5°$$

Hence

$$r_2 = A - r_1$$
$$= 60° - 19.5°$$
$$= 40.5°$$

$$\frac{\sin i_2}{\sin r_2} = 1.50$$

Thus

$$i_2 = 76.9°$$

Thus

$$\text{Total deviation} = 30° - 19.5° + 76.9° - 40.5°$$
$$= 46.9°$$

Further problems

6.5 Two mirrors are parallel and facing each other. If they are 1.0 m long and 200 mm apart, how many reflections of a ray will occur if it is initially incident at an angle of incidence of 30° to the end of one of the mirrors?

6.6 A point source of light is 200 mm below a water–air surface. What is the diameter of the largest circle at the surface through which light can emerge (the air–water refractive index is 1.3)?

6.7 For a prism of apex angle 60° and refractive index 1.50, what is the smallest angle of incidence with which a beam can enter one face of the prism and emerge from the adjoining face without undergoing internal reflection?

6.8 What is the minimum value for the refractive index of a 45°,45°,90° prism if light incident normally on one of the short faces is to be internally reflected at the long side?

6.9 A beam of light is incident on a glass prism having an apex angle of 60° and a refractive index of 1.6. What is the angle of deviation for the light if it passes through the prism at minimum deviation? What is the angle of incidence for this minimum deviation?

6.10 An optical fibre consists of a plastic material having a refractive index of 1.70 coated with glass having a refractive index of 1.50. What is the critical angle for light to be totally internally reflected at the plastic–glass surface?

Refraction by lenses

Notes

The principal axis joins the centres of curvature of the two curved surfaces of a lens. All rays close to and parallel to the axis either converge or diverge from a point called the focus. The distance of the focus from the centre of the lens is called the focal length. The point on the axis in the centre of the lens is called the optical centre.

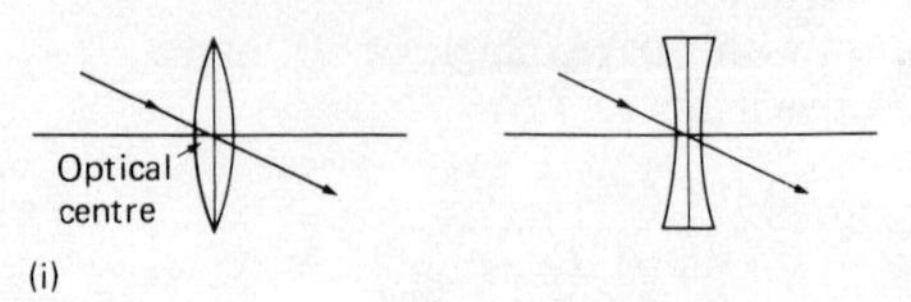

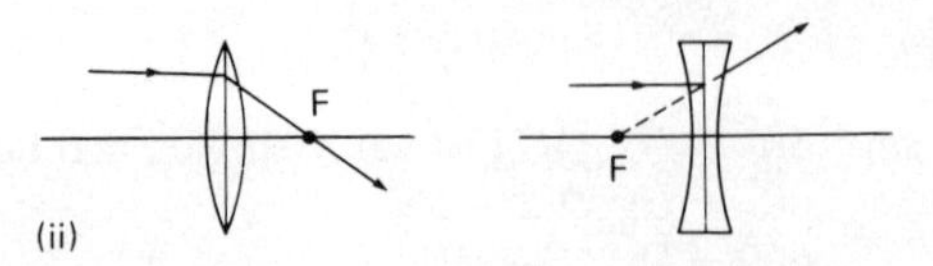

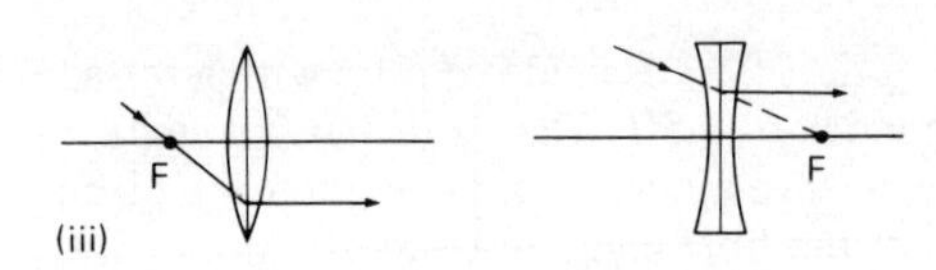

Figure 6.3

For rays of light passing through lenses (*Figure 6.3*):

(i) Any ray passing through the optical centre is undeviated.
(ii) Any ray parallel to the axis passes or appears to pass through the focus after refraction through the lens.
(iii) A ray passing through or proceeding towards the focus is parallel to the axis after refraction.

For the 'real is positive' convention, a real object or image distance is positive and a virtual object or image distance is negative. A real focus through which the light actually passes has a positive focal length and a virtual focus through which the light only appears to pass has a negative focal length. For a thin lens

$$\frac{1}{f} = \frac{1}{v} + \frac{1}{u}$$

where f is the focal length, v the image distance and u the object distance.

The linear magnification is the ratio of the image size to the object size.

$$\text{Linear magnification} = \frac{\text{image size}}{\text{object size}} = \frac{v}{u}$$

In the above equation no signs have been ascribed to any of the quantities.

When light is refracted at a single spherical surface

$$\frac{1}{u} + \frac{n}{v} = \frac{(n-1)}{R}$$

where n is the refractive index of the material and R the radius of curvature. For a thin lens, with two spherical surfaces close together,

$$\frac{1}{f} = (n-1)\left(\frac{1}{R_1} + \frac{1}{R_2}\right)$$

where R_1 and R_2 are the radii of curvature of the two surfaces and the lens is assumed to be in air. The radius is positive for surfaces convex to air and negative for concave surfaces.

The power of a lens is the reciprocal of the focal length. The units used for power are m^{-1}, dioptre or rad/m. All the units are identical.

The term aberration is used when a point object does not give rise to a point image. Spherical aberration occurs by virtue of the shape of a lens. Rays of light passing through regions of a spherical shaped lens have a different focus to those close to the axis. Chromatic aberration occurs because the different constituent colours of white light give different refractive indices and so different focal lengths.

Examples

6.11 An object is situated 22 cm from a concave lens of focal length 16 cm. At what distance from the lens will the image be produced? Will the image be real or virtual, erect or inverted? Draw a ray diagram showing the object and image positions relative to the lens.

Ans Using the real is positive convention:

$u = +22$ cm, $f = -16$ cm

$$\frac{1}{f} = \frac{1}{v} + \frac{1}{u}$$

$$\frac{1}{v} = -\frac{1}{16} - \frac{1}{22}$$

$v = -9.3$ cm

The image is virtual, because v is negative, and erect. *Figure 6.4(b)* shows the ray diagram.

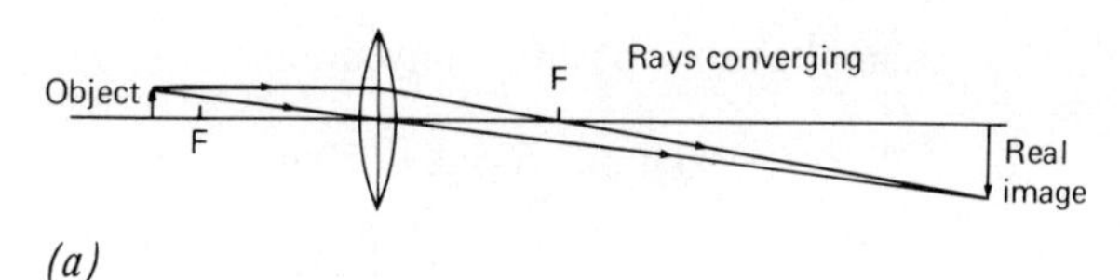

(a)

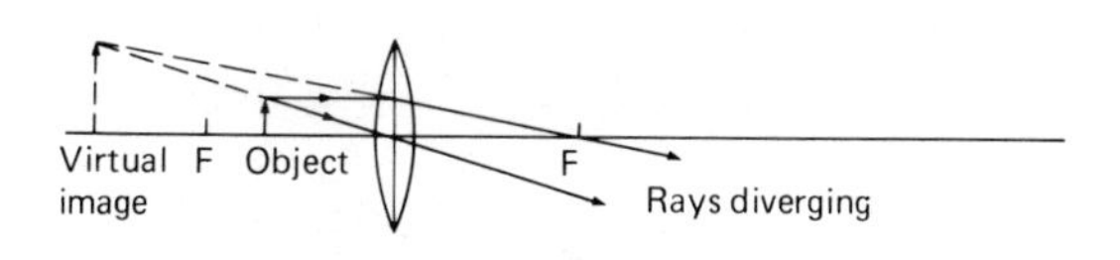

(b)

Object
F Image
F
Rays diverging

(c)

Figure 6.4

6.12 An object 20 mm in height is situated 100 mm from a convex lens having a focal length of 150 mm. What is the size of the image?

Ans Using the real is positive convention:

$u = +100$ mm, $f = +150$ mm

$$\frac{1}{f} = \frac{1}{v} + \frac{1}{u}$$

$$\frac{1}{v} = \frac{1}{150} - \frac{1}{100}$$

$v = -300$ mm

The image is virtual.

$$m = \frac{\text{image size}}{\text{object size}} = \frac{v}{u}$$

Hence

$$\text{image size} = \frac{300}{100} \times 20$$

$= 60$ mm

6.13 A bi-convex lens has one surface with a radius of curvature of 100 mm and the other surface with one of 150 mm. If the lens material has a refractive index of 1.50, and the lens is in air, what is the focal length of the lens?

Ans $\frac{1}{f} = (n-1)\left(\frac{1}{R_1} + \frac{1}{R_2}\right)$

Using the real is positive convention:

$R_1 = +100$ mm, $R_2 = +150$ mm

$$\frac{1}{f} = (1.50 - 1)\left(\frac{1}{100} + \frac{1}{150}\right)$$

$f = 120$ mm

Further problems

6.14 Describe a simple, and very quick, method by which you could determine whether a convex lens has a focal length of 15 cm or 20 cm.

6.15 What are the image distances for the following situations? State whether the images are real or virtual. (a) Object distance 120 mm and convex lens of focal length 60 mm. (b) Object distance 60 mm and convex lens of focal length 120 mm. (c) Object distance 120 mm and concave lens of focal length 60 mm. (d) Object distance 60 mm and concave lens of focal length 120 mm.

6.16 How far is the slide from the lens in a slide projector if the image is produced on a screen 4.0 m from the lens which is of focal length 150 mm? What is the magnification?

6.17 How can the magnification produced by a slide projector be increased?

6.18 A slide projector is to be used 10 m from a screen that is 1.4 m wide. What focal length lens should be used if a 35 mm wide slide is to just fill the width of the screen?

6.19 What is the focal length of a bi-convex lens with radii of 40 mm and a refractive index of 1.5?

6.20 A lens is made of glass of refractive index 1.50. If one face of the lens is convex with a radius of curvature of 200 mm and the other is plane, what is the focal length of the lens in air?

6.21 For two thin lenses of focal lengths f_1 and f_2 in contact, show that the focal length of the combination is given by

$$\frac{1}{f} = \frac{1}{f_1} + \frac{1}{f_2}$$

Optical instruments

Notes

The basic camera consists of a convex lens forming a real image on the photographic emulsion, the lens to emulsion distance being the image distance and the lens to object the object distance.

The nearest point that can be focussed by the unaided eye is called the near point and its distance from the eye is called the least distance of distinct vision. For the average eye this is about 250 mm. The magnifying power of an optical system is defined as the ratio of the angle subtended at the eye by the image formed with the aid of the system to the angle subtended at the unaided eye by the object situated at some specified distance. In the case of the magnifying glass and the microscope this distance is the least distance of distinct vision.

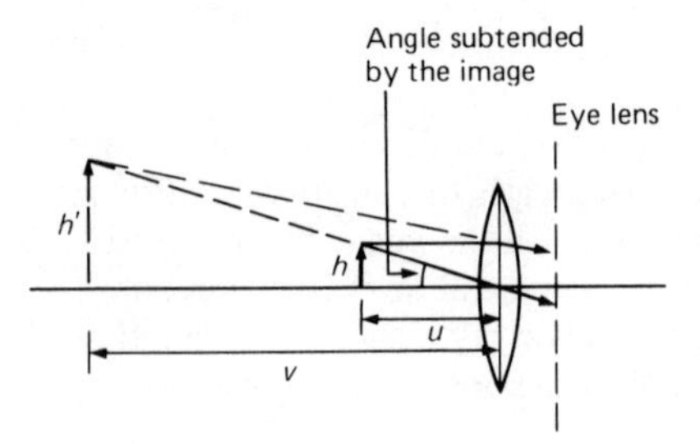

Figure 6.5 The simple magnifying glass

For a simple magnifying glass (*Figure 6.5*), the angle subtended by the image at the eye is h'/v. As

$$h'/h = v/u \text{ and } 1/f = 1/v + 1/u$$

then

$$h'/h = (v/f) - 1$$

Hence

$$\text{Subtended angle} = \frac{h}{v}\left(\frac{v}{f} - 1\right)$$

The angle subtended by the object at the least distance of distinct vision D is h/D. Thus

$$\text{Magnifying power} = \frac{D}{v}\left(\frac{v}{f} - 1\right)$$

This is a maximum when the image is at D. Thus, when

$$v = -D$$

$$\text{Maximum magnifying power} = (D/f) + 1$$

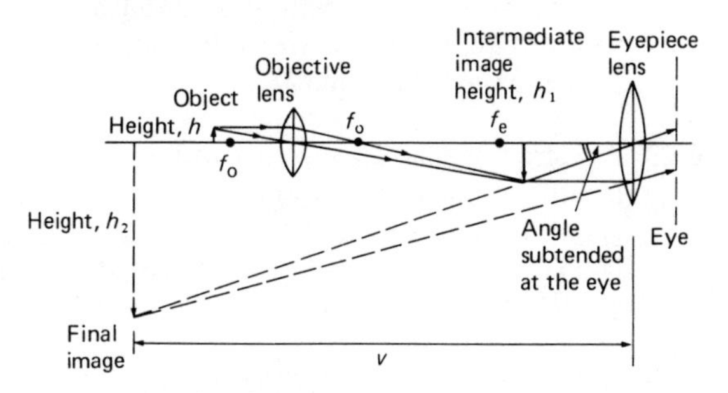

Figure 6.6 Constructional ray diagram for a microscope

For a microscope (*Figure 6.6*), the angle subtended by the image is h_2/v. The angle subtended by the object at D is h/D. Thus

$$\text{Magnifying power} = \frac{h_2}{h} \times \frac{D}{v}$$

The maximum magnifying power is when the final image is at the least distance of distinct vision D (known as normal adjustment), thus

$$\text{Maximum magnifying power} = \frac{h_2}{h}$$

This can be rearranged as

$$\text{Maximum magnifying power} = \frac{h_2}{h_1} \times \frac{h_1}{h}$$

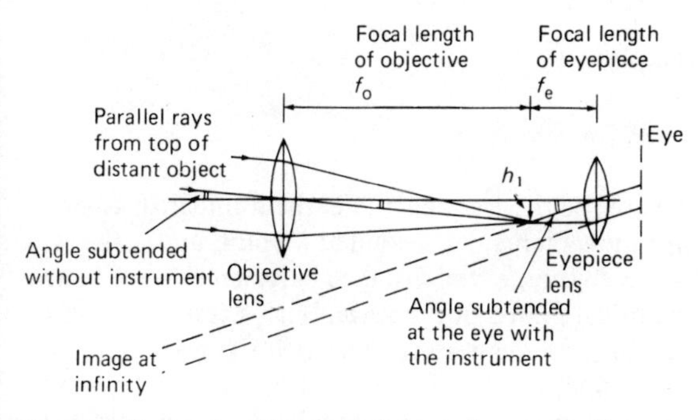

Figure 6.7 Constructional ray diagram for a refracting telescope

This is the product of the magnification of the eyepiece lens and the magnification of the objective lens.

For a refracting telescope (*Figure 6.7*), the angle subtended by the image is h_1/f_e. The angle subtended by the object is h_1/f_o. Thus

$$\text{Magnifying power} = f_o/f_e$$

The separation of the two lenses is $(f_o + f_e)$.

Figure 6.8 shows the basic form of the Newtonian reflecting telescope. The magnifying power is f_o/f_e.

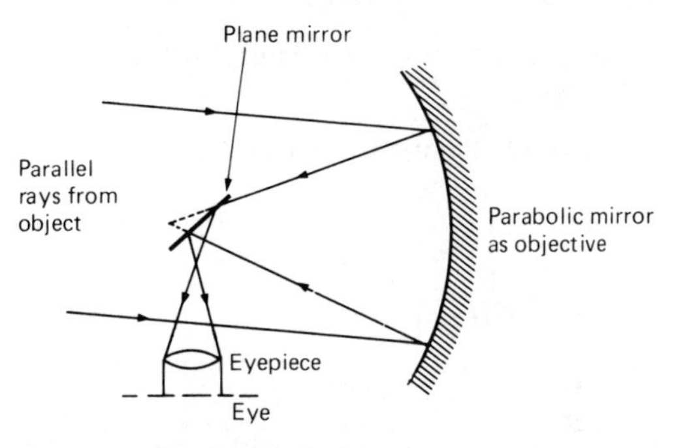

Figure 6.8 The Newtonian reflecting telescope

Examples

6.22 What is the maximum magnifying power of a lens of focal length 50 mm if the least distance of distinct vision is 250 mm?

Ans

$$\text{Maximum magnifying power} = \frac{D}{f} + 1 = \frac{250}{50} + 1 = 6$$

6.23 A microscope has an objective lens of focal length 5.0 mm and an eyepiece of focal length 30 mm. If the lenses are 180 mm apart, what is the magnifying power of the microscope if the final image is produced at the least distance of distinct vision of 250 mm?

Ans For the eyepiece lens, with the real is positive convention

$$f = +30 \text{ mm}$$

and

$$v = -250 \text{ mm}$$

the least distance of distinct vision. Thus

$$\frac{1}{f} = \frac{1}{v} + \frac{1}{u}$$

$$\frac{1}{u} = \frac{1}{30} + \frac{1}{250}$$

$$u = 26.8 \text{ mm}$$

This means that this intermediate image must be

$$180 - 26.8 = 153.2 \text{ mm}$$

from the objective lens. Thus, for the objective lens,

$$f = +5.0 \text{ mm}, \quad v = +153.2 \text{ mm}$$

Thus

$$\frac{1}{u} = \frac{1}{5.0} - \frac{1}{153.2}$$

$$u = 5.2 \text{ mm}$$

The magnification produced by the objective is

$$v/u = 153.2/5.2 = 29.5$$

The magnification produced by the eyepiece is

$$v/u = 250/26.8 = 9.3$$

The overall magnifying power is therefore

$$29.5 \times 9.3 = 275$$

6.24 A simple refracting telescope consists of an objective lens of focal length 450 mm and an eyepiece of focal length 20 mm. What is the magnifying power of the instrument and the distance between the two lenses?

Ans Magnifying power $= \dfrac{f_o}{f_e}$

$$= \frac{450}{20}$$

$$= 22.5$$

The distance between the lenses $= f_o + f_e = 470$ mm

Further problems

6.25 A camera with a lens of focal length 50 mm is used to take a picture of a person 1.6 m tall standing 5.0 m away. What must be the distance between the lens and the photographic emulsion and the size of the image on the picture.

6.26 What is the maximum magnifying power possible with a magnifying glass having a focal length of 100 mm (least distance of distinct vision is 250 mm)?

6.27 A microscope has an objective of focal length 4.0 mm and an eyepiece of focal length 20 mm. What is the magnifying power of the instrument when the object is 5.0 mm from the objective?

6.28 A microscope has an objective of focal length 6.0 mm and the object is placed 8.0 mm from it. If the eyepiece used has a magnification of 5, what is the magnifying power of the instrument?

6.29 The Yerkes refracting telescope has an objective lens of focal length 20 m. If an eyepiece of focal length 6.5 mm is used, what is the magnifying power?

6.30 The objective lens of a telescope has a focal length of 1.00 m. If the distance between the objective and eyepiece is 1.05 m, what is the magnifying power of the instrument?

Properties of waves

Notes

Wave velocity $v = f\lambda$

where f is the frequency, i.e. the number of to-and-fro movements per second at a point, and λ the wavelength, i.e. the distance between two successive identical points in the wave. With a transverse wave the displacement associated with the wave is at right angles to the direction of motion of the wave. With a longitudinal wave the displacement is along the same direction as the direction of wave motion.

Two, or more, waves can move through the same space independently of each other. The displacement at a point where two waves superpose is the sum of the displacements that the individual waves would give. The waves are said to interfere. *Figure 6.9* shows the type of interference pattern that can occur with two point sources of waves of the same frequency and wavelength. The points where two waves always cancel each other out are called nodes and where they reinforce each other antinodes.

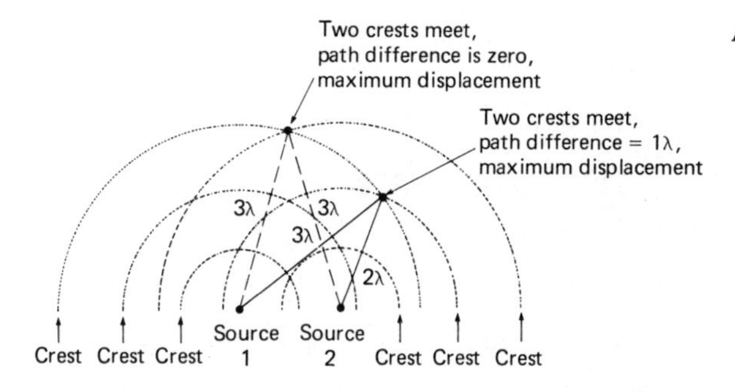

Figure 6.9

Waves can be reflected, the angle of incidence of a wave being equal to the angle of reflection. Waves can be refracted. When refraction occurs there is a change in wavelength.

$$\text{Refractive index} = \frac{\sin i}{\sin r}$$

$$= \frac{\lambda_1}{\lambda_2} = \frac{v_1}{v_2}$$

where λ_1 and λ_2 are the wavelengths in the two 'media' concerned and v_1 and v_2 the wave velocities.

Waves can be diffracted. When waves pass through an aperture that is wide compared with the wavelength then a sharp 'shadow' is produced. When the aperture is comparable with the wavelength the waves 'bend' round the edges of the aperture and into the 'shadow' regions.

Units: wave velocity – m/s, wavelength – m, frequency – s^{-1} or Hz (where 1 Hz is 1 s^{-1}).

Examples

6.31 Complete the following table:

Path difference for two water waves arriving from in-phase sources	*State of water surface*
0	Maximum disturbance
½λ	No disturbance
1λ	?
1½λ	?
2λ	?
5λ	?
5½λ	?

Ans The term 'in-phase' means that the two waves start in exactly the same way from each source, e.g. if one wave starts as a crest then so does the other wave. When the path difference is $n\lambda$, where $n = 0,1,2,3,4,\ldots$ etc., then there is reinforcement and so a maximum displacement. When the path difference is $(n + ½)\lambda$, where $n = 0,1,2,3,4,\ldots$ etc., there is cancellation and so zero displacement. Thus, for the above table maximum disturbance occurs at 0,1λ,2λ,5λ and no disturbance occurs at ½λ, 1½λ, 5½λ.

6.32 A water wave is found to have a wavelength of 4 mm in one depth of water and 2 mm in the next depth. If the wave is incident on this change in depth boundary at an angle of incidence of 30°, what will be the angle of refraction in the different depth water?

Ans

$$\frac{\sin i}{\sin r} = \frac{\lambda_1}{\lambda_2}$$

Hence

$$\sin r = \frac{2 \sin 30°}{4}$$

$$r = 14.5°$$

Further problems

6.33 A water wave in passing from water of one depth to another is found to refract, the angle of incidence being 30° and the angle of refraction 60°. What is the ratio of the wave speeds in the two depths? In which depth of water is the wave speed the greatest?

6.34 Sketch the pattern of nodal lines that are produced when two identical point sources of waves are in phase and separated by a few wavelengths.

6.35 Straight waves of wavelength 2 mm pass through apertures of width (a) 2 mm, (b) 20 cm. What type of behaviour would you expect of the waves after passing through the aperture?

6.36 Two wave pulses are travelling in opposite directions along a rope. Explain what will be observed as the wave pulses meet and cross and how the displacement of the rope at any interval of time can be determined.

Light as a wave motion

Notes

Light can show superposition and diffraction. *Figure 6.10* shows the basic arrangement used to show superposition, the arrangement being known as Young's double slit experiment. For point P on

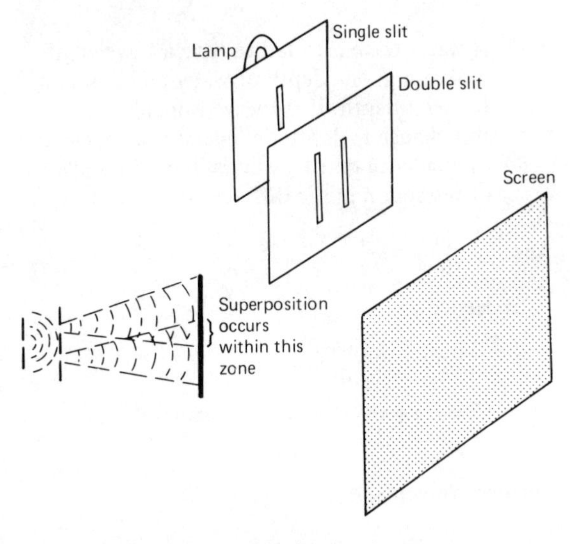

Figure 6.10 Young's double slit experiment

the screen (*Figure 6.11*) there is brightness as it is equidistant from each slit. For point Q there will be brightness if the path difference is 1λ,2λ,3λ, etc. If Q is the first such brightness position after the central point P then

$$S_2Q - S_1Q = \lambda$$

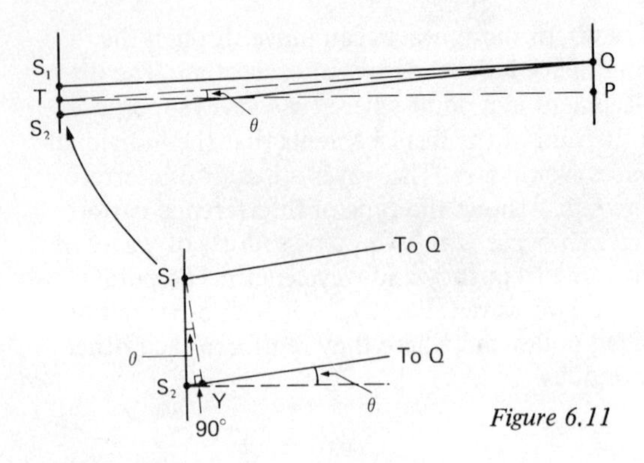

Figure 6.11

Thus

$$S_2Y = \lambda$$

But, to a reasonable approximation,

$$S_2Y/S_1S_2 = \sin\theta$$

and

$$QP/QT = \sin\theta$$

Thus

$$S_2Y/S_1S_2 = QP/QT$$

To a reasonable approximation QT is the distance between the screen and the slits, thus

$$\lambda = \frac{\text{(distance between slits)(distance between maxima)}}{\text{distance of screen from slits}}$$

The light for the double slit experiment must not be from two separate sources but light from one source split into two. This is because light from conventional lamps is not coherent.

Superposition can occur between light reflected from the top and bottom surfaces of a thin film. A phase change of ½λ takes place when the reflection occurs at an interface between a less dense and a more dense medium. Newton's rings is one form of thin film interference (*Figure 6.12*). The central spot is dark even though the film thickness is zero. This is because of the phase change occurring at the air–glass interface. For a dark circular fringe of radius r, Pythagoras' theorem gives

$$R^2 = (R - t)^2 + r^2$$

where t is the film thickness at radius r and R is the radius of curvature of the lens surface. Thus

$$r^2 = 2Rt \quad t^2$$

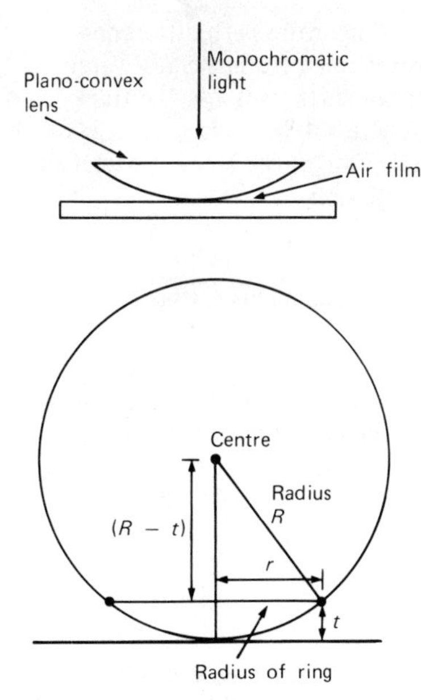

Figure 6.12 Newton's rings arrangement

and as t is much smaller than R this approximates to

$$r^2 = 2Rt$$

For a dark fringe, allowing for the ½λ phase change,

$$2t = m\lambda$$

where $m = 0,1,2,3$, etc. Thus

$$r^2 = Rm\lambda$$

Huygen's principle assumes that every point on a wavefront may be considered as a source of small, secondary wavelets which then spread out in all

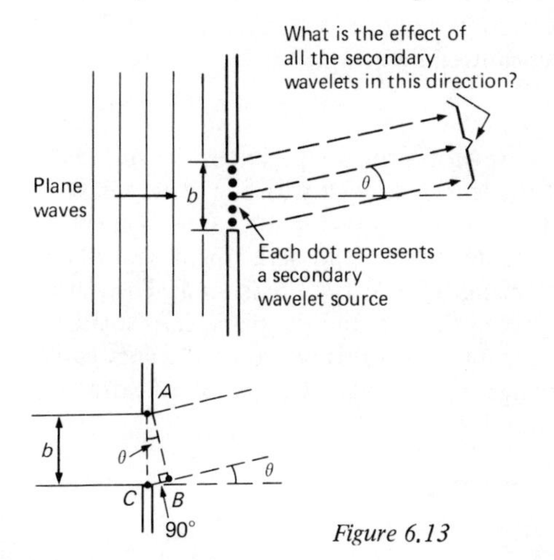

Figure 6.13

directions. The resulting wavefront can be obtained by considering the envelope of all these secondary wavelets. *Figure 6.13* shows this principle applied to diffraction through a single slit. The path difference between the waves coming from a point half way across the slit and from an extreme edge is

$$\tfrac{1}{2}CB = \tfrac{1}{2}b \sin\theta$$

For every point in the lower half of the slit there is a corresponding point in the upper half with this path difference. Thus, for no light at this angle θ we must have

$$\tfrac{1}{2}b \sin\theta = \tfrac{1}{2}\lambda \text{ or } 1\tfrac{1}{2}\lambda \text{ or } 2\tfrac{1}{2}\lambda$$

i.e.

$$\sin\theta = m\lambda/b$$

where $m = 1,2,3,4$, etc. *Figure 6.14* shows how the intensity of light transmitted through the slit depends on the angle θ.

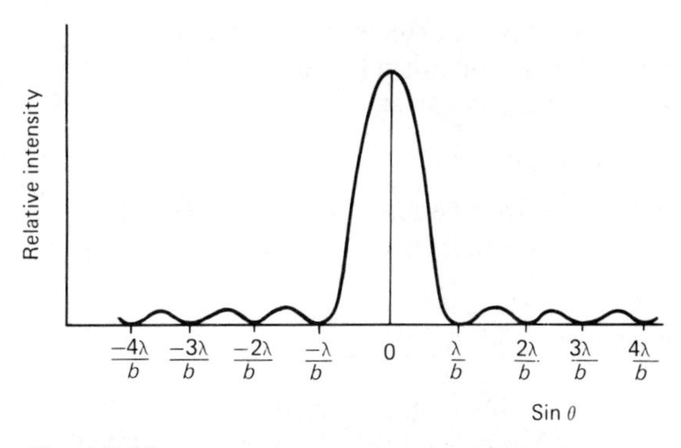

Figure 6.14

If diffraction is through a circular hole instead of a slit then the condition for darkness is

$$\sin\theta = 1.22\, m\lambda/d$$

where d is the diameter of the hole.

Two sources can be resolved if the central maximum from one diffraction pattern falls no nearer than the first node of the other diffraction pattern. This means for a lens of diameter d the minimum angle of resolution θ is given by

$$\sin\theta = 1.22\lambda/d$$

The above criterion is known as the Rayleigh criterion.

The term diffraction grating is used to describe a piece of glass or metal which has a large number of fine, equidistant, closely spaced, parallel lines

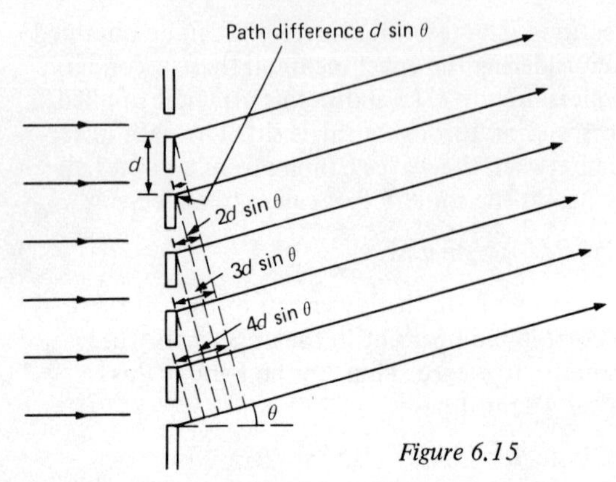

Figure 6.15

ruled on it. The arrangement behaves, for transmission, as a large number of parallel slits (*Figure 6.15*). The path difference between wavelets emerging from similar points in two successive slits is $d \sin \theta$. Reinforcement will occur when

$$d \sin \theta = m\lambda$$

where $m = 0,1,2,3$, etc. When this condition occurs for two slits then it occurs for all slits. Thus, for reinforcement

$$d \sin \theta = m\lambda$$

and $m = 0$ gives the zero order, $m = 1$ the first order, $m = 2$ the second order, etc.

Examples

6.37 What will be the spacing between the fringes in a Young's double slit experiment if light of wavelength 5.0×10^{-7} m is used with slits of separation 0.50 mm a distance of 1.2 m from the screen?

Ans Distance between fringes

$$= \frac{\lambda \times \text{distance of screen from slits}}{\text{distance between slits}}$$

$$= \frac{5.0 \times 10^{-7} \times 1.2}{0.50 \times 10^{-3}}$$

$$= 6.0 \times 10^{-3} \text{ m}$$

6.38 What wavelengths of light will be missing from white light (a) reflected, (b) transmitted almost perpendicularly from a thin water film in air of thickness 320 nm and having a refractive index of 1.33?

Ans (a) For reflected light the path difference between light reflected from the air–water and the water–air surfaces is 2 × the thickness of film. There will be a phase change of ½λ for the reflection at the air–water surface. Thus, for cancellation

$$2t + \tfrac{1}{2}\lambda = (m + \tfrac{1}{2})\lambda$$

where $m = 0,1,2,3$, etc. This is thus

$$2t = m\lambda$$

This is the wavelength in the water, the wavelength in air being $n\lambda$, where n is the refractive index. Thus

$$2t = m\lambda_a/n$$

Hence

$$\lambda_a = 2tn/m$$

and has the values of 851 nm, 426 nm, 284 nm, 213 nm, 170 nm, etc. Only the 851 nm and 426 nm wavelengths lie in the visible part of the spectrum and could therefore be noticed by the eye as being missing. They will, however, be seen in the transmitted light.

(b) For transmitted light the wavelengths missing will be those which give reflections, i.e. those for which

$$2t + \tfrac{1}{2}\lambda = m\lambda$$

$$2t = (m - \tfrac{1}{2})\lambda_a/n$$

These wavelengths are 1700 nm, 567 nm, 340 nm, 243 nm. Only the 567 nm lies in the visible part of the spectrum and could be noticed by the eye as being missing in the transmitted light.

6.39 In a Newton's rings experiment the surface of the lens in contact with the plane surface has a radius of curvature of 1.0 m. What is the radius of the 10th dark ring if light of wavelength 5.9×10^{-7} m is used? The space between the lens and the plane glass surface is occupied by air. How would the answer change if it was filled with water of refractive index 1.3?

Ans $r^2 = Rm\lambda$

$$= 1.0 \times 10 \times 5.9 \times 10^{-7}$$

Hence

$$r = 2.4 \times 10^{-3} \text{ m}$$

With water

$$r^2 = Rm\lambda/n$$

Hence

$$r = 2.1 \times 10^{-3} \text{ m}$$

6.40 A slit of width 1.5×10^{-6} m is illuminated with light of wavelength 5.9×10^{-7} m. At what angle to the line of the slit will the first minimum occur?

Ans $\sin \theta = m\lambda/b$

For $m = 1$

$$\sin \theta = 5.9 \times 10^{-7}/(1.5 \times 10^{-6})$$
$$\theta = 23.2°$$

6.41 A telescope has an aperture of diameter 60 mm. What is the minimum angular separation two stars must have to be resolved, according to Rayleigh's criterion? The light used can be considered to have a wavelength of 500 nm.

Ans Minimum angular separation

$$= 1.22\lambda/d$$
$$= 1.22 \times 500 \times 10^{-9}/(60 \times 10^{-3})$$
$$= 10.2 \times 10^{-6} \text{ radians}$$

6.42 A diffraction grating has 80 000 rulings per metre. If it is illuminated at normal incidence by light of wavelength 5.9×10^{-7} m at what angle will the first order maximum be? What will be the angle between the first and the second order maxima?

Ans $d \sin \theta = m\lambda$

Thus, as

$$d = 1/N$$

where N is the number of rulings per metre, for the first order

$$\sin \theta = 5.9 \times 10^{-7} \times 80\,000$$
$$\theta = 2.705°$$

For the second order

$$\sin \theta = 2 \times 5.9 \times 10^{-7} \times 80\,000$$
$$\theta = 5.417°$$

Thus, the angle between the first and second order is 2.71°.

Further problems

6.43 In a Young's double slit arrangement, the distance between the slits and the screen is 1.0 m and the slits are 4.0 mm apart. What is the fringe separation if light of wavelength 500 nm is used?

6.44 What is the wavelength of the light used in a double slit experiment if the slit separation is 0.80 mm, the slit to screen distance is 1.0 m and the fringes are 0.64 mm apart.

6.45 A layer of oil of thickness 4.5×10^{-7} m floats on the surface of water. If the oil has a refractive index of 1.2 and is viewed normally in white light, for which wavelength of visible light is the reflection the strongest?

6.46 What wavelength of light will be missing from white light reflected almost perpendicularly from a thin film in air or water of thickness 200 nm (refractive index of water is 1.3)?

6.47 What is the thinnest film of refractive index 1.35 with which a glass lens can be coated to give no reflection for a normally incident beam of light of wavelength 6.0×10^{-7} m? What colour will the lens appear by reflected light when illuminated with white light?

6.48 Two glass microscope slides are arranged so that they touch at one end and are separated by a wire 0.48 mm in diameter at a distance of 10 mm from the line of contact. How many dark fringes will appear over this distance when the slides are illuminated normally with light of wavelength 5.9×10^{-7} m?

6.49 The convex surface of a plano-convex lens has a radius of curvature of 1.50 m. When this lens is placed convex side down on a flat glass surface and illuminated normally with light of wavelength 5.9×10^{-7} m, Newton's rings are seen. What is the radius of the 10th ring?

6.50 Describe an experiment by which you could, in the school laboratory, determine the wavelength of the light emitted by a lamp.

6.51 Light of wavelength 600 nm is incident normally on a plane diffraction grating having 600 lines per millimetre. What are the angles of deviation of (a) the first order, (b) the second order maxima?

6.52 A diffraction grating having 500 lines per millimetre is illuminated by white light along the normal to the grating. What is the angular separation between the red and violet ends of the first order spectrum if the wavelengths are taken to be 750 nm and 400 nm?

6.53 Light of wavelength 5.9×10^{-7} m passes through a slit of width 0.5 mm and falls on a screen 1.2 m away. What is the distance between the first and second minima on the screen?

6.54 What is the minimum angle of resolution of the eye if it is assumed to be a circular aperture of diameter 3 mm and the average wavelength of the light is taken to be 5×10^{-7} m?

Polarisation

Notes

Light shows the properties of a transverse wave motion. Light for which the plane of vibration is restricted to just one plane is said to be plane polarised. Light on passing through one sheet of Polaroid is plane polarised. When this plane polarised light is viewed through a second sheet of Polaroid, the intensity of the transmitted light depends on the orientation of the second sheet. At two specific angles 180° apart there is no transmitted light, at angles half way between these two positions the intensity transmitted is a maximum, e.g. at 0° and 180° the intensity is at a maximum while at 90° and 270° there is no transmission.

Polaroid film contains tiny crystals lined up in the same direction. The crystals used have the property of only transmitting the components of the wave displacement in just one particular direction. Another way of producing plane polarised light is by double refraction. Some crystals, e.g. calcite, can produce two refracted rays, each polarised in mutually perpendicular directions.

When unpolarised light is reflected from a surface such as glass or water the light is completely plane polarised at one particular angle of incidence, known as the Brewster angle. This angle occurs when the refracted and reflected rays are at right angles, i.e. when

$$n = \tan i$$

Examples

6.55 How could you determine whether a ray of light was plane polarised or not?

Ans You could pass the ray through a sheet of Polaroid. On rotation of the Polaroid sheet the intensity of the ray should pass through minima and maxima 90° apart if the light is plane polarised.

6.56 At what angle of incidence on an air–glass surface will the reflected light be completely plane polarised (air–glass refractive index is 1.5)?

Ans $\tan i = n = 1.5$

Hence

$$i = 56.3°$$

Further problems

6.57 At what angle of incidence will light reflected from a lake be completely plane polarised (refractive index of the water is 1.3)?

6.58 Rotating one sheet of Polaroid in the path of a beam of light is found to have no effect on the intensity of the transmitted light. If, however, the light is then passed through a second sheet of Polaroid, rotating the second sheet causes the light intensity to change. Explain why the second sheet has this effect and why there is no apparent effect with just one sheet of Polaroid.

Electromagnetic waves

Notes

Electromagnetic waves are transverse wave motions with a speed in a vacuum of about 3×10^8 m/s. The wavelengths of the visible part of the spectrum extend from about 4×10^{-7} m to 7×10^{-7} m. Radiation beyond the red end of the spectrum, i.e. wavelengths greater than 7×10^{-7} m, is called infrared radiation. Radiation of wavelength shorter than the violet end, i.e. wavelengths less than 4×10^{-7} m, is called ultraviolet radiation. *Figure 6.16* shows the full range of the electromagnetic spectrum of waves.

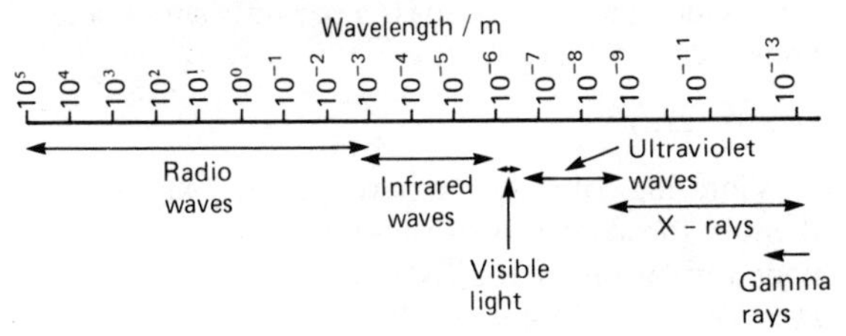

Figure 6.16

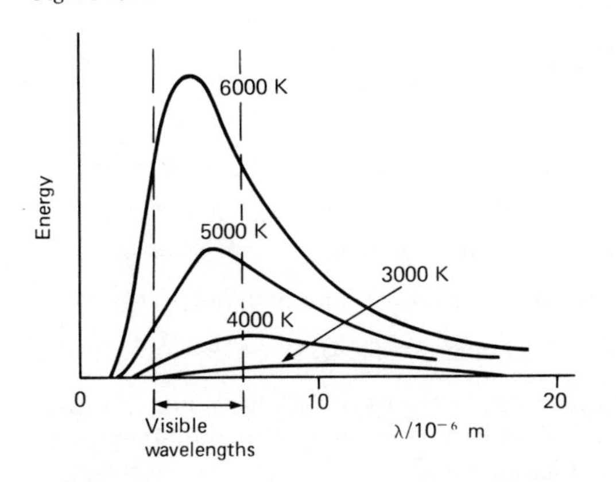

Figure 6.17

Figure 6.17 shows how the energy radiated by a hot body varies with wavelength. The wavelength of maximum energy for a particular temperature is inversely proportional to the absolute temperature of the object. This is called **Wien's law.**

$$\lambda_{max} T = \text{a constant, } 2.8978 \times 10^{-2} \text{ m K}$$

The total energy radiated per second from unit surface of a body is proportional to the fourth power of the absolute temperature; this relationship is known as **Stefan's law** (see chapter 3).

Examples

6.59 The wavelength of maximum energy for the radiation from the Sun is found to be at 4.8×10^{-6} m. What temperature would this indicate for the Sun (Wien's constant is 2.9×10^{-2} m K)?

Ans $\lambda T = 2.9 \times 10^{-2}$

Hence

$$T = \frac{2.9 \times 10^{-2}}{4.8 \times 10^{-6}}$$

$$= 6.0 \times 10^3 \text{ K}$$

6.60 How long does it take a radio signal to travel 200 km from the transmitter to the receiving aerial (velocity of electromagnetic waves is 3×10^8 m/s)?

Ans
$$\text{Time} = \frac{\text{distance}}{\text{speed}}$$

$$= \frac{200 \times 10^3}{3 \times 10^8}$$

$$= 6.7 \times 10^{-4} \text{ s}$$

Further problems

6.61 Describe an experiment which can be used to determine the speed of light.

6.62 What is the wavelength of a radio station broadcasting at 100 MHz (velocity of electromagnetic waves is 3×10^8 m/s)?

6.63 If you were supplied with a microwave source having a wavelength of about 3 cm, how could you determine the wavelength in the laboratory?

6.64 What would be the order of magnitude required for the grating spacings in a diffraction grating for use with microwaves of wavelength about 3 cm?

6.65 Venus behaves as a black body at a temperature of 240 K. What would be the wavelength of maximum energy (Wien's constant is 2.90×10^{-2} m K)?

6.66 Some stars have been found which do not emit visible light but do emit infrared radiation. What can you say about the temperature of such stars?

Sound

Notes

Sound is a longitudinal wave motion. It shows the properties of reflection, refraction, superposition and diffraction. The speed of sound depends on the medium in which the longitudinal wave is travelling.

When a wave pulse travels along a string and is reflected at a fixed end, a phase change occurs such that the fixed end is always a node. With reflection at a free end the reflection is such that the free end is always an antinode. When a continuous wave travels along a string a continuous reflected wave occurs which superposes on the incident wave. The displacement of the string is the vector sum of the displacements that would have been produced by the waves alone. The superposition always gives a node at a fixed end and an antinode at a free end. At whatever time the string is considered certain points are always at rest, i.e. nodes. These nodes are half a wavelength apart. Because these nodes remain stationary the result is said to be a standing or stationary wave.

For a string of length L between two fixed ends, the maximum wavelength is given by

$$L = \lambda/2$$

the next standing wave by

$$L = \lambda$$

the next by

$$L = 3\lambda/2$$

(see *Figure 6.18*). The frequencies associated with these wavelengths are thus $v/2L$ (the fundamental or first harmonic frequency), v/L (the second harmonic frequency or first overtone), $3v/2L$ (the third harmonic or second overtone).

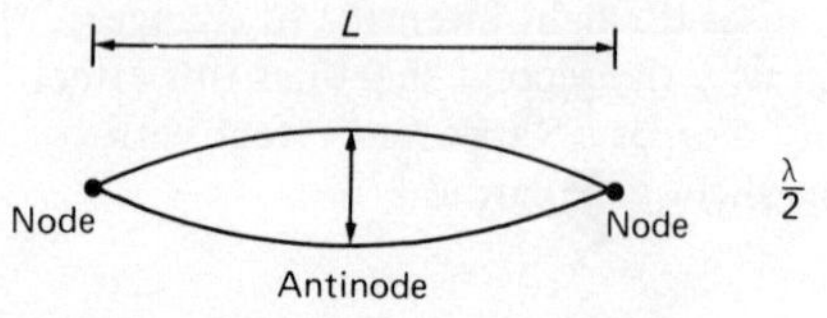

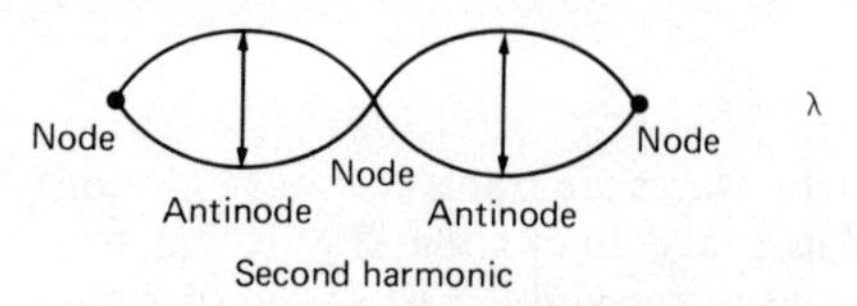

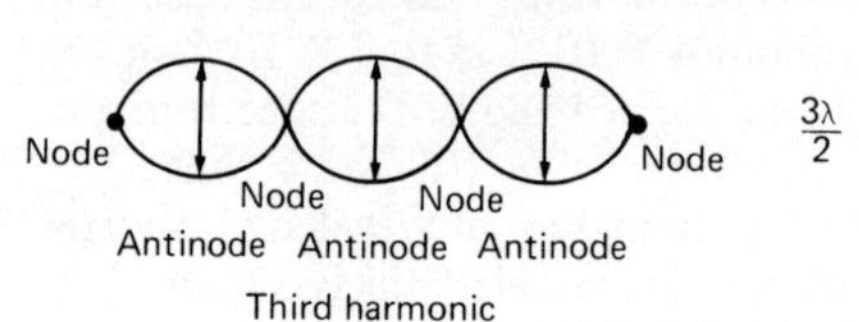

Figure 6.18 Standing waves on strings

The velocity with which a pulse, or a wave, travels along a string of mass μ per unit length and under tension T is given by

$$v = (T/\mu)^{1/2}$$

Vibrating columns can have standing waves in them in the same way as those on strings, i.e. a node displacement at a fixed end and an antinode at an open end (*Figure 6.19*).

$$L = \lambda/4$$

and hence

$$f = v/4L$$

for the fundamental frequency or first harmonic for a tube closed at one end

$$L = 3\lambda/4 \text{ and } f = 3v/4L$$

for the first overtone or third harmonic.

$$L = 5\lambda/4 \text{ and } f = 5v/4L$$

for the second overtone or fifth harmonic. For the tube open at both ends the fundamental frequency or first harmonic is

$$L = \lambda/2 \text{ and } f = v/2L$$

The first overtone or second harmonic is

$$L = \lambda \text{ and } f = v/L$$

The second overtone or third harmonic is

$$L = 3\lambda/2 \text{ and } f = 3v/2L$$

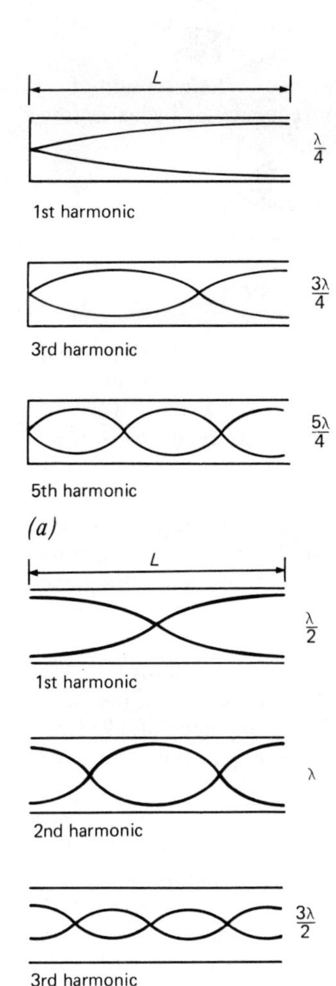

Figure 6.19 Standing wave displacements in tubes; (a) tube closed at one end, (b) tube open at both ends

Beats is the term used to describe the periodic rise and fall in loudness that occurs when two notes of slightly different frequencies but similar amplitubes are sounded together. The two waves superpose to give a wave with an amplitude which varies in a periodic manner, the number of amplitude variations per second being the difference in frequency between the two waves.

The speed of sound through a solid material is given by

$$v = (E/\rho)^{1/2}$$

where E is Young's modulus and ρ the density of the solid. The speed of sound in a gas or liquid is given by

$$v = (B/\rho)^{1/2}$$

where B is the bulk modulus. For an ideal gas

$$B = \gamma p$$

where γ is the ratio of the specific heat capacities at constant pressure and constant volume and p is the pressure. As

$$pV = mRT$$

and

$$m/V = \rho$$

then

$$p/\rho = RT$$

and so

$$v = (\gamma p/\rho)^{1/2}$$
$$= (\gamma RT)^{1/2}$$

The speed of sound in an ideal gas is thus (a) independent of the pressure, (b) directly proportional to the square root of the absolute temperature, (c) dependent on the humidity, which affects both γ and ρ and (d) dependent on any speed the medium might have.

Examples

6.67 What are the first and second harmonic frequencies for a vibrating string fixed at both ends if the string has a length of 0.80 m, a mass per metre length of 0.10 kg/m and under a tension of 60 N?

Ans See *Figure 6.18.* The wavelengths giving the standing waves are when

$$L = \tfrac{1}{2}\lambda \quad \text{and} \quad L = \lambda$$

The first and second harmonic frequencies are thus $v/2L$ and v/L. But

$$v = (T/\mu)^{1/2}$$
$$= (60/0.1)^{1/2}$$

Hence the frequencies are 15.31 Hz and 30.62 Hz.

6.68 What is the fundamental frequency for a tube of length 0.60 m and open at both ends if the speed of sound in the tube is 340 m/s?

Ans The wave giving the fundamental occurs (see *Figure 6.19(b)*) when

$$L = \tfrac{1}{2}\lambda$$

Thus the frequency is

$$v/2L = 340/(2 \times 0.60)$$
$$= 283 \text{ Hz}$$

6.69 What is the frequency of the beats produced when two sounds of frequencies 500 Hz and 510 Hz are sounded together?

Ans Beat frequency = 510 – 500
= 10 Hz

6.70 The speed of a longitudinal wave in a copper rod is 3800 m/s. If the density of the copper is 8900 kg/m^3 what is the value of Young's modulus for the material?

Ans $v = (E/\rho)^{1/2}$

Hence

$$E = \rho v^2$$
$$= 8900 \times 3800^2$$
$$= 1.3 \times 10^{11} \text{ N/m}^2$$

6.71 If the temperature of air is changed from 0 °C to 30 °C by what factor does the speed of sound change?

Ans $v = (\gamma RT)^{1/2}$

Hence

$$v_{30}/v_0 = \left(\frac{303}{273}\right)^{1/2} = 1.05$$

Further problems

6.72 What are the necessary conditions for the establishment of a standing wave?

6.73 Describe an experiment by which you could determine the wavelength in air of the note emitted by a tuning fork.

6.74 If the speed of sound in air is 340 m/s, what would be the length of a pipe, closed at one end, which would resonate to the fundamental frequency of 340 Hz (neglect end correction)?

6.75 What is the fundamental frequency of a violin string of 3.5 × 10^{-4} kg/m and length 0.25 m when stretched under a tension of 20 N? Would the frequency be higher or lower if the string slackens?

6.76 A violin string has a mass of 1.3 g and a length of 360 mm. What tension should be applied to the string if it is to have a fundamental frequency of 440 Hz?

6.77 An organ pipe open at both ends has a length of 1.2 m. If the velocity of sound in air is 340 m/s what is (a) the wavelength of the fundamental note, (b) the frequency of the fundamental note?

6.78 A closed organ pipe has a length of 500 mm. If the velocity of sound in the air in the pipe is 340 m/s what is the fundamental frequency of the pipe?

6.79 Two tuning forks when sounded together are found to give 5 beats per second. If one of the tuning forks has a frequency of 440 Hz, what can be the frequency of the other tuning fork?

6.80 The velocity of sound in air is 331.5 m/s at 0 °C. What will be the velocity at 20 °C?

6.81 If the velocity of sound in mild steel is 5200 m/s and the steel has a density of 7900 kg/m^3, what is the value of Young's modulus for the material?

6.82 Air has a density of 1.29 kg/m^3 at 15 °C under a pressure of 1.0 × 10^5 N/m^2. The ratio of the specific heats for air is 1.4. (a) What is the speed of sound under the above circumstances? (b) What would be the speed of sound if the temperature increased to 20 °C?

Oscillations

Notes

Oscillators which take the same time for each complete oscillation whatever its amplitude are said to be isochronous. If the acceleration of an oscillating object is directly proportional to its distance from a fixed point and is always directed towards that point, then the motion is said to be simple harmonic. Simple harmonic motion is isochronous motion, where

$$\text{Acceleration} \propto -x$$

where x is the displacement from the fixed point. The minus sign indicates that although the acceleration is larger the greater the displacement it is always in the opposite direction, i.e. always directed towards the zero displacement position.

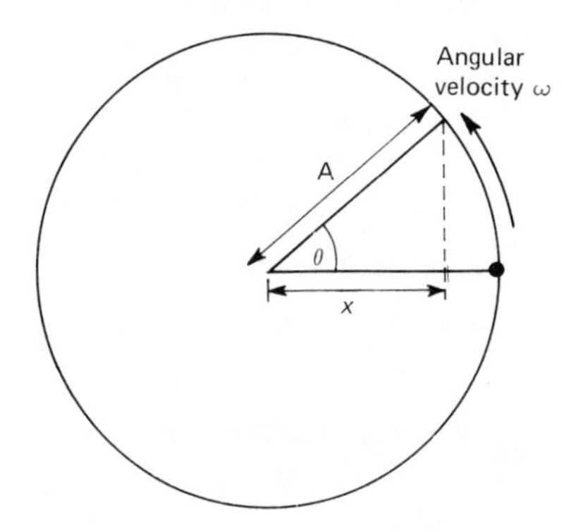

Figure 6.20

Figure 6.20 shows a rotating radius with a constant angular velocity ω. In time t the angle θ covered is

$$\theta = \omega t$$

The projection is

$$x = A \cos \theta$$

Thus

$$x = A \cos \omega t$$

The radius of the circle is A. If we consider the displacement of an oscillating object to be represented by the projection x, then the velocity of the object is $\mathrm{d}x/\mathrm{d}t$. Hence

$$v = \frac{\mathrm{d}x}{\mathrm{d}t} = -A\omega \sin \omega t$$

The maximum velocity is when

$$\sin \omega t = 1, \quad \text{i.e. } \omega t = 90°$$

and the object is passing through the central position.

$$\text{Maximum velocity} = -A\omega$$

and so

$$v = v_{\max} \sin \omega t$$

The acceleration is $\mathrm{d}v/\mathrm{d}t$, hence

$$a = \frac{\mathrm{d}v}{\mathrm{d}t} = -A\omega^2 \cos \omega t$$

Thus

$$a = -\omega^2 x$$

or, as

$$\omega = 2\pi f$$

$$a = -(2\pi f)^2 x$$

$$= -(2\pi/T)^2 x$$

where f is the frequency of oscillation and T the periodic time, i.e. the time for one oscillation.

For an object oscillating on the end of a vertical spring, if we assume that the restoring force F is proportional to the extension x, i.e.

$$F = kx$$

where k is a constant (sometimes called the force constant), then as

$$F = ma$$

$$\text{Acceleration, } a = -kx/m$$

The motion is simple harmonic, thus

$$a = -\omega^2 x$$

and so

$$\omega^2 = k/m$$

Hence, as

$$\omega = 2\pi f = 2\pi/T$$

$$f = \frac{1}{2\pi}\left(\frac{k}{m}\right)^{1/2}$$

$$T = 2\pi\left(\frac{m}{k}\right)^{1/2}$$

For an oscillating simple pendulum (*Figure 6.21*), the horizontal force acting on the pendulum bob is

$$-T \sin \theta = -Tx/L$$

If the vertical acceleration of the bob is very small, $T \cos \theta$ is approximately mg, and for small angles $\cos \theta$ approximates to 1, hence T is approximately mg. Thus

$$\text{Horizontal force} = (-mgx)/L$$

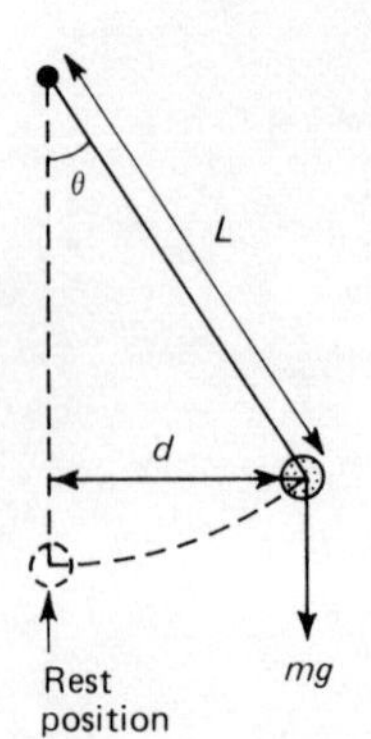

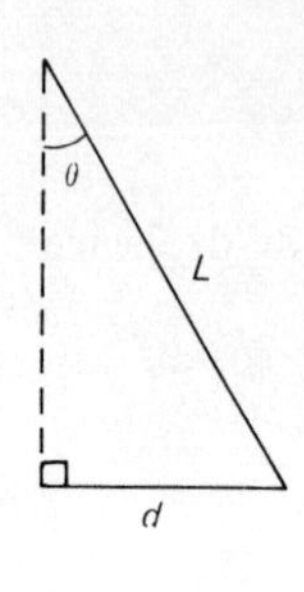

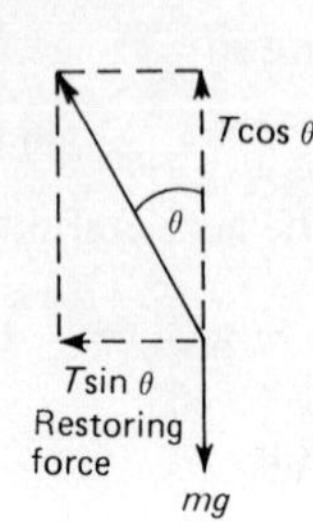

Figure 6.21

As

$$F = ma,$$
$$a = (-mx)/L$$

and so for simple harmonic motion, as

$$a = -\omega^2 x$$

then

$$f = \frac{1}{2\pi}\left(\frac{g}{L}\right)^{½}$$

$$T = 2\pi\left(\frac{L}{g}\right)^{½}$$

The kinetic energy of an oscillating object is $½mv^2$, and thus as

$$v = -A\omega \sin \omega t$$

$$\text{Kinetic energy} = ½mA^2\omega^2 \sin^2 \omega t$$

The maximum value of the kinetic energy is $½mA^2\omega^2$. If it is assumed that no energy leaves the oscillating system, then there is just a transfer back and forth between kinetic and potential energies. The total energy due to the oscillation is $½mA^2\omega^2$.

When an oscillating system is acted on by an external oscillatory force, the amplitude of the oscillator reaches a maximum at some particular frequency of this applied force. When the damping of the oscillator is zero, the maximum amplitude occurs when the driver frequency equals the natural frequency of the oscillator. The system is said to show resonance.

Examples

6.83 An object of mass 100 g is hung from a vertical spring and is found to oscillate with a frequency of 12 Hz. What is the force constant of the spring?

Ans $f = \frac{1}{2\pi}\left(\frac{k}{m}\right)^{½}$

Hence

$$k = 4\pi^2 f^2 m$$
$$= 568 \text{ N/m}$$

6.84 An object of mass 40 g is suspended from a vertical spring and set in oscillation. The frequency of the oscillation is found to be 2.0 Hz and the maximum speed of the object 140 mm/s. What is (a) the amplitude of the motion, (b) the period and (c) the force constant of the spring?

Ans (a) Maximum velocity $= -A\omega = -2\pi fA$

Hence

$$\text{Amplitude}, A = \frac{\text{maximum velocity}}{2\pi f}$$
$$= 11 \text{ mm}$$

(b) Period $= 1/f = 1/2.0 = 0.50$ s

(c) $f = \frac{1}{2\pi}\left(\frac{k}{m}\right)^{½}$

Hence

$$k = 4\pi^2 f^2 m$$
$$= 6.32 \text{ N/m}$$

6.85 A simple pendulum has a length of 1.20 m. What will be its period if it is set in oscillation with a small amplitude at a place where the acceleration due to gravity is 9.81 m/s²?

Ans $T = 2\pi\left(\frac{L}{g}\right)^{½}$

$$= 2\pi\left(\frac{1.20}{9.81}\right)^{½}$$
$$= 2.20 \text{ s}$$

6.86 An object is found to be oscillating with simple harmonic motion, the displacement x being related to time t by the equation

$$x = 0.080 \sin (4\pi t)$$

If all the distances and time represented in the above equation are in metres and seconds, what is (a) the amplitude of the motion, (b) the frequency of the oscillation?

Ans $x = A \sin \omega t$

where A is the amplitude and ω the angular velocity (or $2\pi f$, where f is the frequency).
(a) Amplitude = 0.080 m
(b) $\omega = 4\pi = 2\pi f$; hence $f = 2$ Hz

6.87 What is the maximum kinetic energy, and the maximum potential energy of an object of mass 1 kg oscillating with simple harmonic motion and having an amplitude of 100 mm and a frequency of 5.0 Hz?

Ans Maximum kinetic energy
= maximum potential energy
$= \frac{1}{2}mA^2\omega^2$
$= \frac{1}{2}m4\pi^2f^2A^2$
= 4.95 J

Further problems

Take g to be 9.8 m/s^2.

6.88 How are the following affected for an oscillator performing simple harmonic motion if the amplitude is doubled: (a) the period, (b) the frequency, (c) the maximum velocity, (d) the maximum acceleration and (e) the maximum kinetic energy?

6.89 When an object of mass 200 g is suspended from a vertical spring, the spring extends by 40 mm. The object is then pulled down a further 20 mm from the equilibrium position and released. What is (a) the amplitude, (b) the frequency of the oscillation?

6.90 What is the maximum velocity and the maximum acceleration of the oscillating object described in **6.89**?

6.91 What is the length of a simple pendulum with a period of 1.0 s, if the acceleration due to gravity is 9.8 m/s^2?

6.92 Show that the speed v of an object oscillating with simple harmonic motion is given by

$$v = \pm\omega(A^2 - x^2)^{1/2}$$

where ω is the angular velocity, A the amplitude and x the displacement.

6.93 An object of mass 0.50 kg is suspended from a vertical spring and performs simple harmonic motion with a period of 0.40 s. If the object has a maximum speed of 0.30 m/s what is (a) the force constant of the spring, (b) the amplitude of the oscillation?

6.94 An object is oscillating with simple harmonic motion, having a frequency of 2.0 Hz and an amplitude of 120 mm. What is (a) the maximum acceleration, (b) the maximum velocity and (c) the acceleration and velocity at a displacement of 60 mm from the equilibrium position?

6.95 A cylindrical test tube has a cross-sectional area of 2.0×10^{-4} m^2 and is weighted with lead so that the tube plus contents has a total mass of 40 g. If the tube floats vertically in water, of density 1000 kg/m^3, and is set into vertical oscillations after being pressed down into the water, what is the frequency?

6.96 A pendulum bob has a mass of 50 g and is suspended by a string of length 1.2 m from a fixed support. If the bob is drawn back 200 mm from its rest position and then released, what will be (a) the period, (b) the maximum kinetic energy?

6.97 Explain what is meant by the terms natural frequency and resonance.

Examination questions

6.98 A plane mirror lies at the bottom of a long flat dish containing water, the mirror making an angle of 10° with the horizontal, as shown in *Figure 6.22*.

A narrow beam of monochromatic light falls on the surface of the water at an angle of incidence of θ. If the refractive index of water is 4/3, determine the maximum value

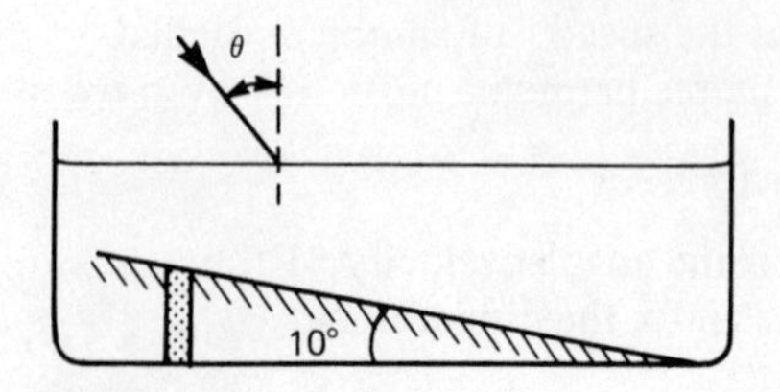

Figure 6.22

of θ for which light, after reflection from the mirror, would emerge from the upper surface of the water.

(AEB)

6.99 A prism *ABC* (*Figure 6.23*) made of glass of refractive index 1.60 has angles $A = C = 45°$; $B = 90°$. A parallel beam of light is incident on the face *AB* and emerges from the face *AC*, the deviation being a minimum. What is this minimum deviation?

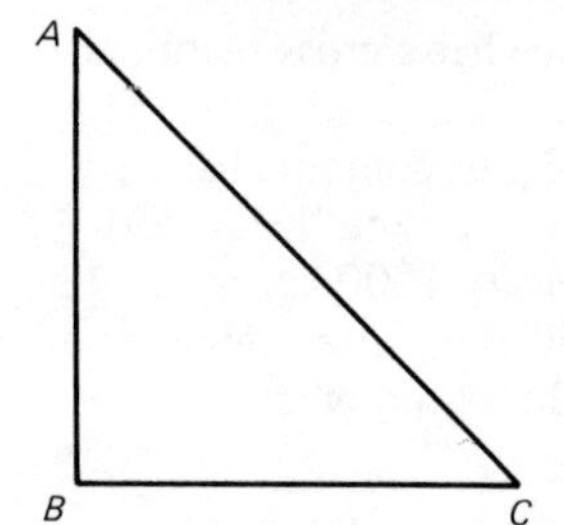

Figure 6.23

Explain, with a diagram, the behaviour of a beam which is incident normally on the face *AB*.

(Southern Universities)

6.100 The deviation of a ray of light in passing through a triangular prism via two adjacent faces depends on the angle of incidence with the first face. Experiment shows that the deviation has a minimum value. Explain why this occurs when the light passes symmetrically through the prism.

Derive an expression for the refractive index of the material of the prism in terms of the refracting angle and the angle of minimum deviation.

Explain how you would use a spectrometer to determine the refractive index of the material of a prism.

State the initial adjustments which are required to be made to the spectrometer before you can make your measurements. Details of experimental procedure are not required.

(University of London)

6.101 A slide projector is required to produce a real image 684 mm wide from an object 36 mm wide. If the distance of the object from the screen is to be 2000 mm, calculate

(i) the distance of the lens from the object,
(ii) the focal length of the lens required.

(AEB)

6.102 A refracting telescope has an objective of focal length 1.0 m and an eyepiece of focal length 2.0 cm. A real image of the Sun, 10 cm in diameter, is formed on a screen 24 cm from the eyepiece. What angle does the Sun subtend at the objective?

(University of London)

6.103 Light from a mercury lamp is incident normally on a plane diffraction grating ruled with 6000 lines per cm. The spectrum contains two strong yellow lines of wavelengths 577 nm and 579 nm. What is the angular separation of the second-order diffracted beams corresponding to these two wavelengths?

(University of Cambridge)

6.104 (a) Draw a labelled diagram showing the optical components of a prism spectrometer.
(b) Describe the adjustments that you would make using the spectrometer with a diffraction grating to ensure that

(i) a parallel beam of light is incident on the grating,
(ii) the grating is normal to the beam of light.

(c) Prove that, for a diffraction grating of which the line spacing is d, and which is placed normally to light of wavelength λ, the angle θ between the normal and the nth order image is governed by the formula

$$n\lambda = d \sin \theta$$

(d) Sodium light of wavelength 5.98×10^{-7} m falls normally on a diffraction grating and the angle between the normal and the first order image is $17° 24'$. Calculate

(i) the number of lines per mm ruled on the grating,
(ii) the angle which the largest order makes with the normal.

(AEB)

6.105 The phenomenon of Fraunhofer diffraction may be demonstrated by illuminating a wide slit by a parallel beam of monochromatic light and focussing the light that passes through the slit onto a white screen. A diffraction pattern may then be observed on the screen.
(a) Sketch the intensity variation in the diffraction pattern as a function of distance across it.
(b) What would happen to the intensity variation if the width of the slit were halved?

(University of Cambridge)

6.106 (a) Describe in detail how you would determine the wavelength of monochromatic light either by using a diffraction grating or by some other method based on optical interference.
(b) Explain, without going into detailed calculations, how the 'Newton's rings' pattern observed in reflected monochromatic light is produced, why the centre of the pattern is (or ought to be) dark, and why the pattern seems to be localised in the air film between the lens and the plane glass plate.
(c) White light transmitted normally through a very thin parallel-sided film of mica of thickness t and refractive index n falls on the slit of a spectrometer, and the continuous spectrum is interrupted by dark lines of which the first three, starting from the violet end of the spectrum, appear at wavelengths 452 nm, 499 nm and 558 nm.

(i) Show that these observations are consistent with the formula

$$2nt = (p + \tfrac{1}{2})\lambda$$

where p is an integer and λ is the wavelength at which a dark line is found.
(ii) What are the wavelengths of the two other dark lines that are seen in the visible region of the spectrum?

(Oxford Local Examinations)

6.107 Give the theory of an experiment to determine the wavelength of yellow light using two narrow slits. Point out any approximations you make.

Why is a third slit usually necessary?

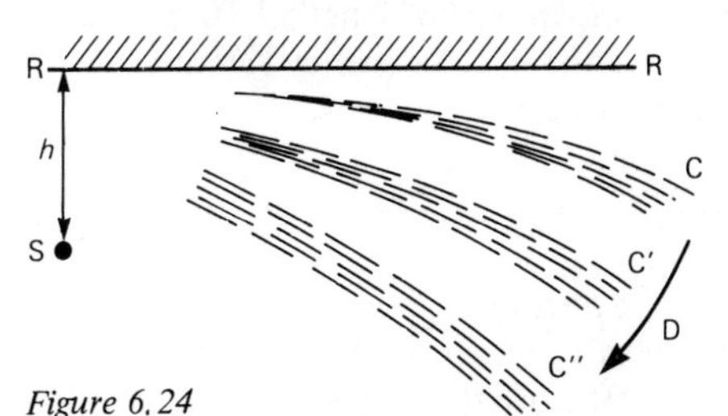

Figure 6.24

A source S (*Figure 6.24*) of continuous waves a distance h from a plane reflector R produces regions of high intensity such as C,C′ and C″. Account for this. When the frequency of S is changed slowly, the regions C,C′ and C″ move in the direction D as shown. Account for this, and deduce whether the frequency has been increased or decreased.

In Appleton's experiment, S was a radio transmitter on the Earth's surface, and R was the Heaviside layer – a reflecting layer in the atmosphere 80 km above the ground. When the wavelength, transmitted slowly, changed from 200 m to 180 m, a receiver on the ground 120 km away from S observed fluctuations in the received signal strength.

Calculate the number of signal strengths maxima observed during this change of frequency.

(University of Cambridge)

6.108 (a) Describe a laboratory experiment to determine the speed of sound in air. Show how you would calculate the speed from the readings you would take.
(b) State the equation relating the speed of sound, v, to the pressure, p, and density, ρ, of the gas through which it is passing. Use this equation to show how the speed of sound depends upon

(i) the pressure of the air at a given temperature,
(ii) the presence of water vapour in the air.

(c) If the speed of sound in air is 336 m s^{-1}, what would be the length of an open organ pipe giving the fundamental frequency of 96 Hz? If this pipe were sounded with another organ pipe of length 2.10 m, what would be the beat frequency (ignore end corrections)?

(AEB)

6.109 A source of sound of frequency 250 Hz is used with a resonance tube, closed at one end, to measure the speed of sound in air. Strong resonance is obtained at tube lengths of 0.30 m and 0.96 m. Find (a) the speed of sound, (b) the end correction of the tube.

(University of Cambridge)

6.110 The displacement y of a particle vibrating with simple harmonic motion of angular speed ω is given by $y = a \sin \omega t$ where t is the time. What does a represent?

Sketch a graph of the velocity of the particle as a function of time starting from $t = 0$ s.

A particle of mass 0.25 kg vibrates with a period of 2.0 s. If its greatest displacement is 0.4 m what is its maximum kinetic energy?

(University of London)

6.111 A certain mass, suspended from a spring, performs vertical oscillations of period T when on Earth. If the system were transferred to the Moon, where the acceleration of free fall is one-sixth of that on Earth, what would be the period?

(University of Cambridge)

6.112 Define simple harmonic motion, and explain what is meant by the amplitude and period of such motion.

Show that the vertical oscillations of a mass suspended by a light helical spring are simple harmonic, and obtain an expression for the period.

A small mass rests on a scale-pan supported by a spring; the period of vertical oscillations of the scale-pan and mass is 0.5 s. It is observed that when the amplitude of the oscillations exceeds a certain value, the mass leaves the scale-pan. At what point in the motion does the mass leave the scale-pan, and what is the minimum amplitude of the motion for this to happen?

(Southern Universities)

Answers

6.5 Ans Horizontal distance between reflections $= 200 \tan 30^\circ = 115$ mm. This gives 8 reflections within the length of the mirrors

6.6 Ans $\sin C = 1/1.3; R/200 = \tan C$; hence diameter $= 481.5$ mm

6.7 Ans See *Figure 6.2*. For no reflection $r_2 = C$; but $r_1 + r_2 = A$ (see **6.4**), therefore $r_1 = A - C$. As $\sin C = 1/1.50$ and $A = 60^\circ$, then $r_1 = 18.2^\circ$. As $\sin i_1 / \sin r_1 = 1.50$ then $i_1 = 27.9^\circ$. This is the smallest angle of incidence

6.8 Ans Minimum value is when $C = 45^\circ$, but $\sin C = 1/{}_{air}n_{prism}$; hence ${}_{air}n_{prism} = 1.41$

6.9 Ans $n = \sin [(A + D_{min})/2] / \sin (A/2)$, hence $D_{min} = 46.3^\circ$ (see **6.2**).

Deviation $= (i_1 - r_1) + (i_2 - r_2)$. For minimum deviation $i_1 = i_2$ and $r_1 = r_2$, thus deviation $= 2i - 2r$, but $A = r_1 + r_2 = 2r$, hence $i = 53.2°$

6.10 Ans See *Figure 6.25*. For the plastic to glass interface, $\sin i_1 / \sin r_1 = {}_{p}n_{g}$. For the

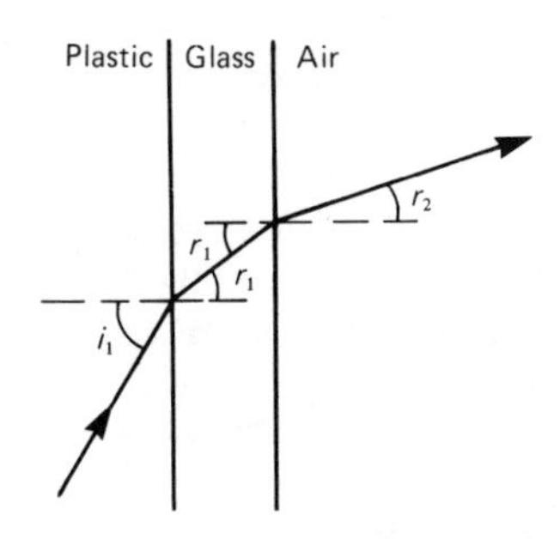

Figure 6.25

glass to air interface $\sin r_1 / \sin r_2 = {}_{g}n_{a}$. But ${}_{p}n_{a}$ would equal $\sin i_1 / \sin r_2$ hence ${}_{p}n_{a} = {}_{p}n_{g}\,{}_{g}n_{a}$. Thus ${}_{p}n_{g} = {}_{p}n_{a} / {}_{g}n_{a}$ $= {}_{a}n_{g} / {}_{a}n_{p} = 1.50/1.70 = 0.882$ and the critical angle $= 61.9°$

6.14 Ans You can hold the lens in front of a piece of paper and determine the distance of the paper from the lens when a clear image of a distance object is produced. The lens to paper distance is then the focal length

6.15 Ans (a) +120 mm, real
(b) −120 mm, virtual
(c) −40 mm, virtual
(d) −40 mm, virtual

6.16 Ans Distance = 156 mm, focal length 25.6 mm

6.17 Ans By increasing the distance between the projector and the screen and so making u smaller to produce a clear image or by using a shorter focal length lens

6.18 Ans $m = 1400/35 = v/u$, hence $u = 250$ mm; $1/f = 1/v + 1/u$, hence $f = 24.39$ mm

6.19 Ans $1/f = (1.5 - 1)(2/40)$, hence $f = 40$ mm

6.20 Ans $1/f = (1.50 - 1)/(1/200)$, hence $f = 400$ mm

6.21 Ans For the first lens $1/f_1 = 1/v_1 + 1/u$. For the second lens the image at v_1 acts as a virtual object and so $1/f_2 = 1/v - 1/v_1$. Hence $1/f = 1/v + 1/v$ $= 1/f_1 + 1/f_2$

6.25 Ans $1/f = 1/v + 1/u$, hence $v = 50.5$ mm; $m = I/1600 = 50.5/5000$, hence $I = 16.2$ mm

6.26 Ans $(D/f) + 1 = 3.5$

6.27 Ans For the eyepiece: $1/20 = 1/u - 1/250$, hence $u = 18.5$ mm. For the objective: $1/4.0 = 1/5.0 + 1/v$, hence $v = 20$ mm. Thus, the magnifying power $= (250/18.5) \times (20/5.0) = 54$

6.28 Ans For the objective: $1/6.0 = 1/8.0 + 1/v$, hence $v = 24$ mm. Thus the magnifying power $= (24/8) \times 5 = 15$

6.29 Ans Magnifying power = 20 000/6.5 = 3077

6.30 Ans $f_e = 1.05 - 1.00 = 0.05$ m; magnifying power = 1.00/0.5 = 20

6.33 Ans Ratio $= \sin 30°/\sin 60° = 0.58$ and so wave speed is greatest in the depth with a 60° refraction angle

6.34 Ans See *Figure 6.26*

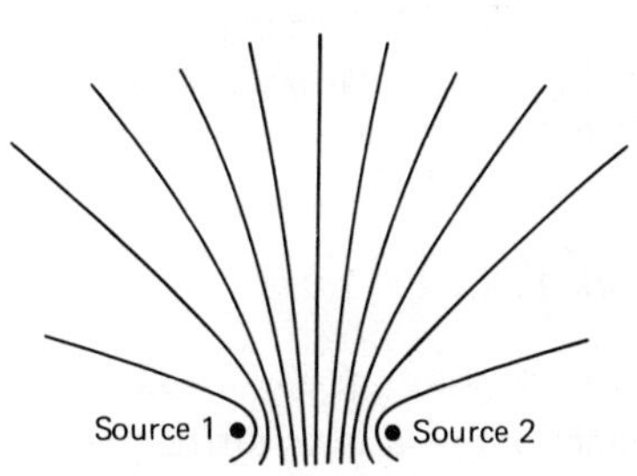

Figure 6.26

6.35 Ans (a) Circular waves spreading out from the aperture
(b) A slight amount of bending into the 'shadow', otherwise straight waves

6.36 Ans When they meet superposition will occur and the displacement of the rope is the vector sum of the separate displacements that would have occurred due to each wave alone. After the meeting the two waves will continue on with the same shape as they had before they met

6.43 Ans Distance between fringes
$= 500 \times 10^{-9} \times 1.0/(4.0 \times 10^{-3})$
$= 1.25 \times 10^{-4}$ m

6.44 Ans $\lambda = 0.64 \times 10^{-3} \times 0.80 \times 10^{-3}/1.0$
$= 5.12 \times 10^{-7}$ m

6.45 Ans $2t = m\lambda_a/n$, there are phase changes at both reflections (see **6.38**). Thus, $\lambda_a = 540$ mm

6.46 Ans $2t = m\lambda_a/n$, thus $\lambda_a = 520$ nm. All other wavelengths are outside the visible spectrum

6.47 Ans $2t = ½\lambda_a/n$, there are phase changes at both reflections (see **6.38**). Thus $t = 1.1 \times 10^{-7}$ m

6.48 Ans A phase change occurs at one reflection, but not the other. For a dark fringe $2t = m\lambda_a/n$. But $t/x = \tan\theta$, where x is the horizontal displacement from the contact line and θ the angle of the wedge. Thus the fringe separation $x_{m+1} - x_m = (6\lambda \tan\theta)/2n$. The number of fringes that can be contained within 10 mm is thus $10 \times 10^{-3} \times 2n/(\lambda \tan\theta)$. But $\tan\theta = d/(10 \times 10^{-3})$, where d is the wire diameter. Thus number of dark fringes $= 1.63 \times 10^3$

6.49 Ans $r^2 = Rm\lambda$, hence $r = 2.97 \times 10^{-3}$ m

6.50 Ans You could describe a Newton's ring experiment or one using a diffraction grating

6.51 Ans (a) $d \sin\theta = m\lambda$; angle of deviation $= 21.1°$
(b) angle of deviation $= 46.1°$

6.52 Ans $d \sin\theta = m\lambda$, hence angular separation $= 10.48°$

6.53 Ans $\sin\theta = m\lambda/b$. To a reasonable approximation $\sin\theta = x/D$, where x is distance along the screen from the zero order and D the distance of slit from screen. Thus, distance $= 1.4 \times 10^{-3}$ m

6.54 Ans Minimum angle $= 1.22\lambda/d = 2.0 \times 10^{-4}$ radians

6.57 Ans $\tan i = n$, hence $i = 52.4°$

6.58 Ans See the notes

6.61 Ans The Fizeau method involving a rotating toothed wheel, or Foucalt's method involving a rotating mirror, or Michelson's method involving a rotating octagonal mirror, could be discussed

6.62 Ans $v = f\lambda$, hence $\lambda = 3$ m

6.63 Ans You could try a Young's double slit type experiment or the superposition occurring between the incident microwaves and the normal reflection from a metal reflector

6.64 Ans $d \sin\theta = m\lambda$. In order for θ to be less than 90° for the first order, d must be more than λ. A reasonable value for d would be 9 cm

6.65 Ans $\lambda_{max} T = 2.90 \times 10^{-2}$, hence $\lambda_{max} = 1.2 \times 10^{-4}$ m

6.66 Ans See *Figure 6.17*; certainly less than 3000 K

6.72 Ans The conditions are that an incident and and a reflected wave travel along the same path, the two waves having the same wavelength, speed and intensity and being continuous, with a constant wavelength

6.73 Ans You could describe a resonance tube experiment in which the tuning fork is sounded over a variable length column of air held in a tube and the length of the tube found at which the first resonance occurs

6.74 Ans $L = \lambda/4 = v/4f = 0.25$ m

6.75 Ans $L = \frac{1}{2}\lambda$, hence as $v = (T/\mu)^{1/2} = f\lambda$, $f = 478$ Hz

6.76 Ans $L = \frac{1}{2}\lambda$, hence as $v = (T/\mu)^{1/2} = (TL/m)^{1/2} = f\lambda$, $T = 362$ N

6.77 Ans (a) $L = \lambda/2$, hence $\lambda = 2.4$ m
(b) $v = f\lambda$, hence $f = 142$ Hz

6.78 Ans $L = \lambda/2$, hence $f = v/\lambda = v/2L = 340$ Hz

6.79 Ans 440 ± 5 Hz

6.80 Ans $v = (\gamma RT)^{1/2}$, hence $v_{20}/v_0 = (293/273)$, thus $v_{20} = 343.4$ m/s

6.81 Ans $v = (E/\rho)^{1/2}$, hence $E = 2.1 \times 10^{11}$ N/m^2

6.82 Ans (a) $v = (\gamma p/\rho)^{1/2} = 329$ m/s
(b) $329 \times (293/288)^{1/2} = 332$ m/s

6.88 Ans (a) No change, (b) no change, (c) doubled, (d) doubled, (e) quadrupled

6.89 Ans (a) Amplitude = 20 mm
(b) $f = (1/2\pi)(k/m)^{1/2}$, where $k = 200 \times 9.8 \times 10^{-3}/(40 \times 10^{-3})$ N/m; hence $f = 2.5$ Hz

6.90 Ans Maximum velocity $= -A\omega = -2\pi fA = -0.31$ m/s. Maximum acceleration $= -\omega^2 x$ when x has its maximum value, i.e. A. Maximum acceleration $= -\omega^2 A = -4.9$ m/s^2

6.91 Ans $T = 2\pi(L/g)^{1/2}$, hence $L = 0.25$ m

6.92 Ans $v = -A\omega \sin \omega t$ (see the notes). But $\sin^2\theta + \cos^2\theta = 1$, hence $v^2 = A^2\omega^2(1 - \cos^2\omega t) = A^2\omega^2[1 - (x/A)^2]$; hence $v = \pm\omega(A^2 - x^2)^{1/2}$

6.93 Ans (a) $T = 2\pi(m/k)^{1/2}$, hence $k = 123$ N/m
(b) Maximum velocity $= A\omega = 2\pi fA$ and so $A = 0.019$ m

6.94 Ans (a) Maximum acceleration $= -\omega^2 A = -18.9$ m/s^2
(b) Maximum velocity $= -A\omega = -1.5$ m/s
(c) Acceleration $= -\omega^2 x = -9.5$ m/s^2, velocity $= \pm\omega(A^2 - x^2)^{1/2}$ (see **6.92**); hence $v = \pm 1.3$ m/s

6.95 Ans For length h immersed, a depression x from this equilibrium position gives a restoring force of upthrust minus weight $= -Ax\rho g$. Hence $a = F/m = -Ax\rho g/m$. Hence as $a = -\omega^2 x$ and $\omega = 2\pi f$, we have $f = (1/2\pi)(A\rho g/m)^{1/2}$; hence $f = 1.1$ Hz

6.96 Ans (a) $T = 2\pi(L/g)^{1/2} = 2.2$ s
(b) Maximum kinetic energy $= \frac{1}{2}mA^2\omega^2 = 0.04$ J

6.97 Ans The natural frequency is the frequency with which a system freely oscillates in the absence of any periodic force. Resonance is said to occur when an oscillator is acted on by a periodic force with a frequency equal to the natural frequency of the oscillator. In the absence of damping the amplitude of the oscillator increases to infinity

6.98 Ans Angle of reflection from the mirror must be $C - 10°$, where C is the critical angle and $\sin C = 3/4$, i.e. C is 48.6°. Thus, the angle of incidence on mirror = 38.6°. This is an angle of refraction in the water of 38.6° − 10° = 28.6°. Thus, $\sin\theta/\sin 28.6° = 4/3$ and so $\theta = 39.5°$

6.99 Ans $n = \sin[(A + D_{min})/2]/\sin(A/2)$, where $A = 45°$ and $n = 1.60$; thus $D_{min} = 30.5°$. For the ray normal to AB there is no deviation on passing through AB, but internal reflection at AC as the angle of incidence is 45° and this is greater than the critical angle of 38.7°. It reflects with an angle of reflection of 45°. It meets BC along the normal and passes undeviated through that surface

6.100 Ans Since a ray directed backwards through the prism will follow the same path, if there is minimum deviation for just one angle of incidence then the angle of incidence at the first face of the prism must be the same as the angle of emergence at the second face. See *Figure 6.2.* Angle of deviation at first face $= (i_1 - r_1)$. Angle of deviation at second face $= (i_2 - r_2)$. Total deviation $D = (i_1 - r_1) + (i_2 - r_2)$. In the quadrilateral *APQR* the angles add up to 360°. Thus angle $PQR = 180° - A$. For the triangle *PQR* the sum of the angles is 180°. Thus $r_1 + r_2 = A$. Minimum deviation occurs when $i_1 = i_2$ and $r_1 = r_2$. Thus $D_{min} = 2i - A$, hence $i = (A + D_{min})/2$ and $2r = A$. Thus $n = \sin i/\sin r = \sin[(A + D_{min})/2]/\sin(A/2)$

The spectrometer would be used to determine the apex angle A by directing the parallel light from the collimator at the prism apex and measuring the angle between the reflections from the sides of the prism, the angular difference between the reflections being twice the apex angle. The minimum deviation angle is determined by rotating the prism, i.e. varying the angle of incidence of the light on the first face of the prism, until the deviation is found to be a minimum. Initially, the telescope should be adjusted so that the eyepiece has the cross-wires in focus and then that parallel light is in focus. The collimator is then adjusted to give parallel light

6.101 Ans Use real is positive convention:
(i) $m = 684/36 = v/u = (2000 - u)/u$.
(ii) Hence $u = 100$ mm; $1/f = 1/v + 1/u = 1/1900 + 1/100$, hence $f = 95$ mm

6.102 Ans Note that the image is not at infinity so the equation derived in the notes cannot be used. For the eyepiece $1/f = 1/2.0 = (1/24) + 1/u$; hence $u = 2.18$ cm. This is the intermediate image and is formed at the focal point of the objective. Image size $= 10 \times 2.18/24 = 0.908$ cm. Angle subtended by object at infinity $= 0.908/100 = 0.009\ 08$ radians

6.103 Ans $d \sin\theta = n\lambda$. Thus angles = 43.82° and 44.01°: thus the difference is 0.19°

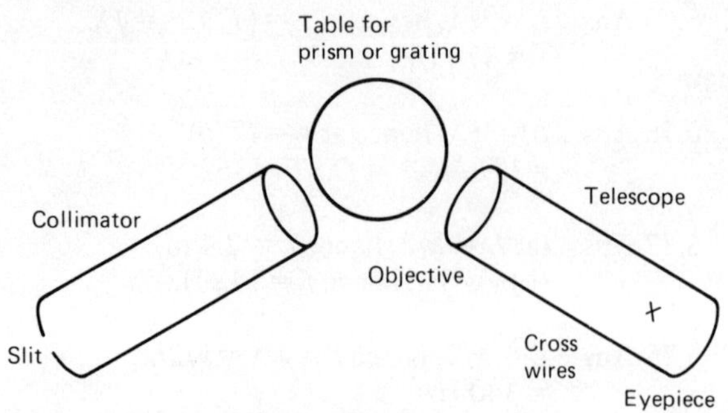

Figure 6.27 Simple spectrometer

6.104 Ans (a) See *Figure 6.27*
(b) (i) The eyepiece of the telescope is focussed on the cross-wires and then the objective adjusted so that a distant object is on focus on the cross-wires. The light from the collimator is then passed into the telescope and the collimator adjusted so that the slit is in focus when viewed through the telescope. (ii) The collimator and telescope are set at right angles to each other and the grating on the table rotated so that a reflected image of the collimator slit is seen in the telescope. The grating is then at 45° to the incident light. The table is then rotated by 45° to bring the grating normal to the light
(c) See the notes
(d) (i) 500 per mm (ii) 63.8°

6.105 Ans (a) See *Figure 6.14*
(b) $\sin\theta = m\lambda/b$. Thus halving b doubles each value of $\sin\theta$. The graph shows the relative intensities between the various maxima. These are unchanged but the central maximum is half the intensity

6.106 Ans (a) See **6.104** for the setting up of a diffraction grating. Measurement of the diffraction angles for light of unknown wavelength with a grating of a known number of rulings per metre enables the wavelength to be calculated
(b) See the notes. The pattern should be dark with zero path difference due to the phase change occurring at reflection at the air–glass interface. See *Figure 6.28*
(c) (i) See **6.38**. (ii) $558(p + ½) = 499(p + 1½) = 452(p + 2½)$; thus

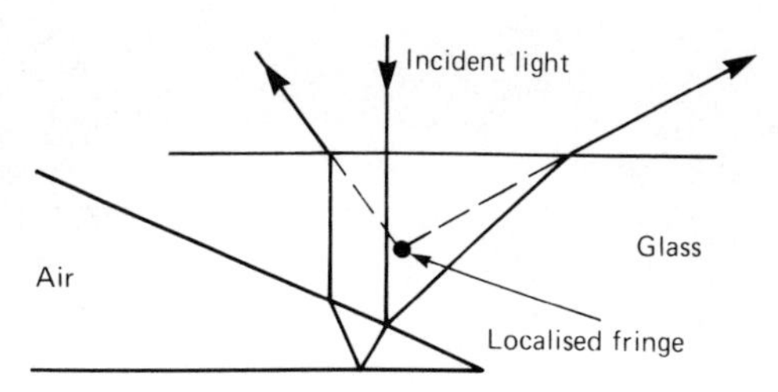

Figure 6.28

$p = 8$. Hence for orders 7½ and 6½, wavelengths are 632 nm and 730 nm

6.107 Ans See the notes for the theory for Young's double slits. The third slit is necessary in order that the light reaching the double slits is coherent. This happens because of diffraction at the single slit used prior to the double slits. With the arrangement shown in *Figure 6.24*, the superposition is occurring between the waves from the source S and its image in the reflector R. The bigger the wavelength in relation to the distance SR the further apart the maxima. Increasing the wavelength means decreasing the frequency. Thus, for the movement D the frequency has been decreased

Distance from image = 200 km. Thus path difference between direct and reflected waves = 200 − 120 = 80 km. Ignoring any possible phase changes $80 \times 10^3/200 = n$ and $80 \times 10^3/180 = n + p$. Hence the number of maxima $p = 44$

6.108 Ans (a) You could describe a direct method in which the time taken for sound to travel to a reflector and back is measured, the equipment being a frequency generator, microphone, reflector, loudspeaker and oscilloscope. An alternative is an indirect method in which the wavelength of a known frequency sound is measured and the speed calculated from $v = f\lambda$. The resonance tube could be used
(b) $v = (\gamma p/\rho)^{1/2}$, (i) $p/\rho = RT$, thus at a given temperature p changes by the same factor as ρ and thus v is independent of pressure. (ii) Water vapour affects both γ and ρ, the value of γ/ρ increasing with increasing water vapour content
(c) $L = \lambda/2 = v/2f = 1.75$ m;
For the 2.10 m pipe $f = 80$ Hz. Hence beat frequency = 16 Hz

6.109 Ans (a) $\frac{1}{4}\lambda = L_1 + e$ and $\frac{3}{4}\lambda = L_2 + e$ where e is the end correction;
$L_2 - L_1 = \frac{1}{2}\lambda$, hence $v = f\lambda = 330$ m/s
(b) $e = 0.03$ m

6.110 Ans a is the amplitude. The graph is minus cosinusoidal with a period of $2\pi/\omega$. Maximum kinetic energy $= \frac{1}{2}mA^2\omega^2$
$= \frac{1}{2} \times 0.25 \times 0.4^2(2\pi/2.0)^2 = 0.197$ J

6.111 Ans No change, the period depends on the mass and the force constant of the spring and has no term involving the acceleration due to gravity

6.112 Ans See the notes. Acceleration $= -\omega^2 x = -(2\pi/T)^2 x$. The maximum acceleration occurs at the maximum displacement, i.e. when $x = A$. The mass will leave the scale-pan when the pan is at its maximum displacement and is just starting to go downwards if the acceleration is greater than the acceleration due to gravity. Minimum amplitude $= 9.8 \times 0.5^2/(4\pi^2) = 0.062$ m

7 Atoms and quanta

Spectra

Notes

Emission spectra in the visible region are produced when light from luminous substances is directly examined with a spectroscope. Absorption spectra are obtained when white light, or some other continuous spread of wavelengths, is passed through the substance concerned before entering the spectroscope.

A continuous spectrum is produced by hot solids and consists of a continuous spread of wavelengths. Line spectra are produced by luminous gases and vapours and consist of a number of discrete wavelengths characteristic of the elements present in the substance. The spectra from luminous molecules consist of several well defined bands of lines and are called band spectra.

Absorption spectra consist of the continuous regions and lines or bands absorbed from white light when it is passed through the substance concerned. The wavelengths absorbed appear as black lines or regions in the white light continuous spectrum. The white light consists of all wavelengths and thus the molecules or atomic lattices through which the white light is passed are subjected to a large range of forces at different frequencies. The forces with frequencies equal to the natural frequencies of the molecules or lattice result in oscillations and so energy at that frequency is extracted from the white light, i.e. absorbed.

Examples

7.1 What are the main characteristics of emission line spectra?

Ans They consist of a number of discrete wavelengths, characteristic of the elements being excited. The pattern of wavelengths emitted is rather like a finger-print for the element concerned; no two patterns are the same.

7.2 Sodium vapour gives an emission line spectrum consisting predominantly of two wavelengths close together in the yellow region. What will be the appearance of the absorption spectrum produced when white light is passed through the sodium vapour before entering the spectroscope?

Ans The spectrum will appear as a continuous range of wavelengths with just two dark lines in the yellow part of the spectrum. The wavelengths of these dark lines will correspond to the wavelengths of the yellow lines in the emission spectrum.

Further problems

7.3 Explain how line spectra are used for the identification of the elements present in a particular substance.

7.4 State two methods by which light can be dispersed into a spectrum.

7.5 What are the differences in origin of line and band emission spectra?

Photoelectricity

Notes

Photoelectricity is the emission of electrons from materials by the action of electromagnetic waves such as light. The maximum kinetic energy of the emitted electrons is independent of the intensity of the light. The higher the frequency of the light the greater the maximum kinetic energy of the electrons. Below a certain frequency a particular material will not emit any electrons, however intense the light. The current produced at any particular frequency above the minimum frequency for which emission occurs increases with the intensity of the light.

Einstein explained the above experimental facts by proposing that light comes in packets, called quanta or photons. The energy E of a photon depends on the frequency f, with

$$E = hf$$

where h is Planck's constant, 6.63×10^{-34} J s. When photons are incident on a material they can cause emission of electrons if the photon energy is greater than the work function W. This is the energy needed to remove an electron from the material. Thus

Maximum kinetic energy of emitted electrons
$$= \tfrac{1}{2}mv^2_{max}$$
$$= hf - W$$

This is known as Einstein's equation for photoelectricity. The greater the intensity of light the greater the number of photons; the energy of each photon is, however, only determined by the frequency of the light.

Examples

Take Planck's constant, h, to be 6.6×10^{-34} J s and the speed of light c to be 3.0×10^8 m/s

7.6 What is the energy of a photon of yellow light with a wavelength in a vacuum of 5.9×10^{-7} m?

Ans
$$E = hf$$
$$= hc/\lambda$$
$$= \frac{6.6 \times 10^{-34} \times 3.0 \times 10^8}{5.9 \times 10^{-7}}$$
$$= 3.36 \times 10^{-19} \text{ J}$$

7.7 A sodium lamp of power 15 W emits photons of wavelength 5.9×10^{-7} m in a vacuum. If all the power is converted into photons, what is the number of photons emitted per second?

Ans As in the previous question the energy of a photon is 3.36×10^{-19} J. Thus, as power is the rate of transfer or transformation of energy,

$$\text{Number of photons per second} = \frac{15}{3.36 \times 10^{-19}}$$
$$= 4.47 \times 10^{19} \text{ s}^{-1}$$

7.8 For potassium the cut-off wavelength in photoelectricity is 5.5×10^{-7} m. (a) If light of wavelength 5.9×10^{-7} m is shone onto a potassium surface will there be any emission of electrons? (b) What is the work function for potassium? (c) What will be the maximum kinetic energy of the electrons emitted when light of wavelength 5.0×10^{-7} m is shone onto a potassium surface?

Ans (a) Maximum kinetic energy of electrons
$$= hf - W$$
At the cut-off frequency the electrons have zero kinetic energy, only having enough energy to just reach the surface. Thus

$$hf = W$$

Hence, the cut-off wavelength is given by

$$hc/\lambda = W$$

If hc/λ is greater than W electrons will emerge from the surface; if hc/λ is less than W no emission will occur. At cut-off

$$\frac{hc}{5.5 \times 10^{-7}} = W$$

At a wavelength of 5.9×10^{-7} m we have

$$W = hc/(5.9 \times 10^{-7})$$

This is less than $hc/(5.5 \times 10^{-7})$ and thus no emission will occur.

(b) $$W = \frac{6.6 \times 10^{-34} \times 3.0 \times 10^8}{5.5 \times 10^{-7}}$$
$$= 3.6 \times 10^{-19} \text{ J}$$

(c) Maximum kinetic energy of electrons
$$= hc - W$$
$$= \frac{6.6 \times 10^{-34} \times 3.0 \times 10^8}{5.0 \times 10^{-7}} - 3.6 \times 10^{-19}$$
$$= 3.6 \times 10^{-20} \text{ J}$$

7.9 When light of frequency 8.0×10^{14} Hz is shone onto a caesium surface emission of electrons occurs. A potential difference of 1.4 V maintained between an electrode above the emitting surface and the caesium is needed to prevent any electrons escaping from the caesium and reaching the electrode.

(a) What is the maximum kinetic energy of the electrons? (b) What is the work function of carsium ($e = -1.6 \times 10^{-19}$ C)?

Ans (a) Energy used to stop electrons
$= Ve$
$= 1.4 \times 1.6 \times 10^{-19}$
$= 2.24 \times 10^{-19}$ J

This is the maximum kinetic energy of the electrons.

(b) Maximum kinetic energy of electrons
$= hf - W$

Hence

$$W = 6.6 \times 10^{-34} \times 8.0 \times 10^{14} - 2.24 \times 10^{-19}$$
$$= 3.04 \times 10^{-19} \text{ J}$$

Further problems

Take Planck's constant to be 6.6×10^{-34} J s, the speed of light to be 3.0×10^8 m/s and the charge on an electron to be -1.6×10^{-19} C.

7.10 How is the maximum kinetic energy of the electrons emitted from a surface related to the frequency of the light falling on it and responsible for the emission?

7.11 What is the minimum frequency of the light that can be used to illuminate a calcium surface having a work function of 4.3×10^{-19} J and still cause the emission of electrons?

7.12 Platinum has a work function of 10.0×10^{-19} J. What will be the maximum kinetic energies of the photoelectrons emitted when the following wavelengths are used to illuminate the surface: (a) 300 nm, (b) 500 nm and (c) 700 nm?

7.13 A stopping potential of 2.0 V is required to prevent electrons emitted from a lithium surface escaping to a collector electrode. If lithium has a work function of 2.3 eV, what is the wavelength of the light used?

7.14 The minimum energy that is needed to remove an electron from a caesium surface is 2.6 eV. What is the longest wavelength photon that can be used to eject an electron from caesium?

7.15 At what rate are photons emitted from a sodium lamp of power 20 W if all the lamp energy is radiated at a wavelength of 590 nm?

7.16 How many photons strike a surface per second if that surface is illuminated by 10 W of yellow light having a wavelength of 590 nm?

Energy levels

Notes

In 1914 J. Franck and G. Hertz carried out experiments in which electrons of different energies were used to hit atoms. They found that at certain electron energies the electrons suffered inelastic collisions with the atoms, i.e. kinetic energy was not conserved and the atoms 'absorbed' energy. At other electron energies the collisions were elastic and kinetic energy was conserved. The absorption of energy is said to produce excited atoms. An atom has an entire series of energy levels at which it can become excited. The excited atom loses its surplus energy by the emission of photons. The photons that can be emitted are only those which allow the excited atom to fall from one energy level to another, or to the ground state where it has no surplus energy. Each photon has a definite energy, the difference between the energy values of the energy levels, and hence a definite frequency given by

$$E = hf$$

The result is a line spectrum when an atom is excited sufficiently to fall through a number of energy levels before reaching the ground state.

The energy needed to completely remove an electron from an atom is called the ionisation energy.

Examples

7.17 *Figure 7.1* shows some of the energy levels for mercury. What wavelengths will be produced by excited atoms having 7.8 eV energy when they revert to the ground state

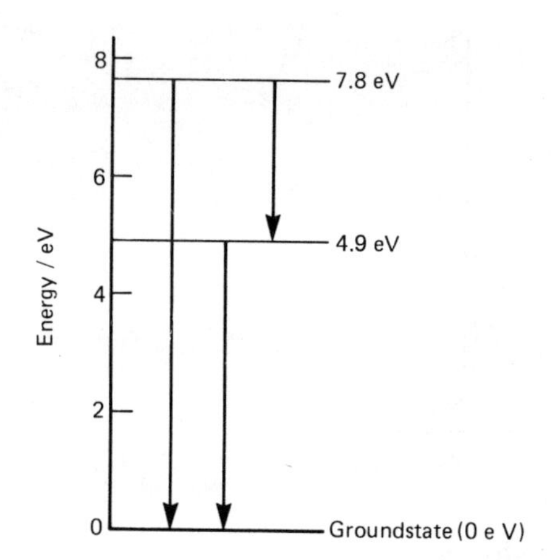

Figure 7.1

by the transitions shown (speed of light is 3.0×10^8 m/s, charge on the electron is -1.6×10^{-19} C and Planck's constant is 6.6×10^{-34} J s)?

Ans The possible transitions between energy levels are from 7.8 eV to the ground state, from 7.8 eV to 4.9 eV and from 4.9 eV to the ground state. The energy lost by a mercury atom in making any of these transitions is converted into photons.

$$E = hf = hv/\lambda$$

Thus for the transition from 7.8 eV to ground state

$$E = 7.8 \times 1.6 \times 10^{-19} = hv/\lambda$$

and so

$$\lambda = 1.6 \times 10^{-7} \text{ m}$$

For the 7.8 eV to 4.9 eV transition

$$E = (7.8 - 4.9) \times 1.6 \times 10^{-19} = hv/\lambda$$

and so

$$\lambda = 4.3 \times 10^{-7} \text{ m}$$

For the 4.9 eV to ground state transition

$$E = 4.9 \times 1.6 \times 10^{-19} = hv/\lambda$$

and so

$$\lambda = 2.5 \times 10^{-7} \text{ m}$$

The spectrum of the mercury when excited to 7.8 eV will consist of just three wavelengths.

7.18 Electrons of energy (a) 2.8 eV, (b) 4.9 eV are used to bombard mercury (see *Figure 7.1*). What will be the energy of the electrons after collisions with mercury atoms?

Ans (a) This collision will be elastic and because of the much greater mass of the mercury atom little kinetic energy will be transferred from the electron to the atom in such a collision. The electron will thus continue with virtually 2.8 eV energy.
(b) This energy coincides with an energy level of mercury and thus the collision will be inelastic. After the collision the energy of the electron will be zero as all its kinetic energy will be 'absorbed' by the mercury atom. Some electrons will not collide with atoms and so will retain their 4.9 eV energy.

Further problems

Take Planck's constant to be 6.6×10^{-34} J s, the speed of light to be 3.0×10^8 m/s, and the charge on an electron to be -1.6×10^{-19} C.

7.19 The lowest excited states of a hydrogen atom are 10.2 eV and 12.1 eV above the ground state. What are the wavelengths that might occur in the spectrum of hydrogen vapour excited to an energy of 12.1 eV if all transitions are possible?

7.20 The first ionisation energy of mercury vapour is 10.4 eV. Explain the significance of this statement?

7.21 The ionisation energy of hydrogen vapour is 13.6 eV. What is the minimum velocity of the electron which can ionise hydrogen (mass of the electron = 9.1×10^{-31} kg)?

7.22 The lowest excited state of caesium is 1.38 eV above the ground state. What wavelength line will be produced by bombarding caesium vapour with electrons having energies of 1.38 eV?

7.23 How many spectrum lines might be produced if a vapour was excited to the fourth of its energy levels above the ground state?

X-rays

Notes

X-rays are electromagnetic radiation having a speed in a vacuum of 3.0×10^8 m/s and wavelengths of the order of 10^{-10} m. *Figure 7.2* shows a modern form of X-ray tube, the X-rays being produced when the electrons bombard a metal target.

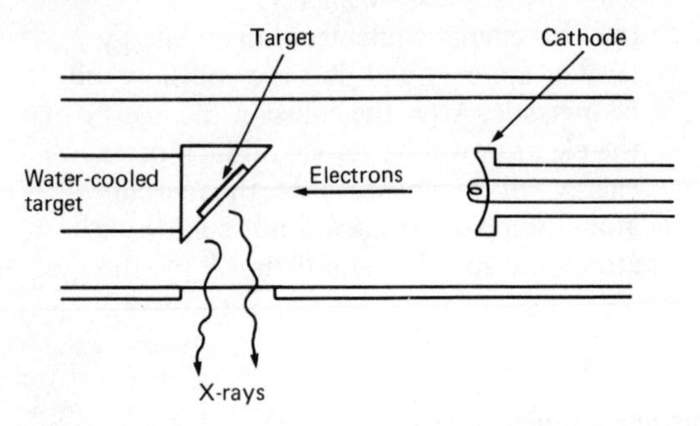

Figure 7.2

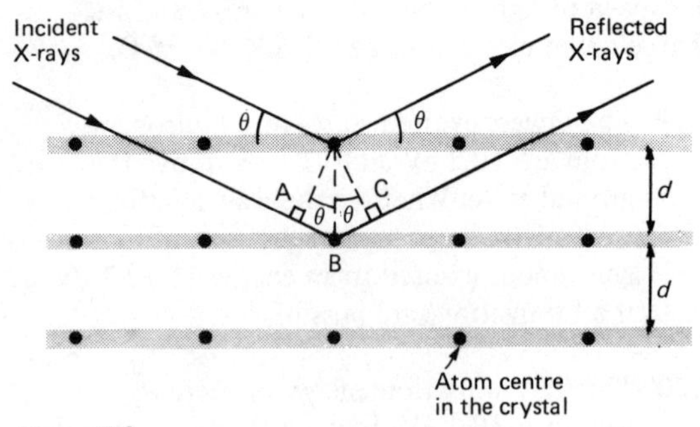

Figure 7.3

X-rays can be diffracted by the regular planes of atoms or ions in a crystal lattice. Planes of atoms act rather like mirrors in reflecting part of the incident beam of X-rays. Each of the planes within a crystal gives rise to reflected beams. Superposition occurs between these beams and reinforcement, i.e. a maximum, occurs when the path difference between the X-rays reflected from successive planes is a whole number of wavelengths. This means (*Figure 7.3*)

$$AB + BC = n\lambda$$

where n is an integer.

$$AB = BC = d \sin \theta$$

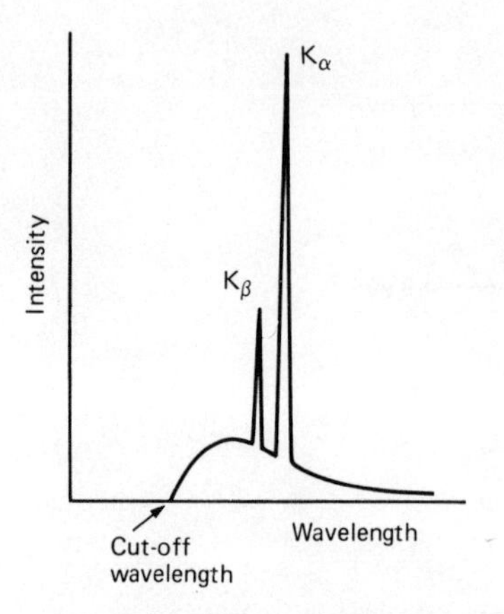

Figure 7.4

Hence for reinforcement

$$2d \sin \theta = n\lambda$$

This equation is known as **Bragg's law.**

Figure 7.4 shows the intensity distribution with wavelength for X-rays produced by the bombardment of a metal target with electrons. The spectrum consists of two parts: a continuous spectrum which is independent of the target material and a line spectrum which depends on the target material used. The continuous spectrum shows a cut-off wavelength. This is produced when an electron hits the target and its entire kinetic energy is converted into a photon, i.e. when

$$eV = \tfrac{1}{2}mv^2$$
$$= hf$$
$$= hv/\lambda$$

where V is the potential difference through which the electron is accelerated. The wavelengths of the line spectra are related to the position of the target element in the Periodic Table and can be used to determine the element's position, i.e. atomic number.

X-rays are photons, like light. When such photons hit electrons scattering occurs which can be explained by considering the collision as one between two particles. This effect is known as the Compton effect.

Examples

7.24 A monochromatic beam of X-rays is found to give a first order Bragg reflection of 15° from a sodium chloride crystal in which the relevant planes were spaced at 0.28 nm. What is the wavelength of the X-rays?

Ans $2d \sin\theta = n\lambda$

$\lambda = 2 \times 0.28 \times 10^{-9} \sin 15°$
$= 1.4 \times 10^{-10}$ m

7.25 Electrons in an X-ray tube are accelerated through a potential difference of 20 kV before they strike the target. What will be the minimum wavelength produced (Planck's constant is 6.6×10^{-34} J s, charge on the electron is -1.6×10^{-19} C and speed of light is 3.0×10^{8} m/s)?

Ans At the cut-off wavelength

$$eV = hv/\lambda$$

Hence

$$\lambda = \frac{6.6 \times 10^{-34} \times 3.0 \times 10^{8}}{1.6 \times 10^{-19} \times 20 \times 10^{3}}$$

$$= 6.2 \times 10^{-11} \text{ m}$$

Further problems

Take Planck's constant to be 6.6×10^{-34} J s, the charge on an electron to be -1.6×10^{-19} C and the speed of light to be 3.0×10^{8} m/s.

7.26 Monochromatic X-rays are incident on a sodium chloride crystal having a spacing between planes of 2.8×10^{-8} m. If the first order Bragg reflection is found to occur at 25° what is the wavelength of the X-rays and the angle of the second order reflection?

7.27 Tungsten is found to give a first order Bragg reflection at 20° for X-rays of wavelength 2.6×10^{-10} m. What is the spacing between the crystal planes responsible for this reflection?

7.28 The spacing between planes of atoms in a calcite crystal is 3.0×10^{-8} m. The first order Bragg reflection occurs for an angle of 15°. What is the wavelength of the X-rays?

7.29 An X-ray tube used for medical diagnosis has a potential difference between the cathode and the target of 25 kV. What is the maximum frequency of the emitted X-rays?

7.30 An X-ray tube is operated at 40 kV. What is the wavelength of the most energetic X-rays emitted?

7.31 Explain the origin of (a) the continuous X-ray spectrum, (b) the line spectrum.

Matter waves

Notes

In 1923 a radical proposal was made by M.L. de Broglie that all matter has wave properties as well as particle properties. The equation he developed was

$$\lambda = h/mv$$

Thus

$$\text{Momentum } mv = h/\lambda$$

It is also valid to say that all waves have particle properties.

Examples

Take Planck's constant to be 6.6×10^{-34} J s, the mass of an electron to be 9.1×10^{-31} kg and the charge on an electron to be -1.6×10^{-19} C.

7.32 What is the de Broglie wavelength for an electron moving with a velocity of 2.0×10^{5} m/s?

Ans $\lambda = h/mv$

$$= \frac{6.6 \times 10^{-34}}{9.1 \times 10^{-31} \times 2.0 \times 10^{5}}$$

$$= 3.6 \times 10^{-9} \text{ m}$$

7.33 If electrons are accelerated through a potential difference of 1.0 kV, what will be their wavelength after the acceleration?

Ans For the acceleration

$$eV = \tfrac{1}{2}mv^2$$

Hence

$$v = \left(\frac{2eV}{m}\right)^{1/2}$$

$$\lambda = h/mv$$

$$= \frac{h}{m}\left(\frac{m}{2eV}\right)^{1/2}$$

$$= \frac{h}{(2meV)^{1/2}}$$

$$= \frac{6.6 \times 10^{-34}}{(2 \times 9.1 \times 10^{-31} \times 1.6 \times 10^{-19} \times 1000)^{1/2}}$$

$$= 3.8 \times 10^{-11} \text{ m}$$

Further problems

Take Planck's constant to be 6.6×10^{-34} J s, the mass of an electron to be 9.1×10^{-31} kg and the charge on an electron to be -1.6×10^{-19} C.

7.34 What is the de Broglie wavelength for electrons having a kinetic energy of 120 eV?

7.35 Sodium ions carrying a charge of $+1.6 \times 10^{-19}$ C are accelerated through a potential difference of 400 V. What is their de Broglie wavelength (mass of sodium ion is $23 \times 1.66 \times 10^{-27}$ kg)?

7.36 What is the wavelength of a proton with a speed of 5.0×10^6 m/s (mass of proton is 1.67×10^{-27} kg)?

7.37 What are the energies of an electron and a photon if each have the same wavelength of 0.40 nm (speed of light is 3.0×10^8 m/s)?

Radiations from radioactive materials

Notes

The main methods used for the detection of radiations from radioactive substances are:

(i) Photographic emulsion. The radiations affect the emulsion so that, after being developed, the emulsion is found to have been blackened by the radiation.

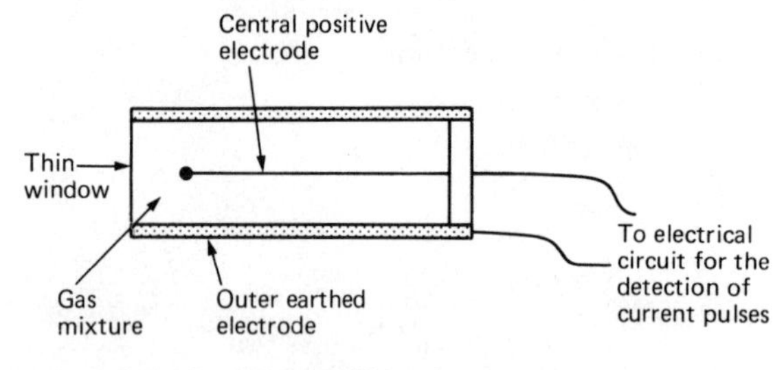

Figure 7.5 The Geiger–Müller tube

(ii) The Geiger–Müller tube (*Figure 7.5*). When radiation enters the tube ionisation occurs in the gas in the tube. The movement of the resulting electrons leads to further ionisations as they collide with other gas atoms. The result is an avalanche of charge. The positive ions produced in the ionisation are prevented from producing a current pulse by, in the case of a halogen tube, dissipating their energy by splitting molecules of bromine that are in the tube. The tube is said to be quenched (there are other quenching mechanisms). To use the Geiger tube the potential difference applied between the electrodes is increased until the count rate becomes reasonably independent of the potential difference. This is the threshold potential difference. For a further 100 to 200 V the count rate is reasonably independent of potential difference. The tube is used in roughly the middle of this region.

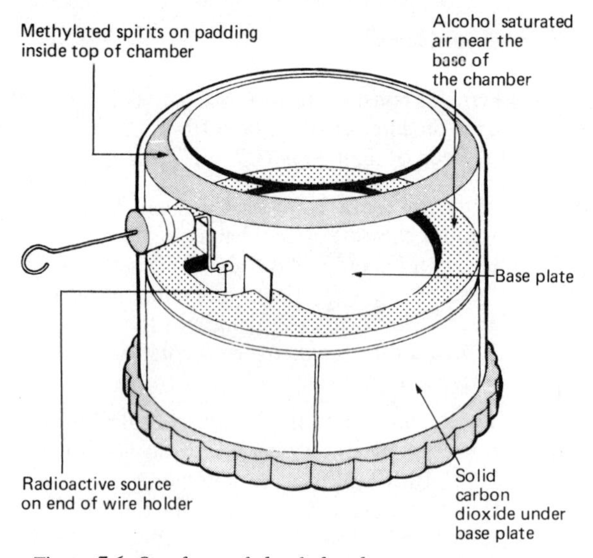

Figure 7.6 One form of cloud chamber

(iii) Scintillation counter. When a material such as zinc sulphide is hit by radiation weak flashes of light can be observed. These flashes can be counted directly, though modern instruments count them electronically.

(iv) Cloud chamber (*Figure 7.6*). Water vapour, or indeed any vapour, will preferentially condense on dust particles or ions. The path of an ionising radiation through a saturated vapour thus shows as a track of condensed vapour, i.e. droplets.

(v) Bubble chamber. The ionising radiation causes a superheated transparent liquid to boil along the trail of ions.

Alpha radiation consists of ionised helium atoms, He^{2+}. The distance such a radiation travels in air at atmospheric pressure, the range, is about 30 to 50 mm. It is unable to penetrate quite small thicknesses of matter. It, being charged, is deflected in electric and magnetic fields from a straight line path. It is intensely ionising. Alpha particles emitted in radioactive decay have well defined energies.

Beta radiation consists of electrons (note that positrons can be emitted). It is able to penetrate greater thicknesses of air and other media than alpha particles. It, being charged, is deflected in electric and magnetic fields from a straight line path. It produces less ions per millimetre of track than alpha particles. The beta radiation produced from radioactive decay is found to contain electrons with a range of kinetic energies from zero to some maximum value.

Gamma radiation is electromagnetic waves of very short wavelength. It is more penetrating than either alpha or beta radiation. It is not deflected by either electric or magnetic fields. It produces much less ions per millimetre of track than either alpha or beta radiation.

Examples

7.38 A Geiger tube with a very thin aluminium foil window is placed in front of a radioactive source. What can you deduce about the nature of the radiation emitted by the source if introducing a piece of aluminium foil 2×10^{-4} mm thick between the source and the tube window is found to (a) have little effect on the count rate, (b) cut the count rate down to just the background level?

Ans (a) The radiation could be beta and/or gamma radiation. Such a thin sheet of aluminium would have virtually no noticeable effect on the count rate with beta radiation and no noticeable effect with gamma radiation.
(b) Alpha radiation would be completely stopped by such a thickness of aluminium, thus the radiation is possibly alpha. Low energy beta radiation would also be stopped.

7.39 An alpha particle in passing through air needs about 30 eV for each ion pair it produces. How many ion pairs would be produced by a 5 MeV alpha particle? If the range of such an alpha particle in air is 65 mm how many ion pairs are produced per millimetre of path?

Ans No of ion pairs produced

$= 5 \times 10^6/30$

$= 1.7 \times 10^5$

No of ion pairs produced per mm

$= 1.7 \times 10^5/65$

$= 2.6 \times 10^3$

7.40 The count rate variation with distance from a small radioactive source is thought to vary as the inverse square of the distance between the end window of a Geiger tube and the source. What type of graph between count rate and distance could be plotted to show an inverse square law?

Ans Plotting the count rate against $1/d^2$ should yield a straight line. There is, however, a problem with this graph in that the distance invariably has a zero error, due to the thickness of the aluminium serving as the Geiger tube window and to the detection position being inside the tube. A better graph to plot is the square root of (1/count rate) against distance. A straight line graph indicates an inverse square law.

Further problems

7.41 A radioactive source is thought to emit beta radiation. What experiment could you do that would establish that the radiation was beta?

7.42 Describe an experiment by which you could determine the range of alpha particles in air.

7.43 How do the cloud chamber tracks of alpha, beta and gamma radiations differ?

7.44 Bismuth 214 emits gamma radiation. This radiation is in the form of several discrete frequencies. The measured wavelength of one of the gamma radiations is 0.016×10^{-10} m. What is the energy of the emitted photon (Planck's constant is 6.6×10^{-34} J s and speed of light is 3.0×10^8 m/s)?

7.45 What would be the radius of curvature of beta particles moving with a velocity of 2.0×10^6 m/s at right angles to a magnetic field of flux density 0.10 Wb/m²? What would be the radius of curvature of alpha particles moving with the same velocity in the same magnetic field (mass of electron is 9.1×10^{-31} kg, mass of alpha particle is 6.7×10^{-27} kg, charge on electron is 1.6×10^{-19} C and charge on alpha particle is $+2 \times 1.6 \times 10^{-19}$ C)?

7.46 What is the evidence for considering alpha particles to be helium nuclei?

The nuclear atom

Notes

The existence of a nucleus, containing the greater part of the mass of an atom, was deduced from the results of the scattering of alpha particles incident on a thin metal foil. Some alpha particles were scattered through angles greater than 90° which could only be explained by the existence of a nucleus.

The alpha scattering experiments also indicated that the nucleus carried a positive charge, the charge being Ze, where Z is the atomic number and e the size of the charge on the electron. The positive charge carriers in the nucleus are called protons. Also in the nucleus are neutrons. These have virtually the same mass as the protons but carry no charge. Atoms thus contain:

(i) Electrons, the number of electrons in the non-ionised atom being equal to the atomic number.
(ii) Protons in the nucleus, the number of protons being equal to the atomic number.
(iii) Neutrons in the nucleus, the number of neutrons plus the number of protons being approximately equal to the atomic mass when reckoned in multiples of the mass of the hydrogen atom (or, more accurately, in terms of a particular isotope of carbon being taken to have the value 12).

Atoms of the same element which have different numbers of neutrons are called isotopes. The unified atomic mass unit (u) is defined as being 1/12th of the mass of the carbon 12 isotope. The atomic mass, on this scale, and the atomic number are written with the chemical symbol of an element as

$$^{\text{atomic mass}}_{\text{atomic number}}\text{Symbol}$$

e.g. $^{35}_{17}\text{Cl}$

The mass spectroscope is used to determine the masses of ions by means of their deflection in magnetic and/or electric fields. In the Bainbridge mass spectrometer (*Figure 7.7*) the ions are

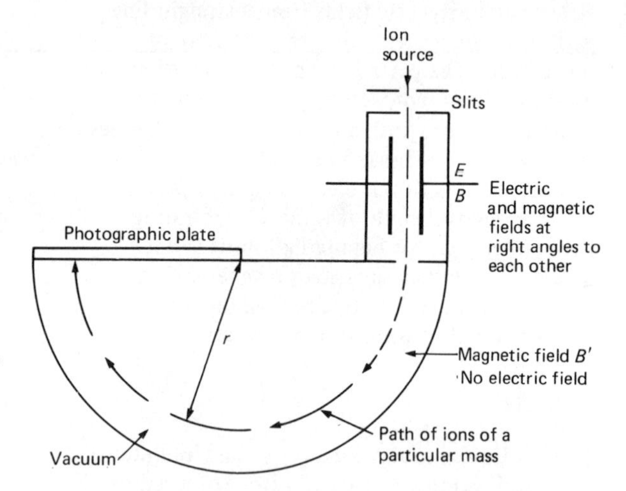

Figure 7.7 The Bainbridge mass spectrometer

passed through a velocity selector so that only ions with the same velocity enter the deflection part of the instrument. The velocity selector consists of an electric and a magnetic field at right angles to each other with only undeviated ions passing on, i.e. when

$$Bqv = Eq$$

or

$$v = E/B$$

In the deflection part of the instrument the ions are deflected into circular arcs by the same magnetic fields

$$Bqv = mv^2/r$$

Thus

$$r = Em/(B^2 q)$$

and m can be determined.

Examples

7.47 What are the number of protons and neutrons in the nucleus of the following atoms: (a) $^{232}_{90}$Th, (b) $^{238}_{92}$U, (c) $^{23}_{11}$Na?

Ans (a) Atomic mass = 232 u, atomic number = 90

Hence

number of protons = 90

Number of protons + number of neutrons = 232

Hence

number of neutrons = 142

(b) Atomic mass = 238 u, atomic number = 92

Hence

number of protons = 92

Number of protons + number of neutrons = 238

Hence

number of neutrons = 146

(c) Atomic mass = 23 u, atomic number = 11

Hence

number of protons = 11

Number of protons + number of neutrons = 23

Hence

number of neutrons = 12

7.48 If a metal foil 4×10^{-7} m thick scatters 1 in 20 000 alpha particles through more than 90°, what is the approximate size of the nucleus? The metal foil atoms can be considered to be about 2.5×10^{-10} m in size and, for simplicity, you can consider them as cubes.

Ans If a layer of atoms is 2.5×10^{-10} m thick then there are $4 \times 10^{-7}/2.5 \times 10^{-10}$ layers of atoms in the foil, i.e. about 16 000 layers. These layers turn back 1 in 20 000 alpha particles. Therefore, just one layer turns back 1 in 20 000 × 16 000, i.e. 1 in 3.2×10^7 particles. The fraction of alpha particles turned back by each layer represents the fraction of the layer that consists of nuclei. Thus, to a reasonable approximation, we can consider that the fraction of the 'target' area of an atom that is nucleus is the same as the fraction of the layer that is nuclei, i.e. $1/(3.2 \times 10^7)$. The 'target' area that is an atom is $(2.5 \times 10^{-10})^2$. Thus, the nuclear size r is given by

$$r^2 = (2.5 \times 10^{-10})^2/(3.2 \times 10^7)$$

and so r is about 4×10^{-14} m.

7.49 What is the distance of closest approach to a nucleus of atomic number 13 (aluminium) for alpha particles having an energy of 5 MeV (charge on the electron is -1.6×10^{-19} C and permittivity of free space ϵ_0 is 8.85×10^{-12} C^2 N^{-1} m^{-2})?

Ans The kinetic energy of the alpha particles is $5 \times 10^6 \times 1.6 \times 10^{-19}$ J. The alpha particles will have their closest approach when all the kinetic energy is converted into potential energy. The potential a distance r from the nucleus is $Q/(4\pi\epsilon_0 r)$. Thus the kinetic energy required to get an alpha particle to a distance r from the nucleus is

$$\text{K.E.} = \frac{13 \times 1.6 \times 10^{-19} \times 2 \times 1.6 \times 10^{-19}}{4\pi 8.85 \times 10^{-12}\, r}$$

Thus, given the above kinetic energy

$$r = 7.5 \times 10^{-15} \text{ m}$$

Further problems

Take the charge on an electron to be -1.6×10^{-19} C.

7.50 How many protons and how many neutrons are present in the nuclei of (a) $^{14}_{6}$C, (b) $^{208}_{82}$Pb, (c) $^{1}_{1}$H?

7.51 Calculate the forces acting on an alpha particle a distance of (a) 10^{-13} m, (b) 10^{-14} m from the centre of a gold nucleus (atomic number of gold is 79, and ϵ_0 is 8.85×10^{-12} $C^2\ N^{-1}\ m^{-2}$).

7.52 What is the closest distance of approach to the centre of a gold nucleus for an alpha particle of energy 5 MeV aimed at gold foil (the atomic number of gold is 79 and ϵ_0 is $8.85 \times 10^{-12}\ C^2\ N^{-1}\ m^{-2}$)?

7.53 What is the radius of the path of $^7_3$Li singly charged ions in a magnetic field of 0.2 Wb/m^2 if the velocity of the ions at right angles to the field is 4×10^5 m/s (the mass of the ion is 1.16×10^{-25} kg)?

7.54 The electric field between the plates of the velocity selector in a Bainbridge mass spectrometer is 1.2×10^5 V/m and the magnetic flux density is 0.50 Wb/m^2. What will be the velocity of the ions entering the deflection part of the spectrometer? Singly charged ions of neon are found to move in a circular path of radius 72.8 mm after passing through the velocity selector and into a magnetic field of flux density 0.50 Wb/m^2. What is the mass of such ions?

Radioactive decay

Notes

The number of alpha particles, beta particles or gamma photons emitted per second by a radioactive element is called the activity of that element. An activity of one bequerel (Bq) is one emission per second. Previously the unit Curie was used, with an activity of 3.7×10^{10} emissions per second equal to 1 Ci (Curie).

The activity of a radioactive element decreased exponentially with time.

$$\text{Activity} = \frac{dN}{dt}$$

i.e. rate at which number of atoms changes.

$$\frac{dN}{dt} = -kN$$

where N is the number of active atoms and k a constant called the decay constant. This integrates to give

$$N = N_0 e^{-kt}$$

where N_0 is the initial number of active atoms at time $t = 0$ and N the number after a time t. Thus

$$\text{Activity} = \frac{dN}{dt}$$

$$= -kN_0 e^{-kt}$$

Both the number of active atoms and the activity decay exponentially with time.

The half life of a radioactive element is the time taken for the activity to decrease by half or the time taken for the number of active atoms to decrease by half. These statements give identical values. Hence

$$\tfrac{1}{2}N_0 = N_0 e^{-kT}$$

where T is the half life. Thus

$$2 = e^{kT}$$

and taking logarithms gives

$$\ln 2 = kT$$

and so

$$T = 0.693/k$$

Radioactive decay is a random matter governed by chance. The constant k can be considered to represent the probability of decay for an atom.

When a radioactive element emits an alpha particle the new element produced has a nucleus with two protons less, i.e. a decrease in atomic number by 2, and a decrease in mass due to the emission of two protons and two neutrons, i.e. a mass 4 u less. When a radioactive element emits a beta particle the new element has an atomic number one higher and a mass not significantly changed. The emission of gamma radiation from an element does not change either the atomic number or atomic mass of that element.

Units: activity - s^{-1} or Bq (1 Bq = 1 s^{-1}), decay constant – s^{-1}, half life - s.

Examples

7.55 The half life of $^{238}_{92}$U is 4.9×10^9 years. (a) What is the decay constant? (b) What is the activity of 1 kg of this isotope (Avogadro's number is 6×10^{23})?

Ans (a) Half life, $T = \dfrac{0.693}{\text{decay constant}}$

Hence

$$\text{Decay constant} = \frac{0.693}{4.9 \times 10^9} = 1.4 \times 10^{-10}\ \text{y}^{-1}$$

(b) Activity $= -kN$

where N is the number of atoms of the isotope in 1 kg. One mole has a mass of 238 g and contains 6×10^{23} atoms. Hence 1 kg contains $6 \times 10^{23}/0.238$ atoms. Hence

$$\text{Activity} = \frac{6 \times 10^{23}}{0.238} \times 1.4 \times 10^{-10}$$
$$= 3.5 \times 10^{14}\ \text{y}^{-1}$$
$$= 1.1 \times 10^{7}\ \text{s}^{-1}$$
$$= 1.1 \times 10^{7}\ \text{Bq}$$

7.56 A sample of $^{32}_{15}\text{P}$ has an activity of 1.6×10^4 Bq. If this isotope has a half life of 14.3 days, how long must elapse before the activity has fallen to 0.4×10^4 Bq?

Ans The activity has to decrease by a factor of 4 to one quarter its initial value. To drop by a half requires 14.3 days. To drop by a further half, i.e. to a quarter, requires another half life. Thus the total time is 28.6 days.

7.57 $^{14}_{6}\text{C}$ has a half life of 5730 years. It is used in radioactive dating. The activity of this isotope in old trees that were felled in a glaciation period was found to be 12½% of that existing in trees felled now. When were the old trees felled by the glaciation?

Ans 12½% is a decrease in activity by a factor of 8. This is achieved after three half lives, i.e. 17 190 years ago.

7.58 The activity of a radioactive isotope was found to be 250 Bq. Two hours later the activity was found to have dropped to 175 Bq. What is the half life of the isotope?

Ans Activity after time $t = -kN_0 e^{-kt}$
Initial activity $= -kN_0$

Thus

Activity after time t = (initial activity) $\times e^{-kt}$

Hence

ln (activity after time t) = ln (initial activity) $- kt$

$$\ln 175 = \ln 250 - 2 \times 3600 \times k$$

Hence

$$k = 0.178\ \text{h}^{-1}$$

$$\text{Half life} = \frac{0.693}{k} = 3.89\ \text{h}$$

7.59 The isotope $^{232}_{90}\text{Th}$ decays by alpha emission. What is the atomic number and mass of the element resulting from this decay?

Ans $^{232}_{90}\text{Th} \rightarrow {}^{4}_{2}\text{He} + {}^{228}_{88}\text{Ra}$

The atomic number is 88 and the mass number 228.

7.60 The isotope $^{232}_{90}\text{Th}$ decays by alpha emission to another element which in turn decays by beta emission to yet another radioactive element. If the total sequence of decays are alpha, beta, beta, alpha, alpha, alpha, alpha, beta, beta, alpha, what is the final element produced and its atomic number and mass?

Ans The atomic number changes are $-2+1+1-2-2-2-2+1+1-2$, i.e. a total atomic number change of -8, to give 82, the element lead. The atomic mass change is $-4-4-4-4-4-4$, i.e. -24, to give 208. The isotope is $^{208}_{82}\text{Pb}$.

Further problems

Take Avogadro's constant to be 6.0×10^{23}.

7.61 The carbon isotope $^{14}_{6}\text{C}$ is radioactive with a half life of 5730 years. What is the activity of 1 g of this isotope?

7.62 What is the activity of 1 g of the isotope $^{226}_{88}Ra$ if it has a half life of 1620 y?

7.63 The isotope iodine 131 has a half life of 8 days. If initially a sample has an activity of 2×10^4 Bq, what will be its activity after (a) 8 days, (b) 16 days and (c) 32 days?

7.64 One gramme of carbon from a living tree is found to have an activity of 720 Bq due to the isotope $^{14}_{6}C$. One gramme of an old piece of wood is found to have an activity of 45 Bq. If the half life of this carbon isotope is 5700 years, what is the age of the old piece of wood?

7.65 A radioactive isotope is found to give an activity of 350 Bq. Thirty minutes later the activity has dropped to 120 Bq. What is the half life of the isotope?

7.66 The isotope strontium 90 has a half life of 28 years. How much mass of this isotope will be needed to give an activity of 5.0×10^4 Bq?

7.67 What are the products of the following radioactive decay processes; give the element, its atomic number and mass: (a) $^{226}_{88}Ra$ decays by alpha emission, (b) $^{14}_{6}C$ decays by beta emission, (c) $^{3}_{1}H$ decays by beta emission and (d) $^{228}_{90}Th$ decays by alpha emission.

The Rutherford–Bohr model of the atom

Notes

The Rutherford model of the atom is that it contains a small central nucleus which contains most of the mass of the atom, i.e. protons and neutrons. Bohr developed this model by proposing:

(i) Electrons can revolve round the nucleus in only certain allowed orbits.
(ii) When an electron is in one of these orbits it does not emit any radiation.
(iii) An electron in an orbit has a definite amount of energy. This means that as only certain orbits are allowed the electrons in an atom can only have certain allowed energies.
(iv) An electron can jump from one orbit to another. When it jumps to a lower energy orbit the energy E given up by the electron is emitted as a photon of frequency f where $E = hf$.

Figure 7.8 shows the Bohr model for hydrogen, it having just one electron, and the way in which electron jumps between orbits gives rise to the spectral lines of hydrogen. The different groups of

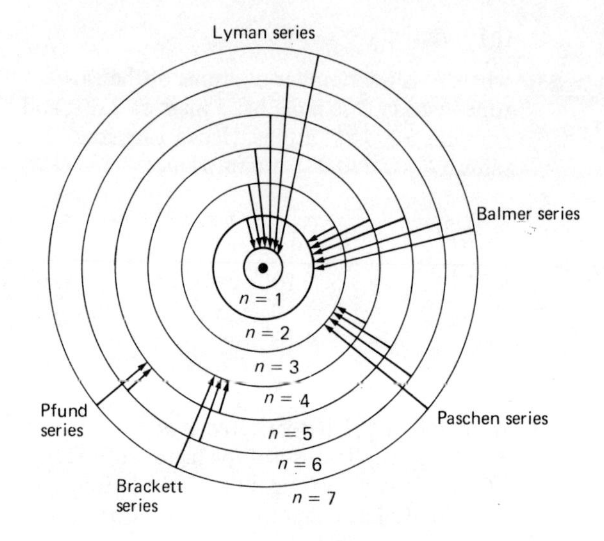

Figure 7.8 The Bohr model of the hydrogen atom

lines are known as the Lyman series, Balmer series, Paschen series, etc. All these series had been found to fit a single equation known as the Rydberg equation.

$$f = cR\left(\frac{1}{m^2} - \frac{1}{n^2}\right)$$

where c is the speed of light, R a constant and m and n are integers. The integer m can have the value 1,2,3, etc., but n is an integer that must be greater than m. When $m = 1$ the Lyman series is produced, $m = 2$ the Balmer series, $m = 3$ the Paschen series, etc. As the energy E of a photon is given by

$$E = hf$$

then

$$E = hcR\left(\frac{1}{m^2} - \frac{1}{n^2}\right)$$

Bohr postulated that this energy arose from a transition between two orbits of energy E_n and E_m, thus

$$E_n - E_m = hcR\left(\frac{1}{m^2} - \frac{1}{n^2}\right)$$
$$= \left(-\frac{hcR}{m^2}\right) - \left(-\frac{hcR}{n^2}\right)$$

Each orbit, i.e. energy level, has thus an energy of $-(hcR/n^2)$ associated with it. *Figure 7.9* shows the resulting energy level picture for hydrogen. The energy of an electron in an atom is defined as being zero when it is just free of the influence of the nucleus, i.e. the atom has become ionised.

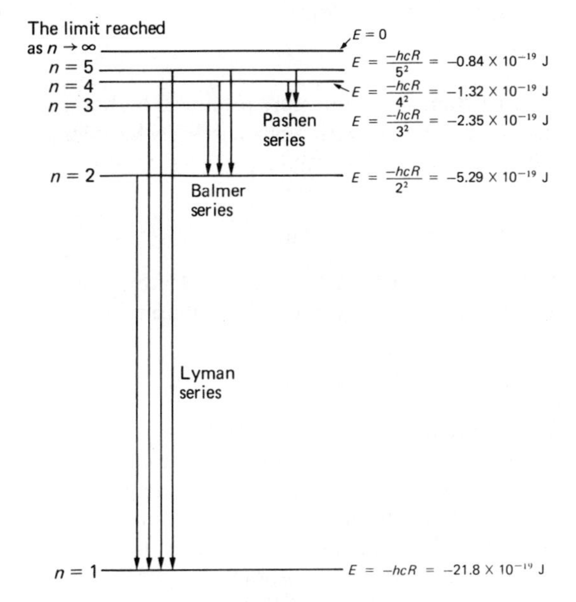

Figure 7.9 The energy levels of hydrogen

The neutral hydrogen atom has just one electron and this is considered to be in the $n = 1$ orbital shell when the atom is not excited. Helium has two electrons and these are both considered to be in the $n = 1$ shell. The $n = 1$ shell is considered to be full when two electrons are in it. The $n = 2$ shell can hold 8 electrons. An atom which has a full shell is considered to be particularly chemically stable, an atom with just one electron in a shell to be particularly reactive. This shell model can give a simple explanation of the Periodic Table.

Examples

7.68 According to the Bohr model, what is the energy of the photon released when an electron jumps from the $n = 3$ level to the $n = 2$ level (use *Figure 7.9*)?

Ans The $n = 3$ level has an energy of -2.35×10^{-19} J. The $n = 2$ level has an energy of -5.29×10^{-19} J. Thus, when the electron jumps from the $n = 3$ level to the $n = 2$ level

$$\text{Energy released} = -2.35 \times 10^{-19} - (-5.29 \times 10^{-19})$$
$$= 2.94 \times 10^{-19} \text{ J}$$

Further problems

7.69 Explain how the Bohr model of the atom explains (a) line spectra, (b) the periodic variation of chemical properties with atomic number.

7.70 What energy is needed to raise the electron in an hydrogen atom from its ground state to the $n = 5$ level (use *Figure 7.9*)?

7.71 What energy photons can be released when a hydrogen atom electron falls from the $n = 3$ level to the $n = 1$ level (use *Figure 7.9*)?

7.72 How does the Bohr atom model explain the origin of the different series of spectrum lines in the hydrogen spectrum?

A wave model for atoms

Notes

A simple one-dimentional introduction to wave mechanics is to consider the motion of a particle trapped between two walls as a standing wave, the wavelength of the particle being given by

$$\lambda = h/mv$$

Only certain wavelengths can 'fit' between the walls and thus only certain particle energies are possible. An atom can be considered as having walls a distance apart equal to the size of the atom and only certain electron waves will fit between the walls and thus only certain energies are possible, i.e. the atom has energy levels.

Examples

7.73 What are the two longest wavelengths an electron can have, and the corresponding energies, if it has to fit in a one-dimensional box having the size of an atom, i.e. 2×10^{-10} m (Planck's constant is 6.6×10^{-34} J s and the mass of the electron is 9.1×10^{-31} kg)?

Ans If L is the size of the box then

$$L = \tfrac{1}{2}\lambda_1 \quad \text{and} \quad L = \lambda_2$$

for standing waves to be produced. Thus

$$\lambda_1 = 4 \times 10^{-10} \text{ m}$$

and

$$\lambda_2 = 2 \times 10^{-10} \text{ m}$$

As

$$E = \tfrac{1}{2}mv^2 \quad \text{and} \quad \lambda = h/mv$$

then

$$E = h^2/(2m\lambda^2)$$

and so

$$E_1 = 1.5 \times 10^{-18} \text{ J}$$

and

$$E_2 = 6.0 \times 10^{-18} \text{ J}$$

This assumes that the only energy the electron has is kinetic energy and that it possesses no potential energy.

Further problems

7.74 Estimate the smallest energy an electron can have and also 'fit' in the nucleus of an atom. The nucleus can be considered to have a diameter of 1.5×10^{-14} m. Assume that the only energy the electron has is kinetic energy.

The electron can be considered to be in a box with walls equal in 'size' to the potential energy that would occur a distance equal to the nuclear radius from a point charge equal to the charge carried by the nucleus. If the kinetic energy of the electron is greater than the potential energy then the electron will escape. Estimate the size of the nuclear wall for the hydrogen atom and consider whether the electron could be contained within the nucleus (Planck's constant is 6.6×10^{-34} J s, the mass of electron is 9.1×10^{-31} kg, the charge on electron is -1.6×10^{-19} C and ϵ_0 is 8.85×10^{-12} F/m).

Transmutation

Notes

The first artificial transmutation was produced by E. Rutherford in 1919 when he bombarded nitrogen with alpha particles.

$$^{14}_{7}\text{N} + {}^{4}_{2}\text{He} \rightarrow {}^{17}_{8}\text{O} + {}^{1}_{1}\text{H}$$

Further transmutations were produced by bombarding other elements with alpha particles. Beryllium, when bombarded with alpha particles, transmutes to give carbon and neutrons.

$$^{9}_{4}\text{Be} + {}^{4}_{2}\text{He} \rightarrow {}^{12}_{6}\text{C} + {}^{1}_{0}\text{n}$$

In 1932 J.D. Cockcroft and E.T.S. Walton used artificially accelerated protons to transmute lithium.

$$^{1}_{1}\text{H} + {}^{7}_{3}\text{Li} \rightarrow {}^{4}_{2}\text{He} + {}^{4}_{2}\text{He}$$

They accelerated the protons to the high energies needed by using a voltage multiplier circuit to produce the large potential difference required.

Another means of producing high potential differences for accelerating particles is the van de Graaff electrostatic generator (*Figure 7.10*). Charge is sprayed onto a moving belt which carries the

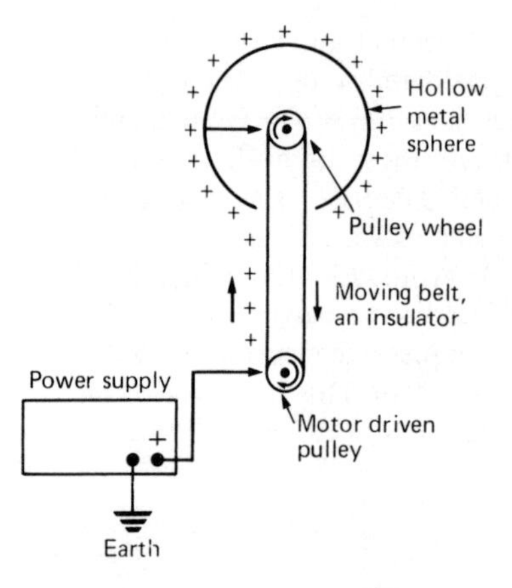

Figure 7.10 The van de Graaff generator

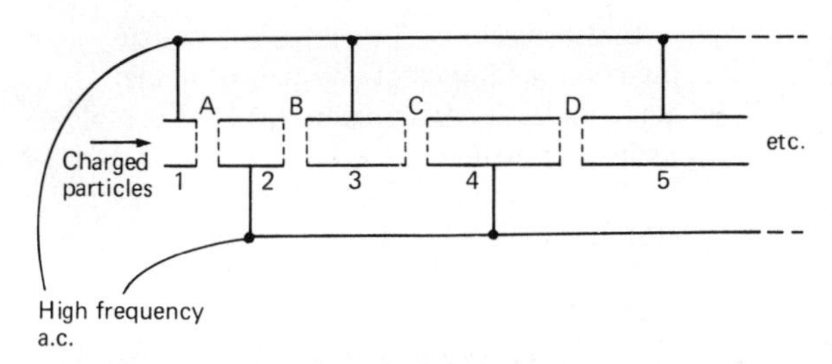

Figure 7.11 The linear accelerator

charge up to a sphere where the charge is removed. The charge on the sphere, a capacitor, thus builds up and hence its potential builds up. Another method is the linear accelerator (*Figure 7.11*). The charged particles pass through a series of electrodes. As the charged particles reach the gap between a pair of successive electrodes the potential difference between the electrodes is so arranged as to accelerate the charged particles. This process is repeated a number of times and the energy of the charged particles is built up. In the cyclotron (*Figure 7.12*) the same principle operates

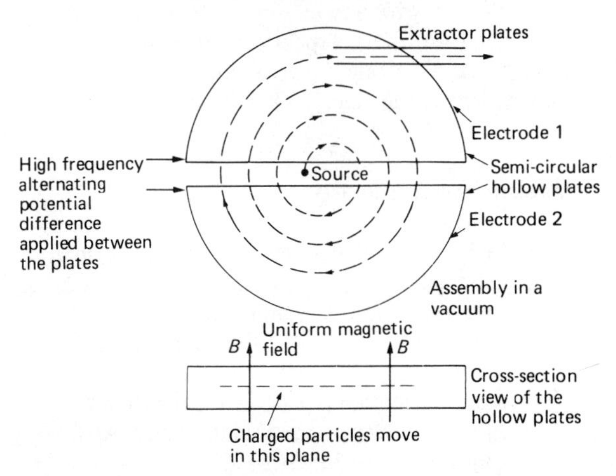

Figure 7.12 The cyclotron

with, however, the charged particles travelling in circular paths due to the presence of a magnetic field.

When the mass of the nucleus of an atom is compared with the sum of the masses of the protons and neutrons in that nucleus they are never found to be equal; the mass of the nucleus is always less than the sum of its constituent particles. This mass loss is called the mass defect. This represents the energy released in forming the nucleus from its constituent particles, being called the binding energy when the quantity of mass is converted into energy by means of the relationship

$$E = mc^2$$

Energy of 931 MeV is equivalent to a mass loss of 1 u.

When $^{235}_{92}U$ is bombarded with neutrons a reaction can occur in which the uranium nucleus splits into two pieces. Thus reaction is known as nuclear fission. When the uranium nucleus splits energy is released in addition to neutrons. This leads to the possibility of a chain reaction with these released neutrons causing further fission reactions.

When two light nuclei combine there is a release of energy due to the mass of the products of the reaction being less than the mass of the reactants. This type of reaction is called fusion.

Examples

7.75 What are the missing components in the following reactions?

(a) $^{10}_{5}B + ^{4}_{2}He \rightarrow ^{1}_{0}n + \ ?$

(b) $^{235}_{92}U + ^{1}_{0}n \rightarrow ^{107}_{43}Tc + 5^{1}_{0}n + \ ?$

(c) $^{118}_{50}Sn + ^{4}_{2}He \rightarrow ^{1}_{0}n + \ ?$

(d) $^{1}_{1}H + ^{127}_{53}I \rightarrow ^{50}_{21}Sc + \ ?$

Ans (a) The atomic mass must be

$$(10 + 4) - 1 = 13$$

The atomic number must be

$$(5 + 2) - 0 = 7$$

The isotope is thus $^{13}_{7}N$.

(b) The atomic mass must be

$$(235 + 1) - (107 + 5) = 124$$

The atomic number must be

$$92 - 43 = 49$$

The isotope is thus $^{124}_{49}In$.

(c) The atomic mass must be

$$(118 + 4) - 1 = 121$$

The atomic number must be 52. Hence, the isotope is $^{121}_{52}Te$.

(d) The atomic mass must be

$$(127 + 1) - 50 = 78$$

The atomic number must be

$$(1 + 53) - 21 = 33$$

Hence the isotope is ${}^{78}_{33}As$.

7.76 What is (a) the mass defect and (b) the binding energy per nucleon for helium ${}^{4}_{2}He$ (mass of a proton is 1.007 28 u, the mass of a neutron is 1.008 87 and the mass of a helium nucleus is 4.0026 u)?

Ans (a) Two protons plus two neutrons gives a total mass of 4.0323 u. This is a mass in excess of that of the helium nucleus by

$$4.0323 - 4.0026 = 0.0297 \text{ u}$$

This is the mass defect.

(b) The mass defect is equivalent to an energy of

$$0.0297 \times 931 = 27.65 \text{ MeV}$$

There are four nucleons, i.e. particles in the nucleus, and so

$$\text{binding energy per nucleon} = 27.65/4 = 6.913 \text{ MeV}$$

Further problems

7.77 What are the missing components in the following reactions?

(a) ${}^{1}_{0}n + {}^{235}_{92}U \rightarrow {}^{141}_{56}Ba + 3{}^{1}_{0}n + \ ?$

(b) ${}^{2}_{1}H + {}^{3}_{1}H \rightarrow {}^{1}_{0}n + \ ?$

(c) ${}^{27}_{13}Al + {}^{4}_{2}He \rightarrow {}^{1}_{0}n + \ ?$

7.78 Derive an equation relating the time taken for a charged particle to complete a semicircle of radius r in a cyclotron, and show that it is independent of the radius of the path. Hence, explain the action of the cyclotron.

7.79 Calculate the mass defects and binding energies per nucleon for the following: (a) ${}^{12}_{6}C$, nuclear mass 12.0000 u, (b) ${}^{56}_{26}Fe$, nuclear mass 55.9349 u, (c) ${}^{238}_{92}U$, nuclear mass 238.0508 u (mass of a proton is 1.007 28 u, the mass of a neutron is 1.008 87 u and 1 u is 931 MeV).

7.80 A nucleus is stable if its mass is less than the combined mass of any pair of nuclei made up by its subdivision. (a) Is ${}^{7}_{3}Li$ stable against a possible break up into ${}^{4}_{2}He$ and ${}^{3}_{1}H$? (b) Is ${}^{5}_{2}He$ stable against a possible break up into ${}^{4}_{2}He$ and ${}^{1}_{0}n$? The atomic mass of ${}^{4}_{2}He$ is 4.002 604 u, of ${}^{3}_{1}H$ is 3.016 049 u, of ${}^{7}_{3}Li$ is 7.016 005 u, of ${}^{1}_{0}n$ is 1.008 665 u and of ${}^{5}_{2}He$ is 5.012 296 u.

7.81 In the nuclear fission of uranium 235, barium 141 and krypton 92 can be formed. What is the energy release in the fission reaction (binding energy per nucleon: barium 141, 8.4 MeV; krypton 92, 8.7 MeV and uranium 235, 7.6 MeV)?

7.82 The mass of the neutron is 1.008 665 u and that of the proton 1.007 825 u. (a) Suppose a neutron decayed to a proton with the emission of an electron. What would be the kinetic energy of the electron? (b) In the beta decay of nuclei the electrons are emitted with any energy up to a maximum, i.e. a continuous spectrum with an upper limit. What might you deduce from this (mass of the electron is 5.48×10^{-4} u)?

Examination questions

7.83 Light of wavelength 0.50 μm incident on a metal surface ejects electrons with kinetic energies up to a maximum value of 2.0×10^{-19} J. What is the energy required to remove an electron from the metal? If a beam of light causes no electrons to be emitted, however great its intensity, what condition must be satisfied by its wavelength (Planck's constant is 6.6×10^{-34} J s and the speed of light is 3.0×10^{8} m/s).

(Southern Universities)

7.84 (a) Explain what is meant by the work function of a metal, ϕ. (b) Draw a diagram of the electrical circuit you would use to find the 'stopping potential' of photoelectrons emitted from a metal surface in a vacuum. Describe how you would determine the value of ϕ and h/e, where h is Planck's constant and e the electronic charge.

(c) The work function of caesium is 1.90 electron volts, equal to 3.04×10^{-19} J. Calculate:

(i) the longest wavelength of light that can cause the emission of photoelectrons from a caesium surface;
(ii) the greatest speed with which electrons are emitted for incident light of wavelength 550 nm

It has been suggested that photocells of some kind might eventually be developed as batteries to make use of solar energy. Supposing that the Sun's radiation is all of a single wavelength, 550 nm:

(iii) estimate the e.m.f. of a caesium cell in sunlight.

With the same oversimplification, the power per unit area received at this wavelength from the Sun can be taken to be about 250 W/m^2.

(iv) Estimate the current per unit area from the caesium surface.
(v) State two difficulties in the way of constructing a practical source of electrical power from such cells.

Take the value of Planck's constant h to be 6.63×10^{-34} J s, that of the electronic charge e to be -1.60×10^{-19} C, that of the electronic mass m_e to be 9.11×10^{-31} kg, and that of the speed of light c to be 3.00×10^8 m/s.

(Oxford Local Examinations)

7.85 Describe the principle of the experiment which established the nuclear model of the atom, explaining how the deduction is made from the observations.

The emission spectrum of the hydrogen atom consists of a series of lines. Explain why this suggests the existence of definite energy levels for the electron in the atom.

By considering the intervals between the energy levels explain the spacing of the lines in the visible hydrogen spectrum.

The ionisation potential of the hydrogen atom is 13.6 V. Use the data below to calculate:
(a) the speed of an electron which could just ionise the hydrogen atom, and (b) the minimum wavelength which the hydrogen atoms can emit.
Charge on an electron = -1.60×10^{-19} C.
Mass of an electron = 9.11×10^{-31} kg.
The Planck constant = 6.63×10^{-34} J s.
Speed of light = 3.00×10^8 m/s.

(University of London)

7.86 (a) What are the chief characteristics of a line spectrum? Explain briefly how line spectra are used:

(i) in analysis for the identification of elements present;
(ii) in astronomy for estimating the component in the line of sight of the velocity of a star relative to the Earth.

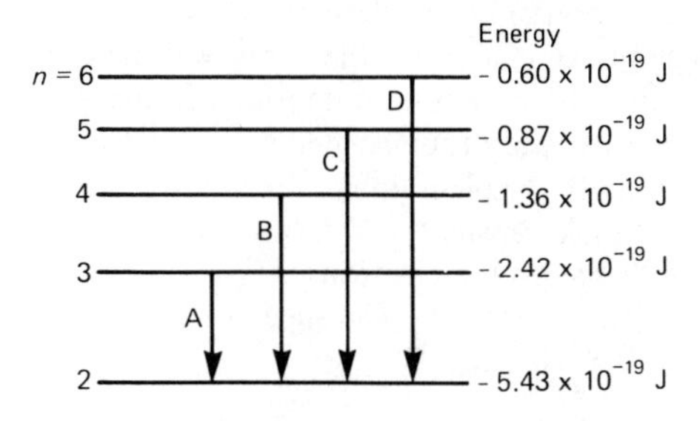

Figure 7.13

(b) *Figure 7.13*, representing the lowest energy levels of the electron in the hydrogen atom, gives the principal quantum number n associated with each, and the corresponding value of the energy, measured in joules.

(i) Calculate the wavelength of the lines arising from the transitions marked A,B,C,D on *Figure 7.13.*
(ii) Show that the other transitions that can occur give rise to lines which are in either the ultraviolet or the infrared regions of the spectrum.

(iii) The level $n = 1$ is the 'ground state' of the unexcited hydrogen atom. Explain why hydrogen in its ground state is quite transparent to light emitted by the transitions A,B,C,D, and also what happens when 21.7×10^{-19} J of energy is supplied to a hydrogen atom in its ground state.

Take the value of the speed of light in vacuum, c, to be 3.00×10^8 m/s, and that of Planck's constant, h, to be 6.63×10^{-34} J s.

(Oxford Local Examinations)

7.87 Draw a diagram, and explain the action, of an X-ray tube.

Explain the presence in X-ray spectra of (a) characteristic wavelengths, (b) a continuous background.

The structure of a particular solid in which the spacing of the atomic planes is 0.30 nm is investigated by X-rays. First order diffraction is observed from these planes at a Bragg angle of 30°. Calculate the X-ray wavelength and find the minimum potential difference across the X-ray tube needed to excite this wavelength. Explain your calculations clearly. The Planck constant, $h = 6.6 \times 10^{-34}$ J s. The charge on the electron, $e = -1.6 \times 10^{-19}$ C. The speed of light $c = 3.0 \times 10^8$ m/s.

(University of Cambridge)

7.88 (a) Explain what is meant by the duality of electrons, and state the relation between the momentum and the wavelength for a mono-energetic beam of electrons.
(b) Electrons are accelerated from rest through a potential difference V. Derive an expression for the wavelength of the electrons.

(JMB)

7.89 What are β and γ rays? Describe the structure and explain the action of a Geiger–Müller tube.

A source emitting both β and γ radiation was placed a fixed distance in front of a Geiger counter. Absorbers of various thicknesses x were placed between the source and the counter and the following readings of counts per second R were obtained:

R/s^{-1}	500	170	125	100	82	70	60
x/mm	0.1	0.3	0.4	0.5	0.6	0.7	0.8

R/s^{-1}	56	53	50	47	45	40
x/mm	0.9	1.0	1.2	1.4	1.6	2.0

Plot a graph of $\lg(R/\text{s}^{-1})$ against x/mm and discuss its shape.

Deduce how many β's were counted each second when the absorber was 0.5 mm thick. State clearly how your result is obtained from the graph.

(University of Cambridge)

7.90 The radioactive isotope $^{218}_{84}\text{Po}$ has a half life of 3 min, emitting α particles according to the equation

$$^{218}_{84}\text{Po} \rightarrow \alpha + ^{x}_{y}\text{Pb}$$

What are the values of x and y?

If N atoms of $^{218}_{84}\text{Po}$ emit α particles at the rate of 5.12×10^4 s^{-1}, what will be the rate of emission after ½ hour?

(Southern Universities)

7.91 (a) A radioactive isotope of strontium, of half-life 28 years, providing a source of beta particles, has been in use for 14 years. If originally 5.0 μg of the strontium isotope were present show graphically, or otherwise, that the amount of this isotope has been reduced to approximately 3.5 μg.

Beta particles emitted from such a source are found to have a *continuous energy spectrum* with an energy maximum that is *characteristic of the source.* Explain what is meant by the italicised phrases.
(b) Describe an experiment to determine the range of beta particles in aluminium and show how you would present your results graphically. Show how you would obtain the range from your graph.

In such an experiment it was deduced from the range that the maximum energy of the beta particles emitted from a strontium source was 2.3 MeV. Would the age of the source have made any difference to this value? Give a reason for your answer.

(University of London)

7.92 (a) What do you understand by half life $T_{1/2}$, and decay constant λ, for a radioactive substance? Deduce the relationship between them.
(b) (i) The first part of the decay series of the artificially produced neptunium ${}^{237}_{93}Np$ involves the following sequence of emissions: $\alpha, \beta, \alpha, \alpha, \beta, \alpha$. Illustrate these changes on a plot of N, the number of neutrons in the nucleus, against Z, the atomic number. Using the Table of elements below, identify (by its symbol) the last element in this portion of the decay series, and make clear which isotope of this element is produced.

Element	Bi	Po	At	Rn	Fr	Ra	Ac	Th	Pa	U
Atomic number	83	84	85	86	87	88	89	90	91	92

(ii) Three other nuclides also initiate decay series. The half lives and abundances are as shown below.

Element	*Half life*/year	*Natural abundance*
${}^{232}Th$	1.4×10^{10}	abundant
${}^{235}U$	7.1×10^{8}	rather rare
${}^{237}Np$	2.2×10^{6}	not found
${}^{238}U$	4.5×10^{9}	abundant

Comment on the age of the Earth in the light of these data.

(University of Cambridge)

Answers

7.3 Ans The wavelengths can be measured and the pattern of wavelengths compared with tables of wavelengths for the elements. An alternative is to match the line spectrum, on a photograph, with standard wavelength photographs

7.4 Ans A prism or a diffraction grating will disperse light

7.5 Ans Line spectra originate from elements, atoms or ions independent of each other. Band spectra originate from molecules

7.10 Ans Maximum kinetic energy $= hf - W$

7.11 Ans $hf = W$, hence $f = W/h = 6.5 \times 10^{14}$ Hz

7.12 Ans Threshold is when $hc/\lambda = W$, i.e. $\lambda = 1.99 \times 10^{-7}$ m. All wavelengths longer than this will cause no emission of electrons, hence the answer to all parts of this question is zero kinetic energy

7.13 Ans $Ve = \frac{1}{2}mv^2 = hf - W$; hence $\lambda = v/f = 3 \times 10^8 \times 6.6 \times 10^{-34}/(2.0 + 2.3) \times 1.6 \times 10^{-19} = 2.88 \times 10^{-7}$m. Note 1 eV $= 1.6 \times 10^{-19}$ J

7.14 Ans $hv/\lambda = W$, hence $\lambda = 4.8 \times 10^{-7}$ m. Note 1 eV $= 1.6 \times 10^{-19}$ J

7.15 Ans $20/(hv/\lambda) = 5.96 \times 10^{19}$ per s

7.16 Ans $10/(hv/\lambda) = 2.98 \times 10^{19}$ per s

7.19 Ans $E = hf = hv/\lambda$; hence wavelengths are 1.21×10^{-7} m, 1.02×10^{-7} m and 6.51×10^{-7} m

7.20 Ans When an atom of mercury is bombarded by, say, an electron having an energy of 10.4 eV then an electron is removed from the mercury atom. The ionisation energy quoted is the energy needed to remove the first electron from the mercury atom

7.21 Ans $\frac{1}{2}mv^2 = E = 13.6 \times 1.6 \times 10^{-19}$ J; hence $v = 2.19 \times 10^6$ m/s

7.22 Ans $E = hv/\lambda$, hence $\lambda = 8.97 \times 10^{-7}$ m

7.23 Ans Ten, but not all of these transitions may, in practice, occur

7.26 Ans $2d \sin\theta = n\lambda$, hence for $n = 1$, $\lambda = 2.4 \times 10^{-8}$ m; for second order $n = 2$ and $\theta = 59°$

7.27 Ans $2d \sin \theta = n\lambda$, hence $d = 3.8 \times 10^{-10}$ m

7.28 Ans $2d \sin \theta = n\lambda$, hence $\lambda = 1.6 \times 10^{-8}$ m

7.29 Ans $eV = hf$, hence $f = 6.1 \times 10^{18}$ Hz

7.30 Ans $eV = hv/\lambda$, hence $\lambda = 3.1 \times 10^{-11}$ m

7.31 Ans See the notes

7.34 Ans $\lambda = h/mv$ and $\frac{1}{2}mv^2 = Ve$; hence $\lambda = 1.1 \times 10^{-10}$ m

7.35 Ans $\lambda = h/mv$ and $\frac{1}{2}mv^2 = Ve$; hence $\lambda = 3.0 \times 10^{-13}$ m

7.36 Ans $\lambda = h/mv = 7.9 \times 10^{-14}$ m

7.37 Ans $E = \frac{1}{2}mv^2 = \frac{1}{2}h^2/(\lambda^2 m)$; $E = hv/\lambda$; energy of electron $= 1.5 \times 10^{-18}$ J, energy of photon $= 5.0 \times 10^{-16}$ J

7.41 Ans An absorption experiment using thin sheets of aluminium foil is probably the simplest method. A better experiment would involve deflection of a beam of the radiation in electric or magnetic fields

7.42 Ans A cloud chamber could be used or an ionisation chamber

7.43 Ans Alpha radiation gives solid straight line tracks, beta radiation gives irregular not so dense tracks and gamma radiation gives small irregular tracks of the electrons knocked out of atoms by the gamma radiation

7.44 Ans $E = hv/\lambda = 1.2 \times 10^{-13}$ J

7.45 Ans $Bev = mv^2/r$, hence $r = 1.1 \times 10^{-4}$ m, for alpha particles $r = 0.42$ m

7.46 Ans Rutherford and Royd trapped alpha particles in a glass tube and found they gave the helium spectrum. Charge to mass values are those of helium nuclei

7.50 Ans (a) 6 protons, 8 neutrons
(b) 82 protons, 126 neutrons
(c) 1 proton, no neutrons

7.51 Ans (a) $F = Q_1Q_2/(4\pi\epsilon_0 r^2)$; hence force $= 3.6$ N
(b) Force = 360 N

7.52 Ans See **7.49**. Closest distance $= 4.4 \times 10^{-14}$ m

7.53 Ans $Bqv = mv^2/r$, hence $r = 1.45$ m

7.54 Ans $Eq = Bqv$, hence $v = E/B = 2.4 \times 10^5$ m/s. $Bqv = mv^2/r$, hence $m = 2.4 \times 10^{-26}$ kg

7.61 Ans $k = 0.693/T = 1.209 \times 10^{-4}$ y^{-1}; $N = 6.0 \times 10^{23}/14$, hence activity $= -kN = 5.18 \times 10^{18}$ y^{-1} $= 1.64 \times 10^{11}$ Bq

7.62 Ans $k = 0.693/T = 4.28 \times 10^{-4}$ y^{-1}; $N = 6.0 \times 10^{23}/226$, hence activity $= -kN = 1.14 \times 10^{18}$ y^{-1} $= 3.60 \times 10^{10}$ Bq

7.63 Ans (a) Activity $= 1 \times 10^4$ Bq
(b) Activity $= 0.5 \times 10^4$ Bq
(c) Activity $= 0.125 \times 10^4$ Bq

7.64 Ans This is 4 half lives, i.e. 22 800 years

7.65 Ans See **7.58**. ln(activity after time t) = ln(initial activity) $- kt$; hence $\ln 120 = \ln 350 - kt$, thus $k = 5.95 \times 10^{-4}$ s^{-1} and so $T = 1165$ s

7.66 Ans Activity $= -kN$, where $k = 0.693/T$ $= 0.0248$ y^{-1}; hence $N = 6.36 \times 10^{13}$ and so, as 90 g contains 6×10^{23} atoms, the mass is 9.54×10^{-9} g

7.67 Ans (a) $^{222}_{86}$Em
(b) $^{14}_{7}$N
(c) $^{3}_{2}$He
(d) $^{224}_{88}$Ra

7.69 Ans See the notes

7.70 Ans $-0.84 \times 10^{-19} - (-21.8 \times 10^{-19})$ $= 20.96 \times 10^{-19}$J

7.71 Ans $n = 3$ to $n = 2$ gives 2.94×10^{-19} J;
$n = 3$ to $n = 1$ gives 19.45×10^{-19} J;
$n = 2$ to $n = 1$ gives 16.51×10^{-19} J

7.72 Ans By electron jumps to energy levels with different values of *n*.

7.74 Ans $\lambda = 3.0 \times 10^{-14}$ m, hence $E = h^2/(2m\lambda^2) = 2.7 \times 10^{-10}$ J. The potential *V* a distance *r* from the centre is $Q/(4\pi\epsilon_0 r)$. Hence the potential energy is $VQ = Q^2/(4\pi\epsilon_0 r) =$ 3.1×10^{-14} J. The kinetic energy of the electron is much greater than the potential energy, so the electron cannot be contained within the nucleus

7.77 Ans (a) $^{92}_{36}$Kr
(b) $^{4}_{2}$He
(c) $^{30}_{15}$P

7.78 Ans $Bqv = mv^2/r$, $t = \pi r/v$ and so $t = \pi m/Bq$. The charged particles are accelerated every time they pass from one dee (hollow semicircular electrode) to the other dee, as the result of a constant frequency alternating potential difference applied between the dees such that $f = 1/2t$

7.79 Ans (a) 0.0969 u, 7.5178 MeV
(b) 0.5205 u, 8.6530 MeV
(c) 1.9140 u, 7.4870 MeV

7.80 Ans (a) Stable
(b) Not stable

7.81 Ans $8.4 \times 141 + 8.7 \times 92 - 235 \times 7.6$ $= 199$ MeV. This assumes that none of the mass emerges as neutrons

7.82 Ans (a) 1.3 MeV
(b) Two particles were emitted, not just the electron. The other particle takes some of the energy and is called the neutrino

7.83 Ans Energy to remove electron $= hv/\lambda$; maximum kinetic energy $= 1.96 \times 10^{-19}$ J. For no emission, hv/λ is equal to or less than 1.96×10^{-19} J

7.84 Ans (a) Work function is the energy needed to bring an electron to the metal surface. Only when the incident light has energy greater than the work function can emission occur

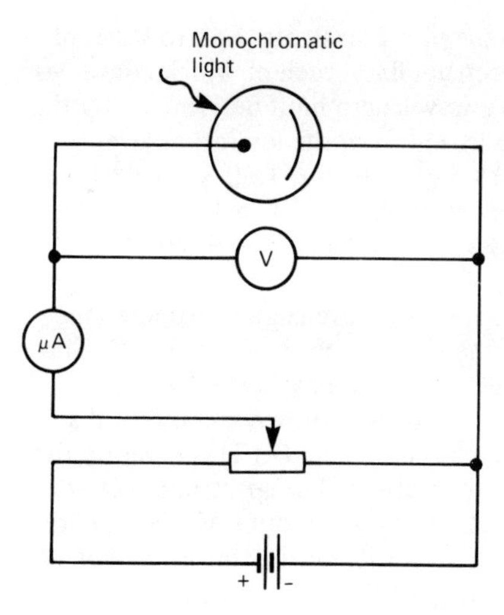

Figure 7.14

(b) See *Figure 7.14*. The voltmeter reading is taken when the current is zero. To obtain this, readings can be taken of current against potential difference and extrapolated to zero current. As $Ve = hf - \phi$ a graph of *V* against *f* has a slope of h/e and an intercept of ϕ/e. The experiment thus needs to be repeated for a number of different frequencies.
(c) (i) $hv/\lambda = \phi$, hence $\lambda = 6.51 \times 10^{-7}$ m. (ii) $\frac{1}{2}mv^2 = hv/\lambda - \phi$, hence $v = 3.51 \times 10^5$ m/s. (iii) The e.m.f. can be considered to be opposite and equal to the potential difference that would have to be applied across the cell to give zero current, i.e. $Ve = hv/\lambda - \phi$, and thus e.m.f. is 0.35 V. (iv) Number of photons hitting 1 m^2/s = $250/hf$. If each photon ejects an electron then charge emerging from 1 m^2/s is $(250/hf)e$. This is a current of 111 A/m^2 and assumes an efficiency of 100%. (v) Sunlight is not always at this 250 W/m^{-2} value. The photocell has a high internal resistance.

7.85 Ans See the notes for the alpha scattering experiment and the discussion of energy levels and spectrum lines. The further the levels from the ground state the closer together they become (see *Figure 7.9*). The interval between successive energy levels thus becomes smaller and smaller. Transitions to the ground state, to the $n = 1$ level,

to the $n = 2$ level, etc. lead to series of spectrum lines, each of which converges to a wavelength limit determined by the transition from the ionisation level.
(a) $\frac{1}{2}mv^2 = 13.6 \times 1.60 \times 10^{-19}$ J, hence $v = 2.19 \times 10^6$ m/s
(b) $E = hv/\lambda$, hence $\lambda = 9.099 \times 10^{-8}$ m

7.86 Ans (a) Discrete wavelengths characteristic of the element producing them. (i) The spectrum of an unknown element is compared with known spectra until a match is obtained. (ii) This relies on the Doppler effect. The spectrum lines are displaced by an amount which depends on the velocity of the star in the line of sight.
(b) (i) $E = hv/\lambda$, thus for A where $E = 3.01 \times 10^{-19}$ J, $\lambda = 6.61 \times 10^{-7}$ m. For B, $\lambda = 4.89 \times 10^{-7}$ m; for C, $\lambda = 4.36 \times 10^{-7}$ m and for D, $\lambda = 4.12 \times 10^{-7}$ m. (ii) The ultraviolet has wavelengths longer than about 3×10^{-7} m. For energy level transitions to give rise to wavelengths at wavelengths greater than this then the energy interval between levels must be less than $hv/\lambda = 6.6 \times 10^{-19}$ m. The infrared has wavelengths longer than about 7×10^{-7} m. The energy interval between levels must be greater than $hv/\lambda = 2.8 \times 10^{-19}$ m for wavelengths less than this wavelength. Thus, for emission in the visible region the energy interval must be between 2.8×10^{-19} J and 6.6×10^{-19} J. All the other transitions possible, with the energy levels given in *Figure 7.13*, are outside these limits. (iii) For a transition to the $n = 2$ level the energy supplied by the light must be at least 16.27×10^{-19} J. None of the photons arising from the A,B,C,D transitions have these energies. The result is that no light is 'absorbed' by the atoms. When 21.7×10^{-19} J is supplied the atom is ionised and loses its electron

7.87 Ans See *Figure 7.2.* (a) Characteristic wavelengths arise from transitions between energy levels in the atom
(b) The continuous background arises from the deceleration of the electrons on hitting the target. $2d \sin\theta = n\lambda$, hence for $n = 1$, $\lambda = 0.30$ nm; $hv/\lambda = Ve$, hence $V = 4.1 \times 10^3$ V

7.88 Ans (a) Electrons have both a wave and a particle characteristic, $\lambda = h/(mv)$
(b) $Ve = \frac{1}{2}mv^2$, hence $\lambda = h/(2Vem)^{1/2}$

7.89 Ans See the notes for the rays and the Geiger–Müller tube. For the absorption experiment, if there was a pure gamma

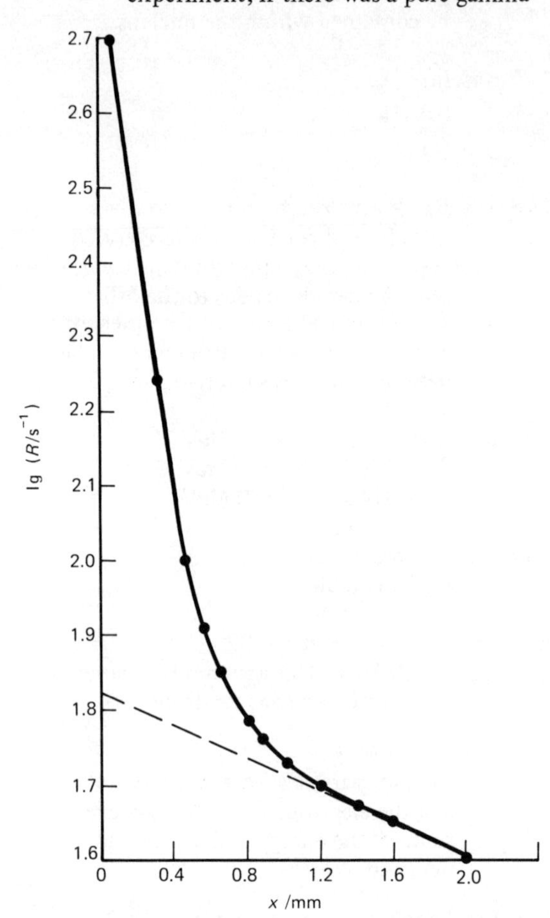

Figure 7.15

source the count rate R would have given a straight line graph of $\lg(R/s^{-1})$ against x/mm because the relationship between R and x is of the form $R = R_0 e^{-\mu x}$, where μ is a constant for the material concerned. The graph, *Figure 7.15*, can be considered to be the summation of two straight line graphs, one for the beta and one for the gamma. The upper x count rate values will be purely for the gamma, thus by extrapolating that end of the graph back the gamma graph can be deduced. The difference between the curve and this extrapolated line is the count rate due to the beta rays. For $x = 0.5$ mm, lg(R) for gamma is about 1.77. Hence R for gamma alone is about 59 s^{-1} and so R for beta is about $100 - 59 = 41$ s^{-1}

7.90 Ans $x = 214, y = 82$. As 30 min is 10 half lives the rate will have decreased by a factor of $1/2^{10}$, i.e. 1024, and so it will be 50 s^{-1}

7.91 Ans (a) $N = N_0 e^{-kt}$, where $k = 0.693/T$. For $t = ½T$ the equation becomes $N = N_0 e^{-0.347} = 0.707\, N_0$. This is the factor by which the mass of the isotope would be reduced. The mass is thus $0.707 \times 5.0 = 3.54$ μg. When beta particles are emitted from the nucleus another particle called the neutrino is also emitted and the energy is shared between the two particles. Hence, the beta particles show a continuous spectrum. The energy maximum occurs when all the energy is taken by the beta particle. The energy for the emission occurs from the mass difference between the nucleus before the event and that of the nucleus after the event plus the mass of the beta particle. The value of this depends on the nucleus concerned.
(b) This could involve a thin-window Geiger tube a fixed distance from the beta source. Different thicknesses of aluminium can then be introduced between the source and the tube and count rate readings taken. A graph of lg(count rate) against thickness will give a straight line, apart from at large thickness, and can be extrapolated to the zero count rate thickness value. The age of the source has no effect on the range, only affecting the count rate. This is because the range is determined only by the energy of the radiation which is in turn only affected by the nuclear change occurring

7.92 Ans (a) See the notes
(b) (i) Alpha emission gives a decrease of 2 neutrons and 2 protons, beta emission gives an increase of 1 proton, and a decrease of 1 neutron (see *Figure 7.16*). Last element is $^{221}_{87}Fr$. (ii) If all the elements were present in significant amounts

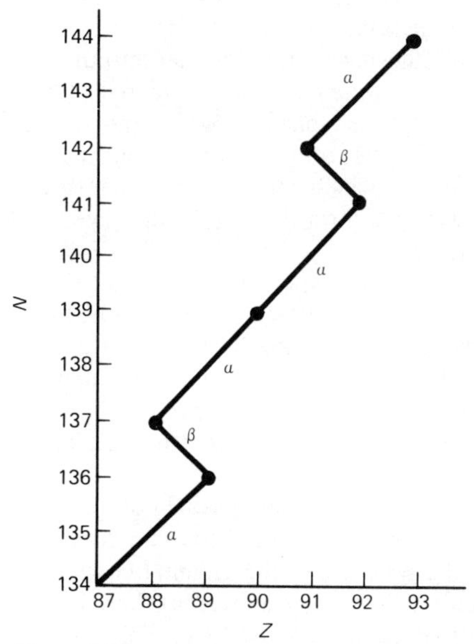

Figure 7.16

when the Earth was formed, the age of the Earth must be greater than 7.1×10^8 years, sufficiently greater for many half lives to have elapsed. The age cannot however be too much bigger as a significant amount of ^{238}U is still around. This would suggest an age of the order of 10^9 to 10^{10} years.

8 Electronic systems

Systems

Notes

The term system is used for an assembly of parts or components which are connected together in some organised way and for which we can identify some general function. A system can be represented by a 'black box' having an input and an output.

A measurement system can be considered to be made up of three linked 'black boxes': a transducer which changes a signal from one form to another; a signal conditioner which converts the signal from the transducer into a form that can be displayed; and the display element.

If a signal from the output of a system is fed back to the input then the system is said to be a feedback system. With positive feedback the signal is fed back in such a way as to enhance the change produced by the system. With negative feedback the signal is fed back in such a way as to reduce the change.

There are many systems that have an output which is determined by the condition of a number of inputs to the system. For the system known as an AND gate, for there to be an output there have to be inputs to each of the two inputs. This can be represented by a truth table:

Input A	B	*Output*
0	0	0
0	1	0
1	0	0
1	1	1

The symbol 0 in the above table represents no input and 1 represents an input. With an OR gate, when either of the two inputs is 1 then the output is 1. With a NOT gate there is an output 1 when there is no input. A NOR gate gives no output when either input is 1. A NAND gate gives no output when both inputs are 1. Systems of this form are known as logic gates.

Examples

8.1 What are the inputs and outputs for the following systems: (a) a d.c. power pack operating off the a.c. mains, (b) an amplifier and (c) a loudspeaker?

Ans (a) Input is the mains a.c. supply; output is the d.c. voltage.
(b) Input is the small signal; output is the bigger signal.
(c) Input is the electrical signal; output is the sound.

8.2 Give an example of a temperature transducer.

Ans An electrical resistance thermometer: input is the temperature signal; output is the resistance change.

8.3 Describe the feedback system of you walking along a straight line.

Ans See *Figure 8.1.* The system is you walking along the line. The output from the system,

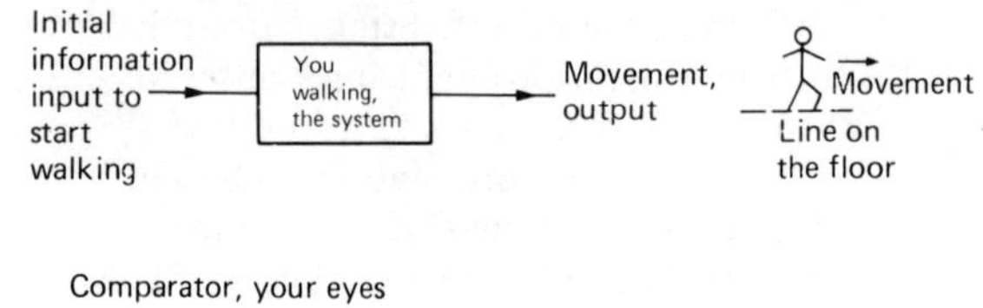

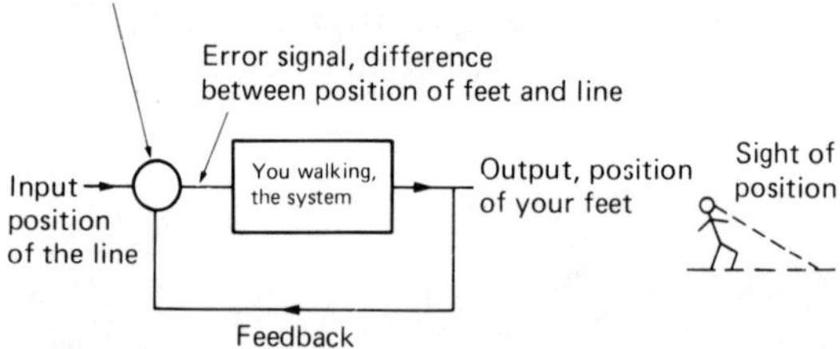

Figure 8.1

which is fed back, is the displacement of your feet relative to the line as perceived by your eyes. The difference between the actual position and the line provides a signal which determines the output from your system. The feedback is negative feedback.

8.4 Draw a simple electrical circuit in which a lamp lights when both of two switches are closed, i.e. an AND gate type system.

Ans See *Figure 8.2.*

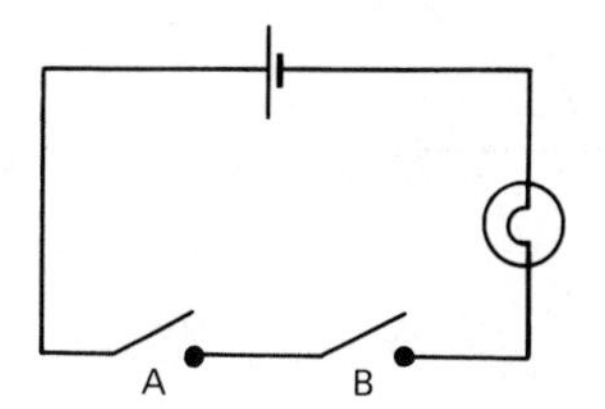

Figure 8.2 The lamp lights when A and B are closed

8.5 State the logic system and specify the truth table for a system where a tape recorder will record the input when both the record and play switches are depressed.

Ans This is an AND gate.

Record switch	*Play switch*	*Output i.e. recording*
On	Off	No
On	On	Yes
Off	On	No
Off	Off	No

Further problems

8.6 What are the inputs and outputs for the following systems: (a) a microphone, (b) an oscillator operated from the mains supply, (c) a Geiger–Müller tube and (d) a car engine?

8.7 Represent the following by a series of connected 'boxes', each box representing an identifiable sub-system: (a) a radio from input at the aerial to sound output, (b) a factory producing canned beans, (c) a smoothed d.c. supply from a mains a.c. input and (d) an assembly for counting radiations from radioactive materials.

8.8 What are the inputs and outputs of the following transducers: (a) a thermocouple, (b) a photoconductive cell and (c) the pick-up of a record player?

8.9 Explain the terms negative and positive feedback and give examples of each form of feedback.

8.10 Describe, by means of a block diagram, the feedback system of a central heating system.

8.11 What type of logic gates are needed for the following: (a) a machine that will operate when the start button is pressed and the machine guard is in place, (b) a lamp that will come on when either of two switches is operated and (c) a burglar alarm that must operate if the infrared beam to a photocell is interrupted?

8.12 Complete the truth tables for the gate systems in *Figure 8.3.*

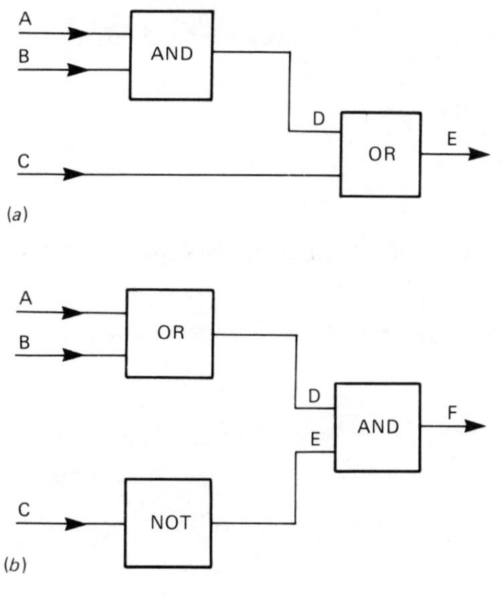

Figure 8.3

Alternating current circuits

Notes

With a capacitor in an alternating current circuit both the current to the capacitor and the potential

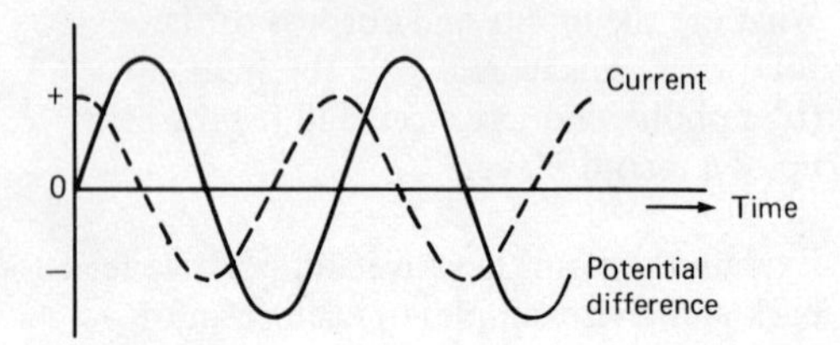

Figure 8.4 Variation of current and potential difference with time for a capacitor

difference across the capacitor vary with time (*Figure 8.4*). The current is said to lead the potential difference by 90° or $\pi/2$. The current and the potential difference can be represented by the rotation of two phasors 90° apart (*Figure 8.5*).

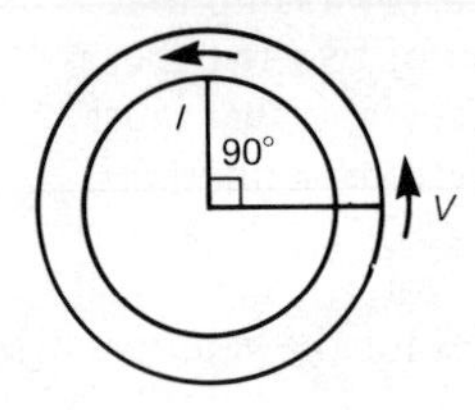

Figure 8.5 Phasors for a capacitor in an a.c. circuit

For an alternating applied potential difference given by

$$V = V_{max} \sin \omega t$$

and as

$$Q = CV$$

the variation of the charge on the capacitor plates with time is given by

$$Q = CV_{max} \sin \omega t$$

But

$$I = dQ/dt = \omega CV_{max} \cos \omega t$$

The maximum current is given by

$$I_{max} = \omega CV_{max}$$

Hence

$$I = I_{max} \cos \omega t$$

and so there is a 90° phase difference between the current and the potential difference. The capacitive reactance X_C is defined as V_{max}/I_{max} (this is the same as V_{RMS}/I_{RMS}). Hence

$$X_C = 1/\omega C$$

With an inductor in an a.c. circuit both the current through and the potential difference across the inductor vary with time (*Figure 8.6*). The current is said to lag the potential difference by 90° or $\pi/2$. The current and the potential difference can be represented by the rotation of two phasors 90° apart (*Figure 8.7*).

When the current through an inductor changes a changing magnetic field is produced and this produces an induced e.m.f. in the wires of the inductor, the direction of the induced e.m.f. being such as to

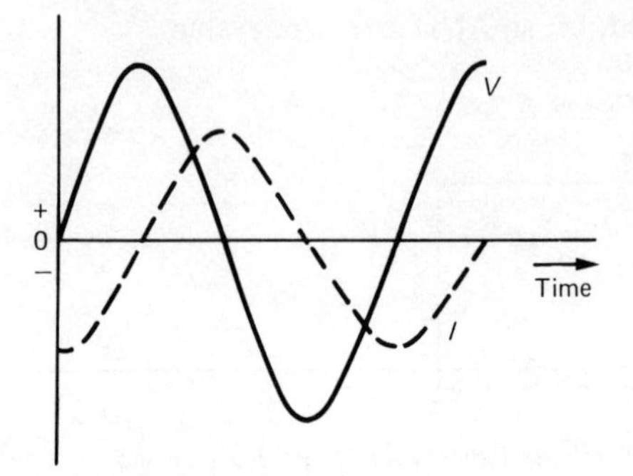

Figure 8.6 Variation of current and potential difference with time for an inductor

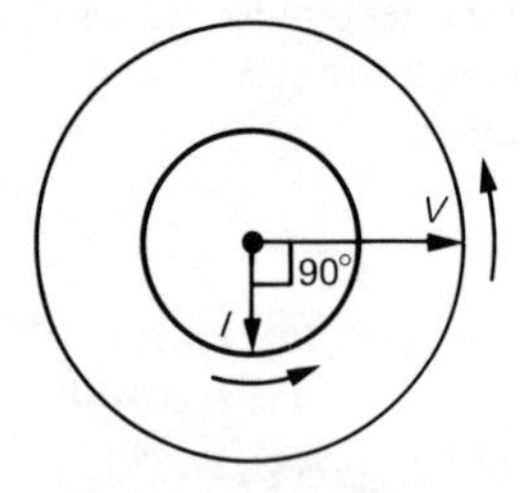

Figure 8.7 Phasors for an inductor in an a.c. circuit

oppose the change producing it. The potential difference across the inductor, when the rate of change of current through it is dI/dt, is LdI/dt. Thus, if the potential difference varies with time according to

$$V = V_{max} \sin \omega t$$

we have

$$LdI/dt = V_{max} \sin \omega t$$

and so by integrating

$$I = (V_{max}/\omega L) \cos \omega t$$

There is thus a 90° phase difference between current and potential difference. The inductive reactance X_L is defined as

$$V_{max}/I_{max} = \omega L$$

A circuit with an inductor and a resistor in series have the same current through each, but the potential differences across each component are not in phase. The potential difference V across the two together can be obtained from a consideration

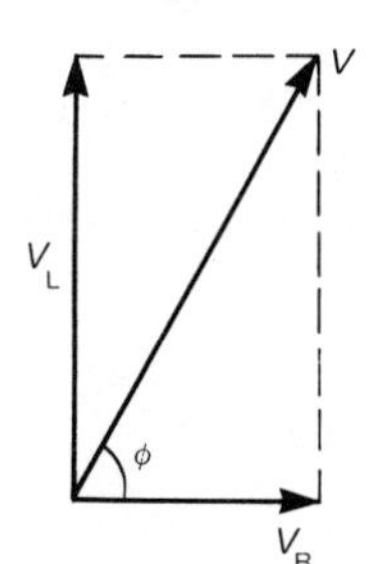

Figure 8.8

of the phasors (*Figure 8.8*). By the use of Pythagoras' theorem

$$V^2 = V_R^2 + V_L^2$$

As

$$V_R = IR \text{ and } V_L = IX_L$$

this can be written as

$$V = I(R^2 + X_L^2)^{1/2}$$

The quantity $(R^2 + X_L^2)^{1/2}$ is called the impedance Z of the circuit. Thus

$$V = IZ$$

The potential difference V leads the current by ϕ, where

$$\tan \phi = V_L/V_R = X_L/R$$

A circuit with a capacitor and a resistor in series will have the same current through each, but the potential differences across each component are not in phase. The potential difference V across the

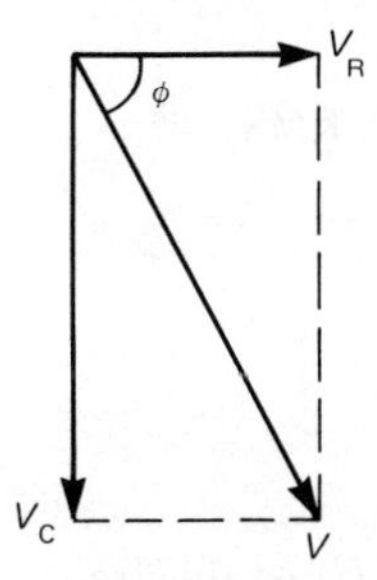

Figure 8.9

two together can be obtained from a consideration of the phasors (*Figure 8.9*). By the use of Pythagoras' theorem

$$V^2 = V_R^2 + V_C^2$$

As

$$V = IR \text{ and } V_C = IX_C$$

this can be written as

$$V = I(R^2 + X_C^2)^{1/2}$$

The quantity $(R^2 + X_C^2)^{1/2}$ is called the impedance Z of the circuit. Thus

$$V = IZ$$

The potential difference V lage the current by ϕ, where

$$\tan \phi = V_C/V_R = X_C/R$$

For a circuit with an inductor, a capacitor and a resistor, all in series, the current through each will be the same, but the potential differences across

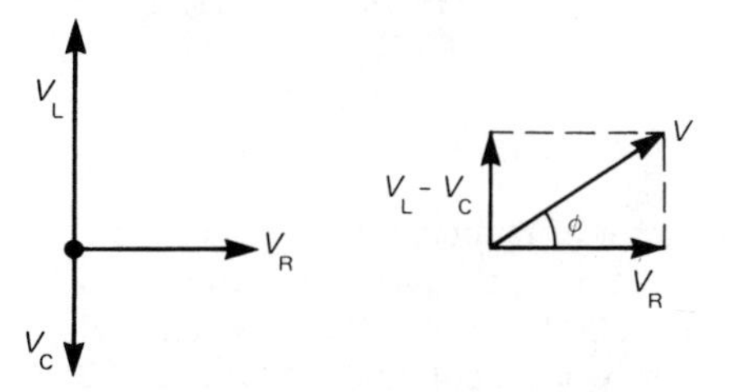

Figure 8.10

each component are not in phase. *Figure 8.10* shows the phasors. Hence

$$V^2 = V_R^2 + (V_L - V_C)^2$$

As

$$V_R = IR,\ V_C = IX_C \text{ and } V_L = IX_L$$

this can be written as

$$V = I\,[R^2 + (V_L - V_C)^2]^{1/2}$$

The quantity $[R^2 + (V_L - V_C)^2]^{1/2}$ is called the impedance Z of the circuit. Thus

$$V = IZ$$

The potential difference across the three components leads the current by ϕ, where

$$\tan \phi = (V_L - V_C)/V_R$$
$$= (X_L - X_C)/R$$

The circuit impedance of the *LCR* circuit varies with frequency, the circuit impedance being a minimum when

$$X_L = X_C \text{ and so } Z = R$$

Then there is no phase difference between the current and the potential difference, i.e. $\phi = 0$. When

$$X_L = X_C$$

we have

$$\omega L = 1/\omega C$$

and so

$$f^2 = 1/(4\pi^2 LC)$$

This is the resonant frequency.

Units: current – A, potential difference – V, reactance – Ω, impedance – Ω, ω – s^{-1} or Hz, frequency – s^{-1} or Hz, capacitance – F, inductance – H.

Examples

8.13 An inductor of inductance 2 H is in series with a resistor. If the pair have a total resistance of 1000 Ω, what is the impedance of the circuit when a potential difference of 3 V RMS and frequency 50 Hz is applied to the two components? What is the RMS current?

Ans $Z = (R^2 + X_L^2)^{½}$

$$X_L = 2\pi f L$$

Hence

$$Z = 1180\ \Omega$$

$$V_{RMS} = ZI_{RMS}$$

Hence

$$I_{RMS} = 2.5\ \text{mA}$$

8.14 A capacitor of capacitance 8 μF is in series with a resistor of resistance 2 kΩ. What are the potential differences, RMS, across each component if the potential difference across the combination is 4 V RMS at a frequency of 50 Hz?

Ans $Z = (R^2 + X_C^2)^{½}$

where

$$X_C = 1/(2\pi f C)$$

Hence

$$Z = 2040\ \Omega$$

As

$$V = IZ$$

the current is 1.96 mA. Hence the potential difference V_R across the resistor is

$$V = IR = 3.92\ \text{V RMS}$$

and that across the capacitor is

$$V = IX_C = 0.78\ \text{V RMS}$$

8.15 An inductor of inductance 2 H, a capacitor of capacitance 4 μF and a resistor of resistance 500 Ω are in series. The series arrangement of components is connected across an alternating voltage supply of 6 V RMS and frequency of 1 kHz. What is the RMS current through the arrangement? If the frequency were to be changed, at what frequency would the current be a maximum and what would be the value of this current?

Ans $Z = [R^2 + (X_L - X_C)^2]^{½}$

where

$$X_L = 2\pi f L$$

and

$$X_C = 1/(2\pi f C)$$

Hence

$$Z = 1.25 \times 10^4\ \Omega$$

Thus, as

$$V = IZ$$

the current is 4.79×10^{-4} A RMS.

Maximum current occurs when

$$X_L = X_C$$

i.e. when

$$f^2 = 1/(4\pi^2 LC)$$

The frequency is thus 56.3 Hz. At this frequency the current is given by

$$V = IR$$

and so is 1.2×10^{-2} A RMS.

Further problems

8.16 What are the reactances of the following components when each are separately connected into an a.c. circuit where the frequency is 100 Hz: (a) an 8 μF capacitor, (b) a 1000 μF capacitor, (c) a 2 H inductor and (d) a 50 mH inductor?

8.17 A 2.0 μF capacitor has a reactance of 20 Ω. What must be the frequency of the supply to which it is connected?

8.18 An inductor of inductance 5 H is connected to a supply of 120 V RMS at 50 Hz. What is the RMS current through the inductor?

8.19 A resistor of resistance 100 Ω and an inductor of inductance 0.2 H are connected in series to a 240 V RMS 50 Hz supply. What is (a) the impedance of the circuit, (b) the RMS current?

8.20 A tuning circuit in a radio has a coil of inductance 2×10^{-5} H in series with a variable capacitor. To what capacitance should the capacitor be adjusted if the series arrangement is to be 'in tune' for a frequency of 10^6 Hz?

8.21 A resistor of resistance 200 Ω, a capacitor of capacitance 1 μF and an inductor of inductance 0.5 H are connected in series to an alternating current supply of 12 V RMS. To what frequency should this supply be adjusted if the current in the circuit is to be a maximum? What is the value of this maximum current?

8.22 High impedance a.c. voltmeters are connected in parallel with each of the following components: a resistor of resistance 300 Ω, a capacitor of capacitance 2 μF and an inductor of inductance 500 mH. If the components are in series and the potential difference across the entire arrangement is 12 V RMS at 50 Hz, what are the readings of the voltmeters?

8.23 Explain the terms reactance, impedance and phase angle.

Power in a.c. circuits

Notes

The instantaneous power developed by a component is IV, where I and V are the current and potential difference values at the same instant of time. Where the current and potential difference are in phase, i.e. for a pure resistance,

$$\begin{aligned}\text{Instantaneous power} &= IV \\ &= (I_{max} \sin \omega t)(V_{max} \sin \omega t) \\ &= I_{max} V_{max} \sin^2 \omega t\end{aligned}$$

The mean power is the mean value of the above expression, and as the mean value of $\sin^2 \omega t$ is ½,

$$\begin{aligned}\text{Mean power} &= \tfrac{1}{2} I_{max} V_{max} \\ &= \frac{I_{max} V_{max}}{2^{1/2} \times 2^{1/2}} \\ &= I_{RMS} V_{RMS}\end{aligned}$$

With an inductor, having no resistance, the current lags the potential difference by 90°.

$$\begin{aligned}&\text{Instantaneous power} \\ &= (-I_{max} \cos \omega t)(V_{max} \sin \omega t) \\ &= -I_{max} V_{max} \cos \omega t \sin \omega t\end{aligned}$$

The mean value of $\cos \omega t \sin \omega t$ is zero, thus the mean power is zero. *Figure 8.11* shows the instantaneous power variation with time. During that

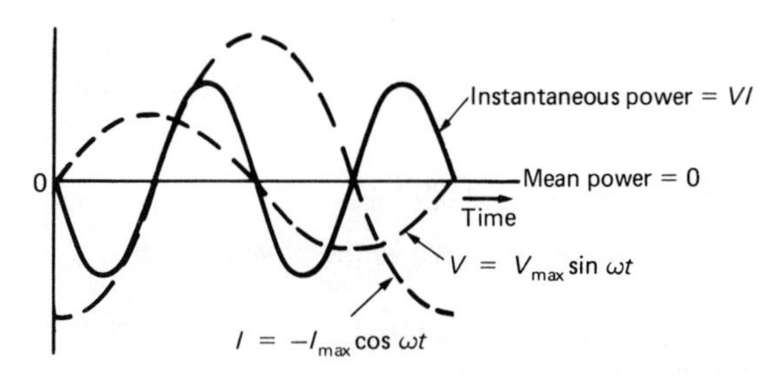

Figure 8.11 Power produced by an alternating current through a pure inductor

part of the cycle for which the power is positive, power is drawn from the source and energy is stored in the magnetic field of the inductor. When the power is negative the inductor is returning the energy stored in its magnetic field back to the source. The net result is an average power of zero.

With a capacitor, having no resistance, the current leads the potential difference by 90°.

Instantaneous power
$= (I_{max} \cos \omega t)(V_{max} \sin \omega t)$
$= I_{max} V_{max} \cos \omega t \sin \omega t$

The mean value of $\cos \omega t \sin \omega t$ is zero, thus the mean power is zero. *Figure 8.12* shows the instantaneous power variation with time. When the power is positive energy is taken from the source to charge the capacitor and so store energy in the

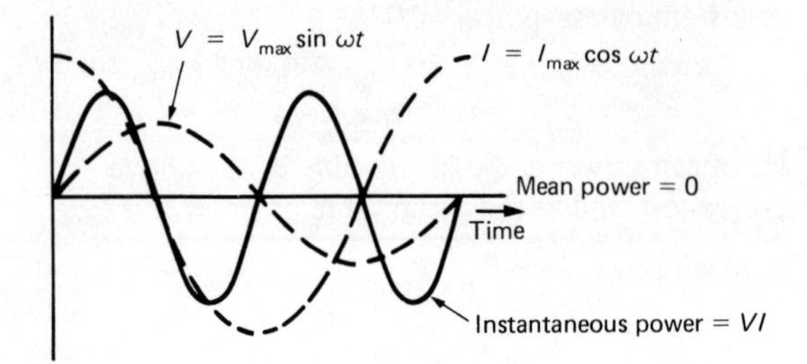

Figure 8.12 Power produced by an alternating current through a pure capacitor

electric field of the capacitor. When the power is negative the capacitor is discharging and returning the energy to the source. The net power is thus zero.

In general, where there is a phase difference of ϕ between the current and the potential difference,

Instantaneous power
$= [I_{max} \sin (\omega t - \phi)](V_{max} \sin \omega t)$
$= I_{max} V_{max} \sin (\omega t - \phi) \sin \omega t$

The mean value of $\sin (\omega t - \phi) \sin \omega t$ is $\frac{1}{2} \cos \phi$. Thus

$$\text{Mean power} = \tfrac{1}{2} I_{max} V_{max} \cos \phi$$

The term $\cos \phi$ is known as the power factor.

Units: current – A, potential difference – V, power – W.

Examples

8.24 What is the power dissipated when an alternating current of 2.0 A RMS passes through a resistor of resistance 10 Ω?

Ans Power $= I_{RMS} V_{RMS}$
$= I^2_{RMS} R$
$= 40$ W

8.25 What is the power dissipated when an alternating current of 2.0 A RMS passes through an inductor, having negligible resistance, of inductance 2 H?

Ans The situation is artificial in that no inductor can have zero resistance. However, for the situation specified the power will be zero (see the notes).

8.26 What is the power dissipated when an alternating current of 2.0 A RMS and frequency 50 Hz passes through an inductor having an inductance of 2 H and a resistance of 200 Ω?

Ans For a resistor in series with an inductor

$$\tan \phi = X_L / R$$

Thus as $X_L = 2\pi f L$

$$\tan \phi = \frac{2\pi f L}{R}$$

$$\phi = 72.3°$$

The impedance is given by

$Z = (R^2 + X_L^2)^{1/2}$
$= 659$ Ω

Thus, as

$V_{RMS} = I_{RMS} Z$
Power $= \frac{1}{2} V_{max} I_{max} \cos \phi$
$= V_{RMS} I_{RMS} \cos \phi$
$= I^2_{RMS} Z \cos \phi$

Thus

Power $= 2.0^2 \times 659 \cos 72.3°$
$= 800$ W

A simpler way, however, to consider the problem is to consider there to be zero power dissipated in the inductor, having no resistance, and all the power to be dissipated in the resistance part that we can consider to be effectively in series with the inductor. Thus

Power $= I^2 R$
$= 4 \times 200$
$= 800$ W

Further problems

8.27 What is the mean power dissipated when there is an alternating potential difference of 4.0 V RMS across a resistor of resistance 20 Ω?

8.28 A capacitor has an alternating potential difference between its plates. Explain why there is no mean power taken from the source of the alternating voltage?

8.29 An alternating voltage of 5.0 V RMS and a frequency of 50 Hz is applied across a capacitor of capacitance 8 μF and a resistor of resistance 500 Ω in series. What is the mean power dissipated?

8.30 An inductor connected to the 240 V RMS mains supply of frequency 50 Hz has a resistance of 20 Ω and a reactance of 10 Ω. What is (a) its impedance, (b) the mean power supplied?

8.31 A capacitor of capacitance 2 μF, an inductor of inductance 0.5 H and a resistor of resistance 200 Ω are connected in series. The inductor has a resistance of 100 Ω. An alternating potential difference of 12 V RMS is applied across the series arrangement. (a) At what frequency will the current be a maximum? (b) At the resonant frequency, what will be the power dissipated? (c) What would be the power dissipated at a frequency of 100 Hz?

Pulses and *CR* and *LR* circuits

Notes

When square voltage pulses are passed through a circuit containing a capacitor in series with a resistor, the potential difference across the capacitor rises as a result of the capacitor being charged during the rising part of the square pulse. If the length of the pulse is large enough the capacitor can become fully charged. This depends on the length of the pulse and the time constant, i.e. *CR*, of the circuit. *Figure 8.13* shows the form of

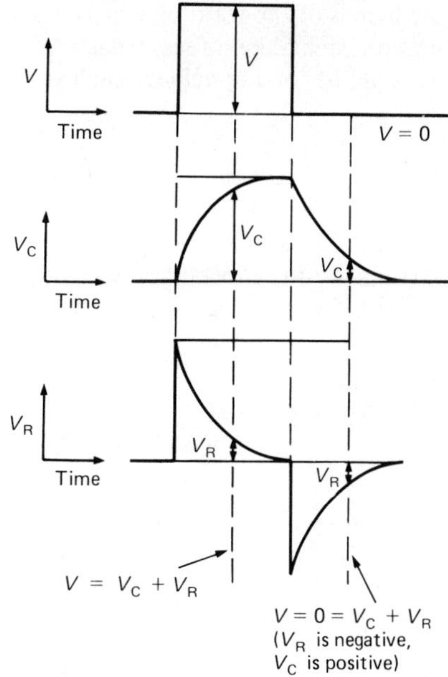

Figure 8.13

the pulses across the components when the capacitor becomes fully charged. With smaller values of *CR* the potential difference across the resistor is reasonably proportional to the rate at which the potential difference across the input varies with time. For this reason the circuit in this condition is known as a differentiating circuit. If the time constant is long with respect to the length of the voltage pulse then the potential difference across the capacitor is the integral, about the mean, of the input pulse and hence is known as an integrating circuit.

With an *LR* circuit, when the time constant *L*/*R* is small compared with the length of the input pulse then the potential difference across the inductor is a reasonable approximation to the differential of the input pulse and hence the circuit is known as a differentiating circuit. When the time constant is large, the potential difference across the resistor is a reasonable approximation to the integral of the input potential difference to the circuit, the circuit in this condition being known as an integrating circuit.

Examples

8.32 From a square pulse voltage input to a circuit an output potential difference is required which is of the form of sharp, short duration, spike-like pulses. What type of circuit could be used to achieve such an output?

Ans A differentiation circuit can be used. *Figure 8.14* shows the effect of differentiating a square pulse. For this a *CR* circuit would need a very small time constant when compared with the length of the input pulse and the output would be taken as the potential difference across the resistor.

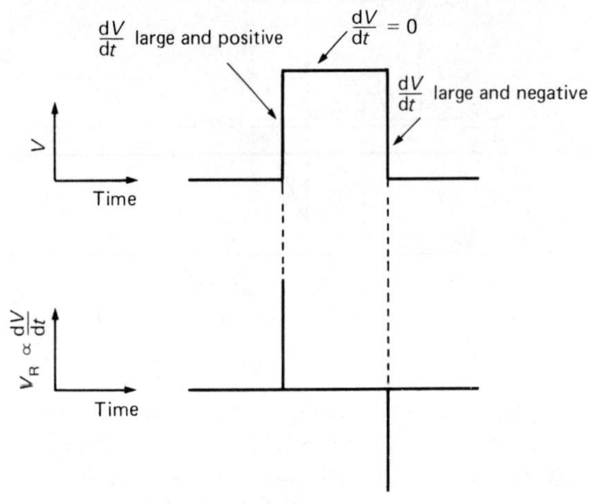

Figure 8.14 The differentiation of a square pulse

Further problems

8.33 Describe how the potential difference across a capacitor and a resistor vary with time when a square voltage pulse is passed through the two components in series for time constants both shorter and longer than the length of the input pulse.

Rectification

Notes

Figure 8.15 shows the basic form of a diode valve. Electrons are emitted from a filament when it is heated, or from an electrode when it is heated by means of a filament. This emission is known as thermionic emission. The higher the temperature the greater the rate of emission of electrons.

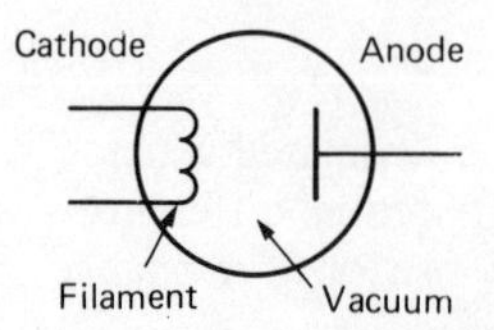

Figure 8.15 The basic diode valve

Opposite the emitting surface is an electrode, known as the anode. When this is made positive with respect to the emitter a current flows across the valve. When it is made negative with respect to the emitter no current flows. *Figure 8.16* shows the current/potential difference relationship. The more positive the anode is made the greater the current is, until the saturation value is reached when all the electrons are gathered as fast as they are produced at the emitter.

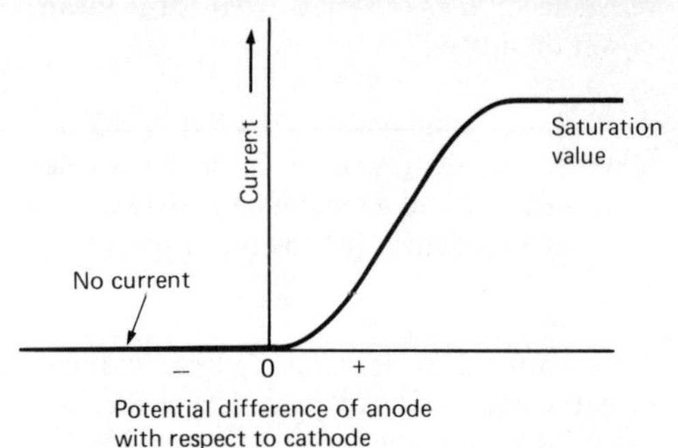

Figure 8.16 Current/potential difference relationship for a diode valve

If an alternating potential difference is applied across a diode valve, a current only occurs when the potential difference is such as to make the anode positive with respect to the emitter. The diode is said to rectify the alternating current (*Figure 8.17*).

A junction between copper and cuprous oxide has the property of offering a very much greater resistance to a current in one direction than in

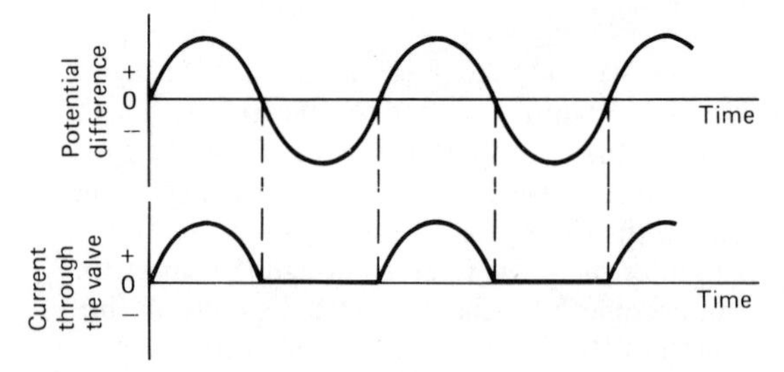

Figure 8.17 Half-wave rectification

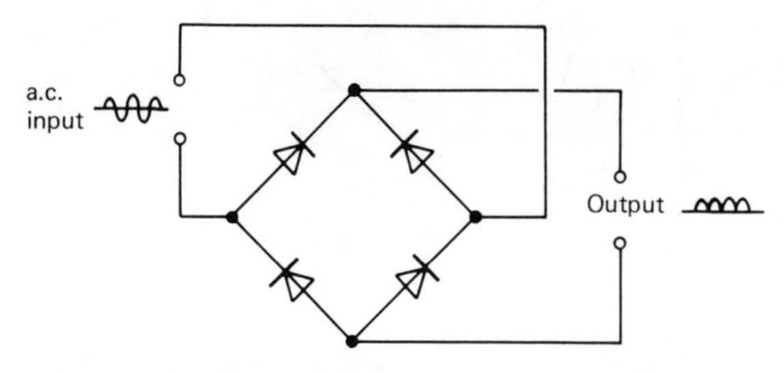

Figure 8.18 Bridge rectifier

the other. Such a junction is known as a metal rectifier. *Figure 8.18* shows a bridge circuit used to produce a full-wave rectifier.

A junction between p- and n-type semiconductor materials also has the property of a high resistance for current flow in one direction through it and a much lower resistance for the opposite direction (*Figure 8.19*). The component is known as a semiconductor junction diode.

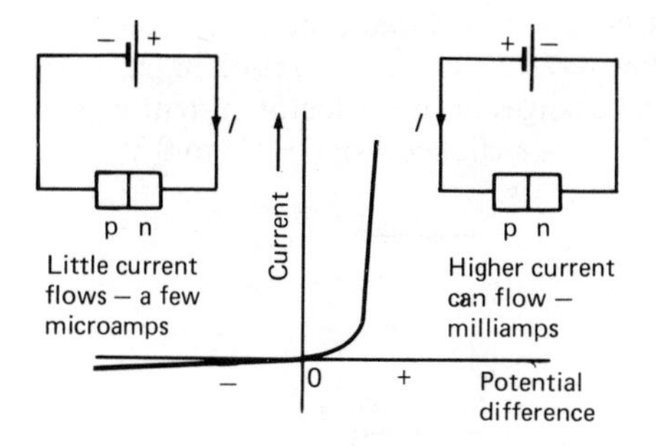

Figure 8.19 The characteristic of a p–n junction

Examples

8.34 What is the form of current/potential difference relationship required for any device that will rectify an alternating current?

Ans It must have a high resistance when one side of the device is positive with respect to the other side and a low resistance for the reverse of this potential difference. When conducting the I/V graph should be linear to eliminate distortion.

8.35 A junction diode is connected across the terminals of a galvanometer. What would be the effect of this on the readings given by that meter when used in d.c. circuits?

Ans With the diode connected in one way across the terminals the application of a d.c. voltage will result in the junction having a very high resistance. This will have little effect on the readings given by the meter in that the percentage of the current that passes through the meter will be considerably greater than the probably negligible percentage passing through the diode. With the diode connected the other way round, or the polarity of the d.c. supply reversed, the diode will have a comparatively low resistance and so will act as a shunt to the meter and take a significant percentage of the current. The meter reading will thus be reduced. Because the current/voltage relationship for a junction diode connected in the forward-biased direction is not linear, the shunting effect will vary with the applied current. As the characteristic is reasonably exponential a log-scale galvanometer will be produced.

Further problems

8.36 Explain how the diode valve is able to rectify alternating current.

8.37 What would be the effect on the value of the saturation current of increasing the temperature of the filament in a diode valve?

8.38 A junction diode is placed in series with a galvanometer in a d.c. circuit. How will this affect the readings given by the instrument?

The transistor

Notes

The transistor can consist of a slice of p-type semiconductor material between pieces of n-type material, or alternatively, a slice of n-type semiconductor material between pieces of p-type material (*Figure 8.20*). Transistors can be connected into circuits in a number of ways. *Figure 8.21* shows the common emitter form of connection. For this arrangement, a variation in the base current results in a much larger change in the collector current for a constant potential difference between the collector and emitter.

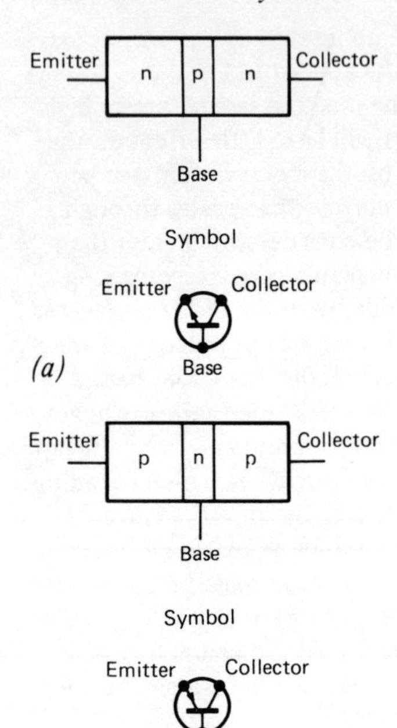

Figure 8.20 (a) The n–p–n transistor; (b) the p–n–p transistor

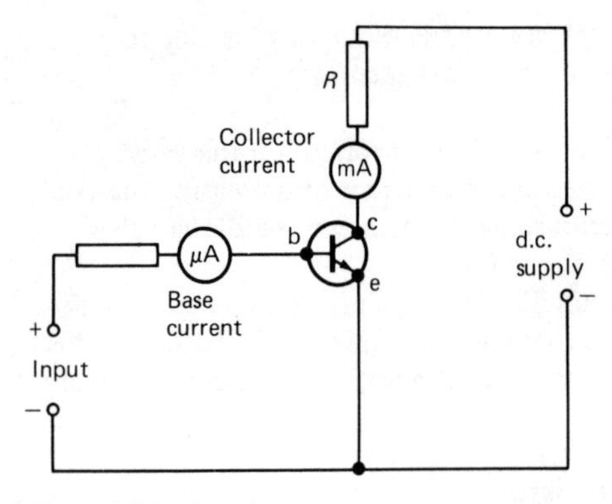

Figure 8.21

Examples

8.39 *Figure 8.22* shows how the collector current depends on the base current for a transistor connected in the common emitter way. If the base current changes from 20 to 30 μA, by how much will the collector current change?

Ans The collector current changes from about 4 mA to 6 mA, i.e. a change of 2 mA. The change is about 200 times greater than the change in the base current.

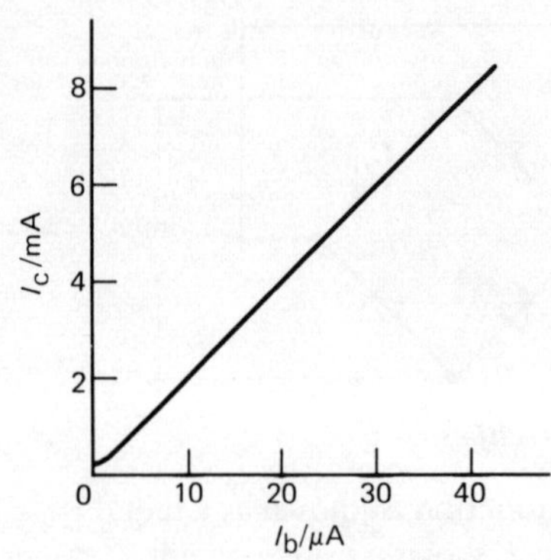

Figure 8.22

Further problems

8.40 For the circuit shown in *Figure 8.23* it is found that when the input is connected to +6 V the output is almost 0 V and, conversely, when the input is at 0 V the output is almost 6 V. Explain this action in terms of the changes in the collector current when the input is switched from +6 V to 0 V.

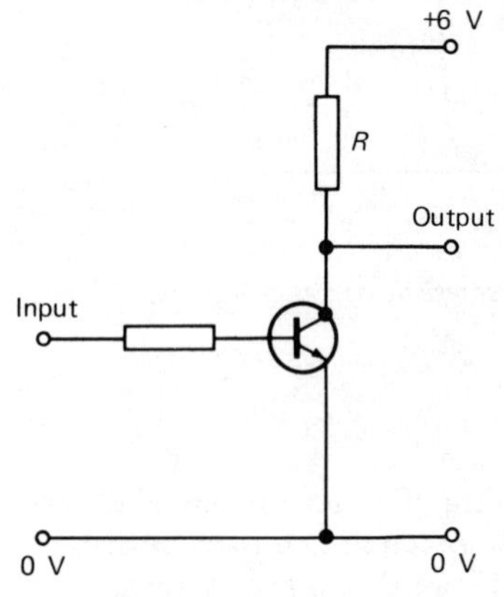

Figure 8.23

8.41 With a transistor connected as in *Figure 8.23*, when the potential difference between the base *b* and the emitter *e* changes from 0.60 V to 0.62 V then the collector current is found to change from 2.0 mA to 4.0 mA, the load resistor *R* having the value 1.0 kΩ. What is (a) the change in the potential difference across the load resistor, (b) the change in the potential difference between the collector and the emitter, if the d.c. supply connected across the load resistor *R* and the transistor is constant and (c) the voltage gain, i.e. the change in the output potential difference between the collector and the emitter divided by the change in the input potential responsible for the change?

Conduction in solids

Notes

Materials can be put into sets according to their resistivities, conductors having resistivities of the order of 10^{-8} Ω m, semiconductors of the order of 1 to 100 Ω m, and insulators of the order of 10^{10} to 10^{20} Ω m. For semiconductors and insulators the resistivity decreases with an increase in temperature. For conductors there is usually an increase in resistivity with an increase in temperature. For metals (conductors) there is usually about one charge carrier for every atom and no change occurs with an increase in temperature. For semiconductors there is about one charge carrier per million atoms; the number, however, increases with an increase in temperature. For insulators there are hardly any charge carriers; the number does, however, increase with an increase in temperature.

In a metal the atoms each have loosely held electrons. At room temperature, about 300 K above absolute zero, these electrons have broken free of the atoms and a general drift in one direction under the action of an electric field is possible, i.e. a current can flow when a potential difference is applied. Germanium and silicon at room temperature have gained sufficient energy for just a few electrons to break free. When an electron breaks free it leaves a 'hole', i.e. a vacancy, into which another electron can move. When a potential difference is applied to a pure semiconductor the electrons move in one direction and the holes in the opposite direction. The introduction of small amounts of impurities into germanium or silicon can markedly affect their conductivity. With an n-type semiconductor, atoms having five electrons in their outer shell have been introduced into a material, e.g. germanium, which has four electrons in its outer shell. The result is an introduction of more electrons and hence the conduction occurs more by electrons than holes. With a p-type semiconductor a material having only three electrons in its outer shell is used, the result being a surplus of holes and hence conduction predominantly by holes.

The junction diode is a junction between n-type and p-type semiconductors. When this junction is produced the electrons in the n-type material close to the junction can 'drop' into the holes in the p-type material just the other side of the junction. The resulting charge movement leads to the n-type material acquiring a positive charge and the p-type material a negative charge. A potential difference is produced across the junction (*Figure 8.24*). When

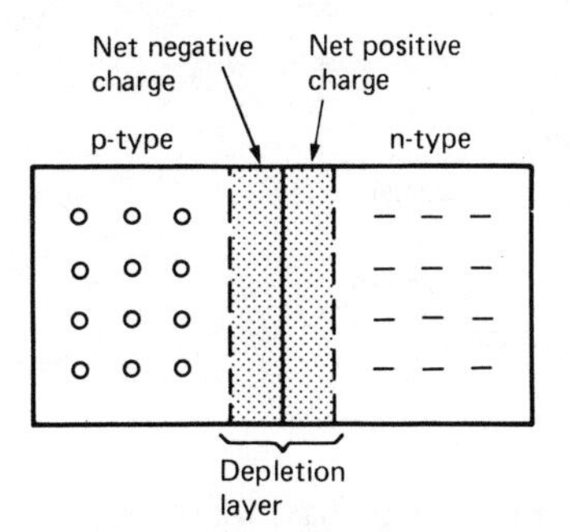

Figure 8.24

a battery is connected across the p–n junction a current can flow when the positive side of the battery is connected to the p-side of the junction, but virtually no current flows if it is connected to the n-side. In one case the battery potential difference is reinforcing the potential barrier at the junction and so no current occurs; in the other case the battery potential difference cancels the potential barrier and so charge moves through the barrier region (*Figure 8.25*).

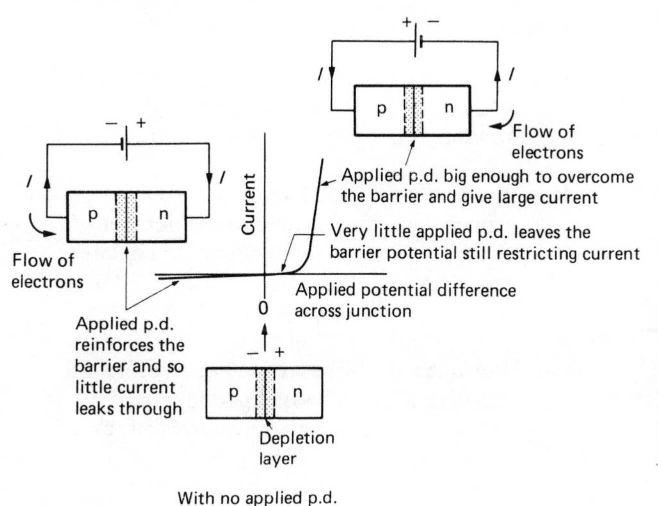

Figure 8.25

A transistor can be considered as being two p–n junctions back to back with the region between the junctions being very narrow. To obtain transistor action the emitter–base junction is forward biased and the collector–base junction reverse biased (*Figure 8.26*).

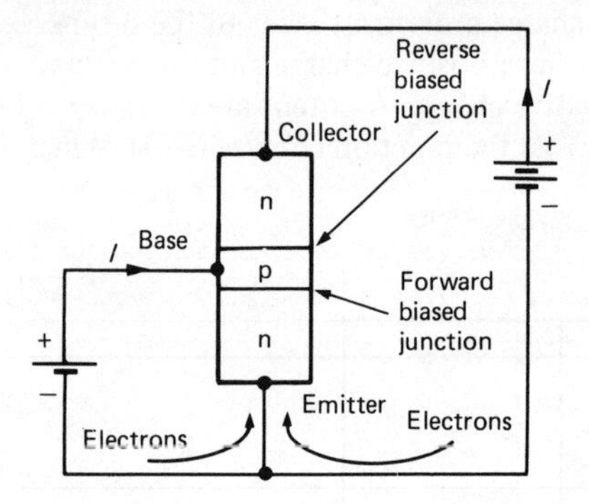

Figure 8.26 Common emitter connection

The forward bias of the emitter–base junction means that electrons flow through into the base. Once in the base these electrons do not have far to travel, because the base is only a thin layer, before they are accelerated towards the positively charged collector. The result is a current in the base-collector circuit despite it being reverse biased. Because the circuit is reverse biased it has a high resistance, the output current is thus flowing in a circuit with high resistance, the input current being in the lower resistance forward biased circuit.

Examples

8.42 Given two pieces of semiconductor material, what experiment could be used to determine whether it is a p-type or n-type semiconductor?

Ans The Hall effect, see chapter 4, enables the sign of the charge, as well as the number of charge carriers per unit volume, to be determined.

8.43 Which has the higher resistance – a forward biased or a reverse biased p–n junction? Explain the terms forward biased and reverse biased.

Ans A forward bias is when the potential difference applied across the p–n junction is such as to decrease or cancel the barrier potential and so permit a current. The reverse bias is when the applied potential difference reinforces the barrier potential and so virtually no current occurs. The reverse bias has high resistance and the forward bias low resistance.

Further problems

8.44 Antimony is used to dope germanium. Is the resulting material a p- or n-type semiconductor (germanium has 4 electrons in its outer shell, antimony has 5 electrons in its outer shell)?

8.45 Describe, with reference to the motion of holes and electrons, how a barrier potential is produced at a p–n junction.

8.46 The resistivity of a pure semiconductor, such as germanium, decreases as the temperature is increased. What causes this decrease?

8.47 Explain how a junction diode can be used for the rectification of alternating current, using the concepts of hole and electron movement in the explanation.

8.48 Describe, with reference to the motion of holes and electrons, how a transistor operates.

Electronic systems

Notes

The performance of amplifiers can be described as that they have a small signal as an input and generally give a larger version of the input signal as an output. The term 'gain' is used to relate the output and input,

$$\text{Gain} = \frac{\text{amplifier output}}{\text{amplifier input}}$$

Often the input signal has to be biased, i.e. be superimposed on a steady d.c. signal, in order that the output is both bigger than the input and has least distortion.

A sine-wave oscillator can be considered to be an amplifier with a feedback loop via an L–C circuit. The L–C circuit will resonate at just one particular frequency, thus only this frequency is effectively fed back into the amplifier. The feedback has to be positive feedback.

With negative feedback a proportion of the signal is fed back in such a way as to reduce the change producing it. The effect of such feedback with an amplifier is to reduce the input to the amplifier and hence the gain. The stability of the amplifier is, however, improved. This means that the gain will not fluctuate so much when changes, such as changes in temperature, occur.

If A is the gain of the amplifier then

$$A = \text{output}/\text{input}$$

If a fraction β of the output is fed back then the modified input becomes

$$\text{Original input} - \beta \times \text{output}$$

But the output must still be related to the input by the gain A, thus

$$\text{Output} = A \times \text{modified input}$$

Thus

$$\text{Output}(1 + \beta A) = A \times \text{input}$$

Gain with feedback is given by

$$A_f = \frac{\text{output}}{\text{original input}}$$

$$= \frac{A}{1 + \beta A}$$

The term operational amplifier was originally used to describe a type of amplifier which had a very high gain and was designed for use in analogue computers. They are, however, much more widely used now. When the operational amplifier is used as an amplifier, negative feedback is invariably used to produce stability. *Figure 8.27* shows an operational amplifier wired as (a) an inverting amplifier, (b) a non-inverting amplifier. With the inverting amplifier, any signal applied between the inverting input (marked –) and the common line appears at the output as an amplified version of the input, but 180° out of phase with it. With the non-inverting input (marked +), any signal applied between this input and the common line appears at the output as an amplified version of the input and in-phase with it.

With the inverting amplifier, when the input V_i tends to drive the potential at point X positive, then point Y becomes negative, the amplifier is

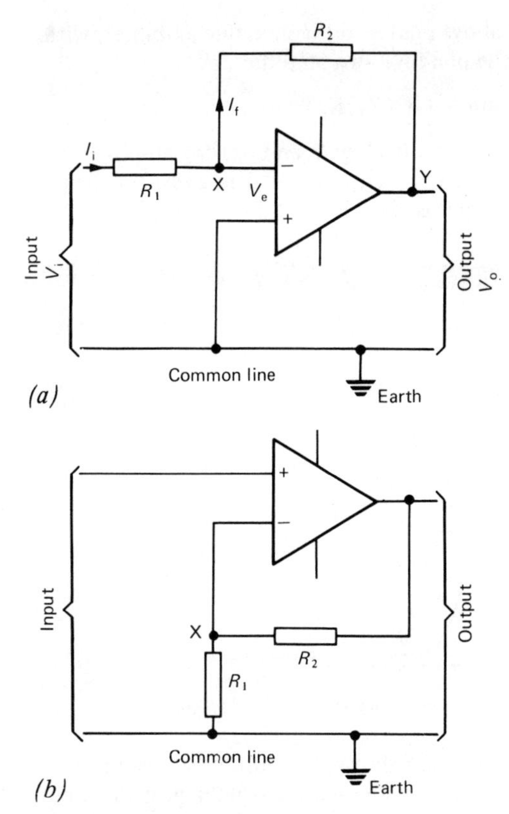

Figure 8.27 (a) An inverting amplifier; (b) a non-inverting amplifier

inverting, and so a potential difference develops between X and Y which causes a current to flow. As the gain is high this means that no matter how the input changes, the effect of the feedback will be to keep the potential at X at a very low level. This point is thus referred to as the virtual earth. A similar situation occurs with the non-inverting amplifier.

The operational amplifier has a very high impedance and draws little current. Thus, the input current I_i is effectively the same as I_f. If X is at earth potential then the potential difference across R_1 is V_i and so

$$I_i = V_i/R_1$$

Also

$$I_f = -V_0/R_2$$

Hence

$$V_0/V_i = -R_2/R_1$$

$$= \text{the amplifier gain}$$

The above relates to the inverting amplifier, with, for the non-inverting amplifier,

$$\text{Gain} = 1 + (R_2/R_1)$$

If the feedback with an inverting amplifier is taken via a capacitor rather than a resistor then the output is given by

$$V_o = -(1/RC) \int_0^t V_i \, dt$$

An operational amplifier integrator has been produced.

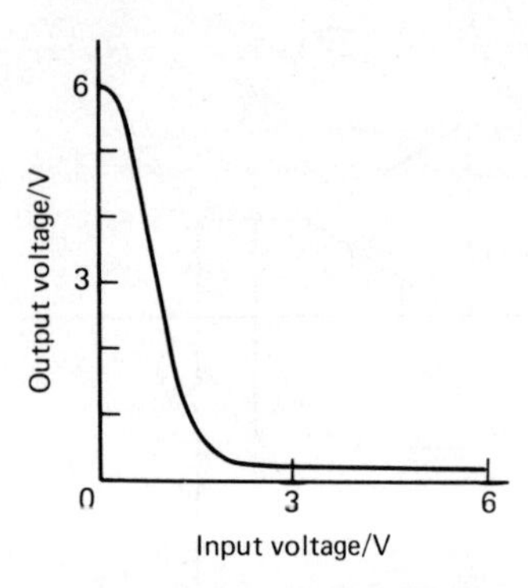

Figure 8.28

Figure 8.28 shows the form of behaviour of a circuit that can be used for switching. If the input is 0 V then the output is 6 V. If the input rises to 6 V then the output switches to 0 V. The circuit behaves as a NOT gate. Other gates can be produced and gate systems combined to enable a wide range of switching and control operations to be achieved.

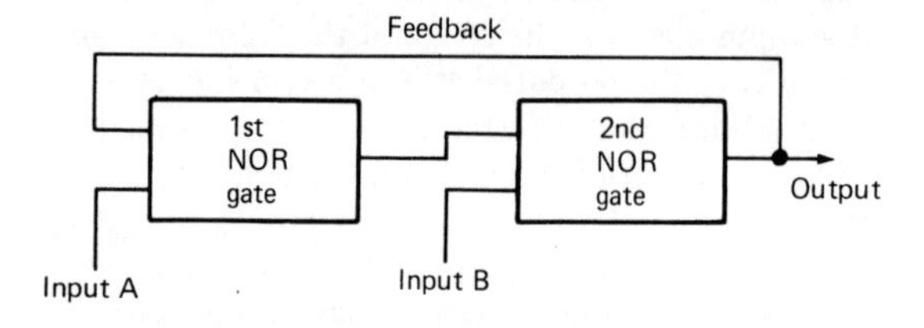

Figure 8.29 A bistable circuit

A bistable circuit is one which can exist in either of two stable states. It can be produced from two NOR gates (*Figure 8.29*).

Examples

8.49 Suppose you wish to use the system that gave *Figure 8.28* as an amplifier. What would be the bias voltage needed? What would be the maximum alternating input that could be amplified without significant distortion? What would the gain of the amplifier in the non-distorted region?

Ans The bias voltage must be such as to allow input signals to arrive at about the middle of the straight line portion of the graph. This would indicate a bias of about +0.8 V. The maximum peak to peak value of input signal possible is about 0.5 V to about 1.1 V. This is a maximum value for a.c. of about 0.3 V. The gain is about 10.

8.50 An amplifier has a gain of one hundred. What is the gain of the amplifier if negative feedback is used to feed back 0.05 of the output?

Ans

$$\text{Gain with feedback} = \frac{A}{1 + \beta A}$$

$$= \frac{100}{1 + 0.05 \times 100}$$

$$= 17$$

8.51 What is the required value for the feedback resistor if a gain of 200 is to be achieved for an operational amplifier with negative feedback and connected as an inverting amplifier. The resistor in series with the input has a value of 5.00 Ω.

Ans

$$\text{Gain} = -\frac{R_2}{R_1}$$

Hence

$$R_2 = 200 \times 500$$
$$= 100 \text{ k}\Omega$$

Further problems

8.52 *Figure 8.30* shows the circuit of a simple sinusoidal oscillator. Explain the functions of each component in the circuit.

8.53 Operational amplifiers can be used as inverting or non-inverting amplifiers. Explain the significance of these terms.

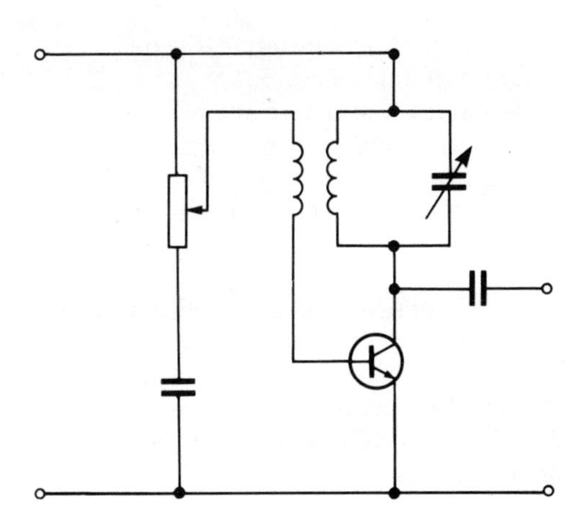

Figure 8.30

8.54 Why is negative feedback used with amplifiers?

8.55 An operational amplifier is to be used with negative feedback to give a gain of 20, the output signal being in phase with the input signal. If the resistor in series with the input has a value of 1 kΩ, what must be the value of the feedback resistance?

Examination questions

8.56 A 500 Ω resistor and a capacitor C are connected in series across the 50 Hz a.c. supply mains (*Figure 8.31*). The RMS potential

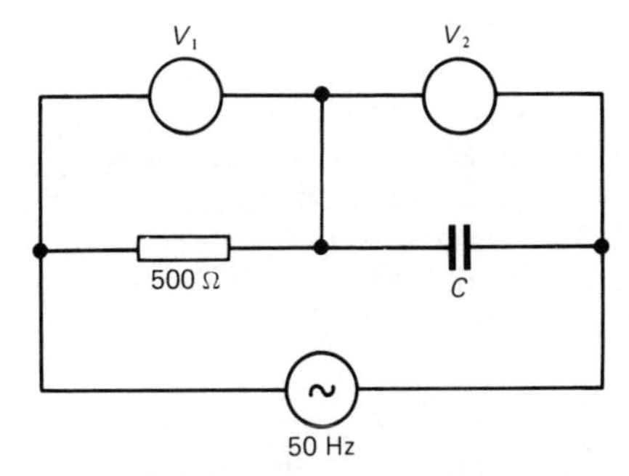

Figure 8.31

differences recorded on high impedance voltmeters V_1 and V_2 are:

$V_1 = 120$ V; $V_2 = 160$ V.

(a) What is the current flowing in the circuit?
(b) What is the power taken from the supply mains?
(c) What is the capacitance of C?

(Southern Universities)

8.57 (a) Define the impedance of a coil carrying an alternating current. Distinguish between the impedance and resistance of a coil and explain how they are related.

Describe and explain how you would use a length of insulated wire to make a resistor having an appreciable resistance, but negligible inductance.
(b) Outline how you would determine the impedance of a coil at a frequency of 50 Hz using a resistor of known resistance, a 50 Hz a.c. supply and a suitable measuring instrument. Show how to calculate the impedance from your measurements.
(c) A coil of inductance L and resistance R is connected in series with a capacitance C and a variable frequency sinusoidal oscillator of negligible impedance. Sketch a graph showing qualitatively how the current in the circuit varies with the applied frequency and account for the shape of the curve.

Sketch on the same axes the curve you would expect for a considerably larger value of R, the values of L and C remaining unchanged, taking care to indicate which curve refers to the larger value of R.

(JMB)

8.58 An inductor and a capacitor are connected one at a time to a variable-frequency power source. State how, and explain in non-mathematical terms, why the current through the inductor and the capacitor varies as the frequency is varied.

A circuit is set up containing an inductor, a capacitor, a lamp and a variable-frequency source with the components arranged in series. Explain why, as the frequency of the supply is varied, the lamp is found to increase in brightness, reach a maximum and then become less bright. Explain why the inductor is heated by the passage of the current while the capacitor remains cool.

(University of London)

8.59 A sealed box with two external terminals (*Figure 8.32*) is known to contain a resistor and a capacitor connected in some way. When a potential difference of 32 V RMS at a frequency of 100 Hz is applied across the terminals of the box a current of 1.0 mA RMS flows in it. If the

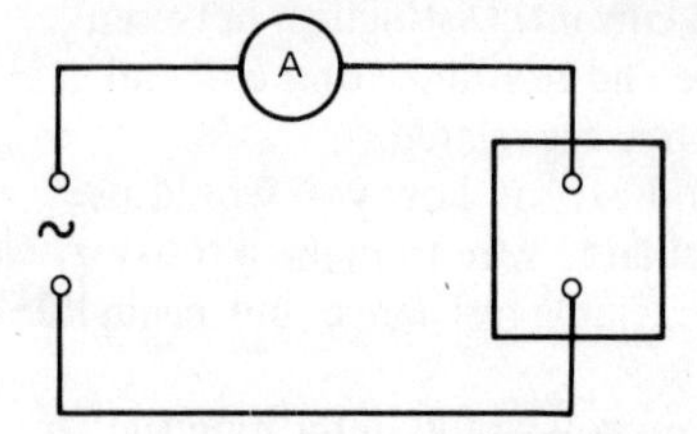

Figure 8.32

frequency of the applied potential difference is gradually increased, the current at first rises and then reaches a steady value of 2.0 mA RMS no matter how high the frequency. There is no current when a d.c. power supply is used instead.

What is the most likely arrangement of the two components inside the box? Explain.

Calculate the resistance of the resistor.

(University of London)

8.60 What is meant by a semiconductor? Explain how the conductivity of such a material changes with (a) temperature, (b) the presence of impurities.

Describe the structure of a solid state diode, explaining the nature of the semiconducting materials from which it is made. Explain the action of the diode in rectifying an alternating current.

The diagrams (*Figure 8.33*), show simple forms of transistor voltage amplifiers using (i) a p–n–p transistor, and (ii) an n–p–n transistor. Choose one of these circuits and explain the functions of the components R_1, R_2, C_1 and C_2. State which circuit you are considering.

(University of London)

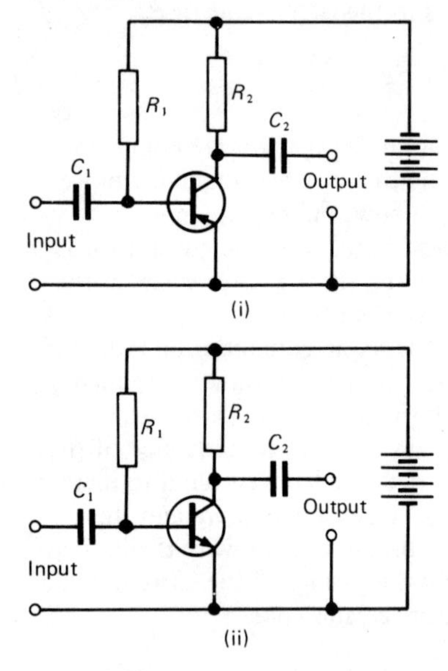

Figure 8.33

8.61 (i) Draw the input and output characteristics of a junction transistor in common emitter connection.
(ii) Draw a circuit diagram for a common emitter transistor amplifier suitable for amplifying small audio-frequency signals and explain the operation of the circuit.
(iii) Explain why a silicon transistor is normally used in preference to a germanium transistor for the above amplifier.

(AEB)

8.62 Give an account of the mechanism of the flow of electric current in n-type and p-type semiconductors.

Explain what is meant by the Hall effect and show how it can be used to distinguish between n-type and p-type semiconductors.

What are the important differences between (a) the resistivities of metals and semiconductors, (b) the variation of these resistivities with temperature?

(Southern Universities)

Answers

8.6 Ans (a) Input is the sound; output is the electrical signals
(b) Input is the mains supply a.c.; output is the alternating signal of the required frequency

(c) Input is the radiations; output is the voltage pulses
(d) Input is petrol; output is the rotation of a shaft

8.7 Ans See *Figure 8.34*

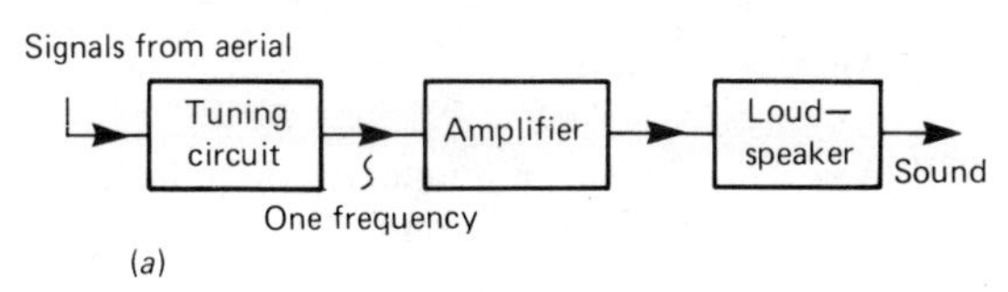

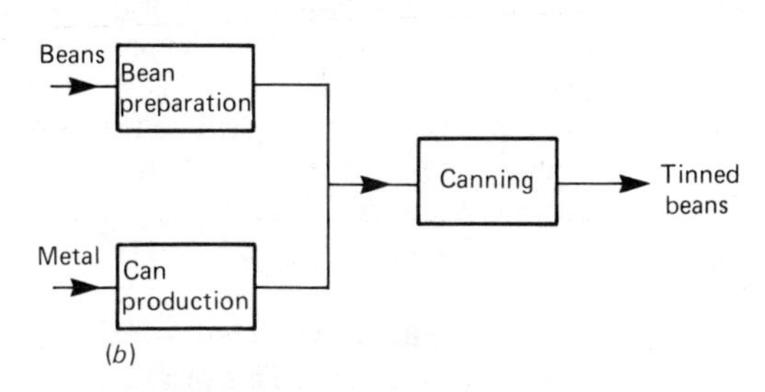

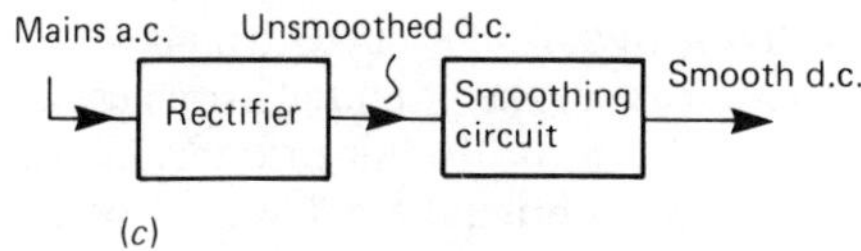

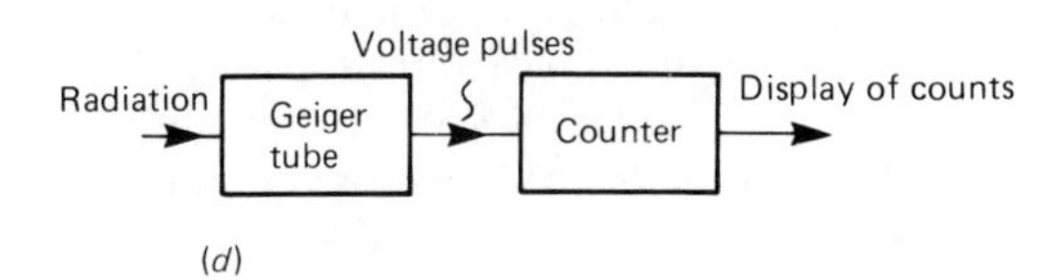

Figure 8.34 (a) A simple radio; (b) simple representation of bean canning; (c) smoothed d.c. supply; (d) radiation counter

8.8 Ans (a) Input is the temperature; output is the e.m.f.
(b) Input is the light; output is the change in resistance
(c) Input is the movement; output is the voltage signals

8.9 Ans See the notes. An example of negative feedback is a thermostat-controlled central heating system; an example of positive feedback is higher prices means more pressure for higher wages, which means higher prices, which means higher wages, etc. or a resonance phenomenon

8.10 Ans See *Figure 8.35*

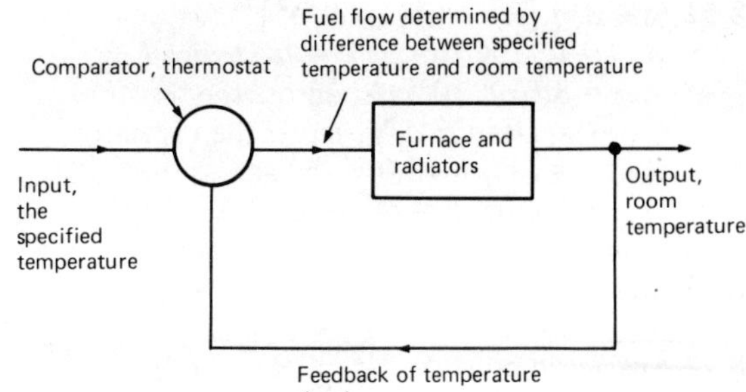

Figure 8.35

8.11 Ans (a) AND
(b) OR
(c) NOT

8.12 Ans (a)

A	B	C	D	E
0	0	0	0	0
1	0	0	0	0
0	1	0	0	0
1	1	0	1	1
0	0	1	0	1
1	0	1	0	1
0	1	1	0	1
1	1	1	1	1

(b)

A	B	C	D	E	F
0	0	0	0	1	0
1	0	0	1	1	1
0	1	0	1	1	1
1	1	0	1	1	1
0	0	1	0	0	0
1	0	1	1	0	0
0	1	1	1	0	0
1	1	1	1	0	0

8.16 Ans (a) $X_C = 1/(2\pi fC) = 199\ \Omega$
(b) $X_C = 1/(2\pi fC) = 1.59\ \Omega$
(c) $X_L = 2\pi fL = 1260\ \Omega$
(d) $X_L = 2\pi fL = 31.4\ \Omega$

8.17 Ans $X_C = 1/(2\pi fC)$, hence $f = 3980$ Hz

8.18 Ans $X_L = 2\pi fL$ and thus as $V = IX_L$, the RMS current is 0.076 A

8.19 Ans (a) $Z = (R^2 + X_L^2)^{1/2}$ where $X_L = 2\pi fL$; hence $Z = 118\ \Omega$
(b) $V = IZ$, hence $I = 2.0$ A

8.20 Ans $f^2 = 1/(4\pi^2 LC)$, hence $C = 1.27 \times 10^{-9}$ F

8.21 Ans $f^2 = 1/(4\pi^2 LC)$, hence $f = 225$ Hz; $V = IR$, hence $I = 0.06$ A

8.22 Ans $Z = [R^2 + (X_L - X_C)^2]^{1/2}$ $= 1.47 \times 10^3\ \Omega$; $V = IZ$, hence I $= 8.16 \times 10^{-3}$ A. Across the resistor, $V = IR = 2.45$ V, across the capacitor, $V = IX_C = 13.0$ V and across the inductor, $V = IX_L = 1.28$ V

8.23 Ans See the notes

8.27 Ans Mean power $= V^2_{RMS}/R = 0.8$ W

8.28 Ans See the notes

8.29 Ans Tan $\phi = X_C/R$, hence $\phi = 38.5°$; $Z = (R^2 + X_C^2)^{1/2} = 639\ \Omega$. Mean power $= (V^2_{RMS}/Z) \cos \phi = 0.0306$ W

8.30 Ans (a) $Z = (R^2 + X_L^2)^{1/2} = 22.4\ \Omega$
(b) Tan $\phi = X_L/R$, hence $\phi = 26.6°$.
Mean power $= (V^2_{RMS}/Z) \cos \phi$ $= 2.30 \times 10^3$ W

8.31 Ans (a) $f^2 = 1/(4\pi^2 LC)$, hence $f = 159$ Hz
(b) Power $= V^2_{RMS}/R = 0.48$ W

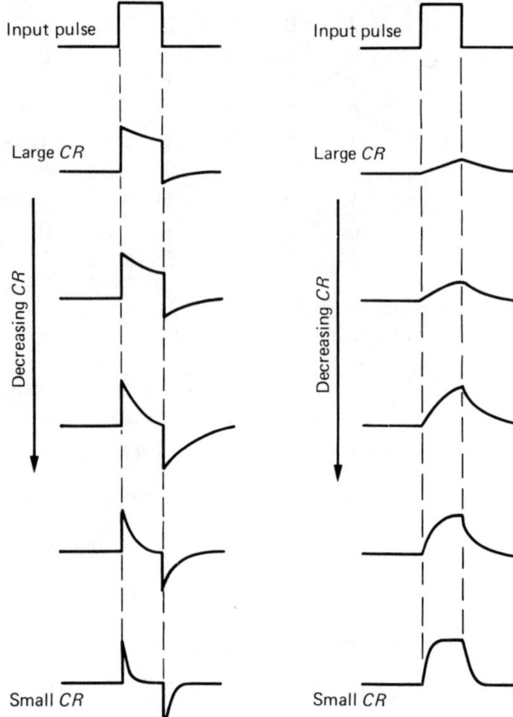

Figure 8.36

(c) Tan $\phi = (X_L - X_C)/R$, hence $\phi = 58.1°$; $Z = [R^2 + (X_L - X_C)^2]^{1/2}$ $= 568\ \Omega$; hence mean power $= (V^2_{RMS}/Z) \cos \phi = 0.134$ W

8.33 Ans See *Figure 8.36*

8.36 Ans See the notes

8.37 Ans The higher the temperature the higher the saturation current

8.38 Ans With the diode connected one way there will be virtually no current detected by the instrument. With the diode the other way round the situation will be the same as having a resistor in series with the instrument and the current will just be reduced. See **8.35**

8.40 Ans When the input is connected to +6 V the base current rises. This results in an increase in collector current. This means an increase in the potential drop across the load resistor. In this case this increase is such as to bring the potential drop to almost 6 V and so the output to 0 V. When the input is connected to 0 V the base current is low and so is the collector current. This means very little potential drop across the load resistor and so the output is almost at the 6 V value.
$V_R + V_{CE} = 6$ V

8.41 Ans (a) Change in potential difference across $R = (4.0 - 2.0) \times 10^{-3} \times 1000$ $= 2.0$ V
(b) The potential difference between the collector and the emitter must therefore change by 2.0 V as the d.c. supply is constant
(c) Voltage gain $= 2.0/0.02 = 100$

8.44 Ans n-Type

8.45 Ans See the notes

8.46 Ans An increase in the number of charge carriers available for conduction

8.47 Ans See the notes

8.48 Ans See the notes

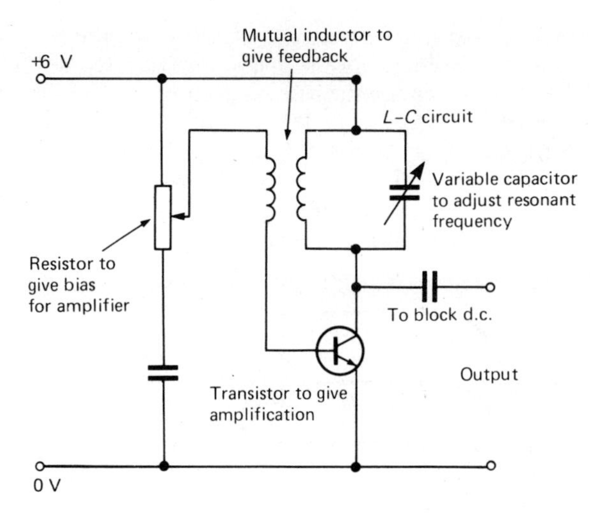

Figure 8.37

8.52 Ans See *Figure 8.37*

8.53 Ans See the notes

8.54 Ans See the notes

8.55 Ans Gain = $1 + (R_2/R_1)$, hence $R_2 = 19$ kΩ

8.56 Ans (a) $V_1 = RI$, hence $I = 0.24$ A RMS
(b) Mean power = $V_{RMS} I_{RMS} \cos\phi$, with $V^2_{RMS} = 120^2 + 160^2$ and $\tan\phi = X_C/R = V_2/V_1$. Hence $\phi = 53.1°$ and power = 28.8 W or alternatively just power = $I^2 R$ = 28.8 W
(c) Tan 53.1° = $X_C/500$, hence $X_C = 666$ Ω and so $C = 4.78 \times 10^{-6}$ F

8.57 Ans (a) The impedance can be defined as being the maximum value of the potential difference across the coil divided by the maximum current value, or the root mean square potential difference divided by the root mean square current. The potential difference maximum of an inductor does not occur at the same time as the maximum value of the current. Thus, if the coil is considered to have inductance the impedance will not be the same as the resistance, as this is the ratio of the potential difference and current at the same instance of time. The impedance depends on the frequency of the alternating current, the resistance does not. The relationship between impedance Z and R is $Z = (R^2 + X^2_L)^{1/2}$. A non-inductive resistance can be made by doubling the wire back on itself before winding it round a bobbin (*Figure 8.38*). The magnetic field produced by the current in any

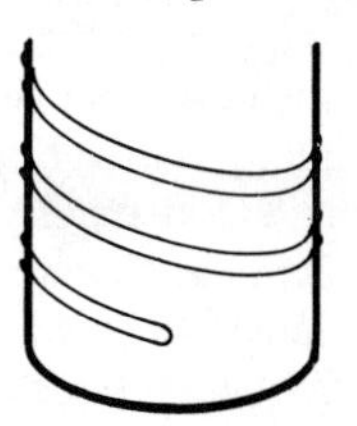

Figure 8.38

part of the wire is thus virtually cancelled by the oppositely directed current in the neighbouring wire.
(b) See *Figure 8.39*. A double beam oscilloscope is used to compare the potential differences across the resistor and the inductor. The maximum current can then

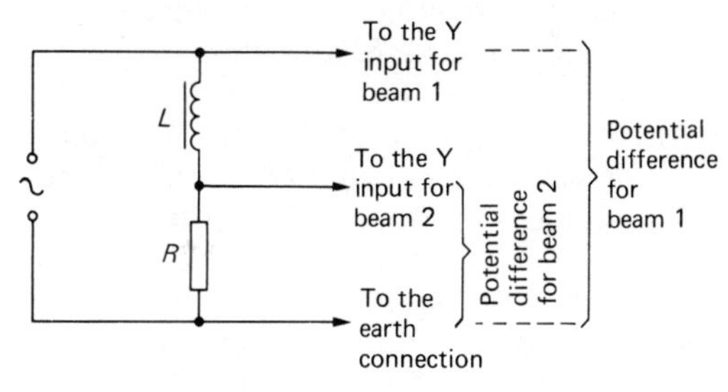

Figure 8.39

be calculated from the maximum potential difference V_R across the resistor, being V_R/R. The impedance of the coil is thus the maximum potential difference across it, V_L, divided by the maximum current.

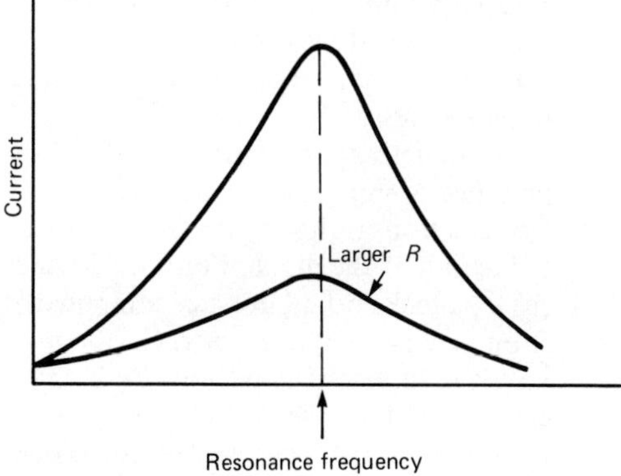

Figure 8.40

(c) See *Figure 8.40*, the current showing a maximum at the resonance frequency given by $f^2 = 1/(4\pi^2LC)$

8.58 Ans The reactance of an inductor is $2\pi fL$, increasing as the frequency is increased. This is because the greater the rate of change of current the greater the induced e.m.f. in the inductor which opposes the changing current. The reactance of the capacitor is $1/(2\pi fC)$ and decreases as the frequency increases. This is because the higher the frequency the shorter the time between the charging and discharging of the capacitor. With the *LCR* circuit resonance occurs at a particular frequency and so the current peaks at this frequency. The inductor gets warm because it has both resistance and inductance and the passage of current through a resistor results in the dissipation of power. No mean power is dissipated with a pure inductor or capacitor

8.59 Ans $Z = V_{RMS}/I_{RMS} = (R^2 + X_C^2)^{1/2}$ if the resistor and capacitor are in series. This would seem likely in view of there being no current when d.c. is used. At 100 Hz the impedance is $32/(1.0 \times 10^{-3}) = 32 \times 10^3\ \Omega$. At the steady value $Z = 32/(2.0 \times 10^{-3}) = 16 \times 10^3\ \Omega$. The higher the frequency the smaller the the value of X_C; eventually it will become insignificant when compared with R. This would appear to be the steady value. Thus R would be expected to be $16 \times 10^3\ \Omega$

8.60 Ans See the notes. A semiconductor has a resistivity intermediate between that of conductors and insulators.
(a) Conductivity increases with an increase in temperature
(b) Conductivity increases with the presence of impurities which add extra free electrons or holes. See the notes for the action of the diode. For both (i) and (ii) C_1 blocks off all but a.c. components from reaching the transistor. C_2 allows only a.c. to reach the output. R_1 is to ensure that the base is always positive for n–p–n. R_2 is the load resistor, the potential difference from the battery being shared between the transistor and this resistor. For the p–n–p transistor R_1 is to ensure the base is negative

8.61 Ans (i) See *Figure 8.41*. (ii) See *Figure 8.42*. See the notes for the explanation of the transistor as an amplifier. (iii) Silicon is less affected by changes in temperature

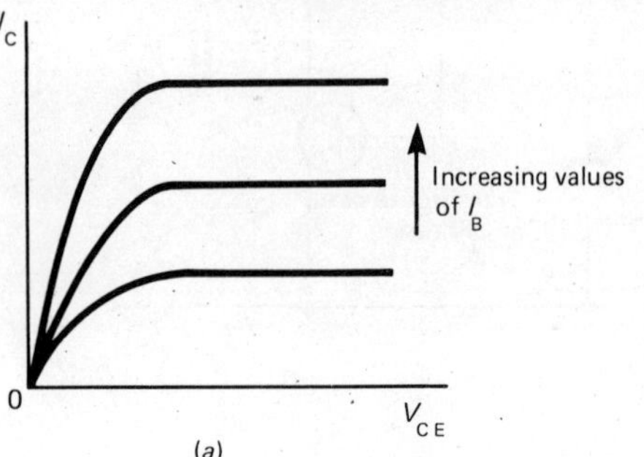

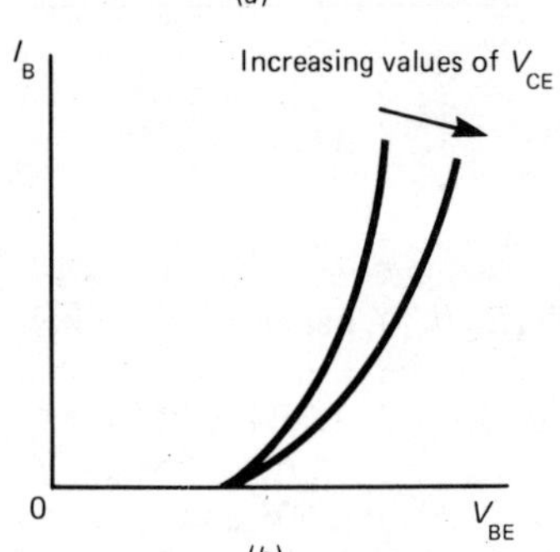

Figure 8.41 (a) Output characteristic; (b) input characteristic

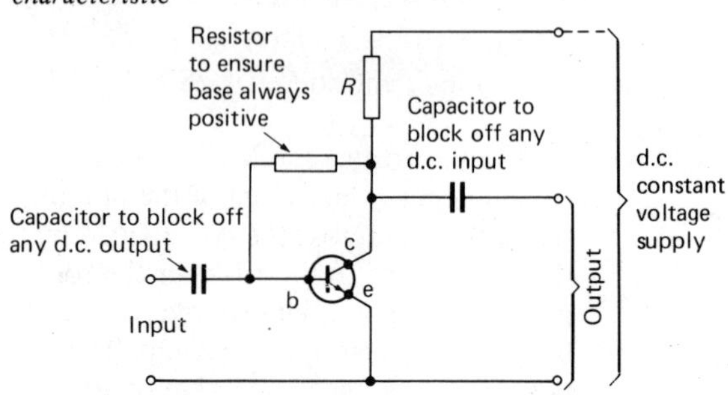

Figure 8.42

8.62 Ans See the notes in this chapter for the mechanism of current and in chapter 4 for the Hall effect.
(a) Metals have resistivities of the order of 10^8 less than semiconductors
(b) The resistivity of a metal normally increases with temperature while that of semiconductors decreases